Ecological Studies
Analysis and Synthesis

Edited by

W.D. Billings, Durham (USA) F. Golley, Athens (USA)

O.L. Lange, Würzburg (FRG) J.S. Olson, Oak Ridge (USA)

H. Remmert, Marburg (FRG)

Volume 86

Ecological Studies

Jurek Kolasa
Steward T.A. Pickett
Editors

Ecological Heterogeneity

Contributors
T.F.H. Allen, J.J. Armesto, J.P. Barry, S. Brand, H. Caswell,
P. Chesson, J.E. Cohen, R.K. Colwell, P.K. Dayton,
J.A. Downing, R.H. Gardner, T.W. Hoekstra, B. Jackson,
P.A. Keddy, C. Loehle, M.J. McDonnell, R.P. McIntosh,
B.T. Milne, S. Naeem, R.V.O'Neill, C.D. Rollo,
M. Shachak, M.G. Turner

With 72 Illustrations

Springer-Verlag
New York Berlin Heidelberg London
Paris Tokyo Hong Kong Barcelona

Jurek Kolasa
Department of Biology
McMaster University
Hamilton, Ontario L8S 4K1
Canada

Steward T.A. Pickett
Institute of Ecosystem Studies
The New York Botanical Garden
Mary Flagler Cary Arboretum
Millbrook, NY 12545
USA

Cover illustration: Haida Dog Salmon: Skaagi by Bill Reid 1974.

Library of Congress Cataloging-in-Publication Data
Ecological heterogeneity/Jurek Kolasa, Steward T.A. Pickett,
    editors.
        p. cm.—(Ecological studies; V. 86)
    Includes bibliographical references and index.
    ISBN 0-387-97418-0
    1. Ecological heterogeneity.   I. Kolasa, Jurek.   II. Pickett,
Steward T., 1950–   .   III. Series.
QH541.15.E24E28   1991
574.5'24—dc20   90-45476

Printed on acid-free paper.

Typeset by Asco Trade Typesetting Ltd
Printed and bound by Edwards Brothers, Inc., Ann Arbor, MI.
Printed in the United States of America.

9 8 7 6 5 4 3 2 1

ISBN 0-387-97418-0 Springer-Verlag New York Berlin Heidelberg
ISBN 3-540-97418-0 Springer-Verlag Berlin Heidelberg New York

Preface

An attractive, promising, and frustrating feature of ecology is its complexity, both conceptual and observational. Increasing acknowledgment of the importance of scale testifies to the shifting focus in large areas of ecology. In the rush to explore problems of scale, another general aspect of ecological systems has been given less attention. This aspect, equally important, is heterogeneity. Its importance lies in the ubiquity of heterogeneity as a feature of ecological systems and in the number of questions it raises—questions to which answers are not readily available.

What is heterogeneity?
Does it differ from complexity?
What dimensions need be considered to evaluate heterogeneity adequately?
Can heterogeneity be measured at various scales?
Is heterogeneity a part of organization of ecological systems? How does it change in time and space?
What are the causes of heterogeneity and causes of its change?

This volume attempts to answer these questions. It is devoted to identification of the meaning, range of applications, problems, and methodology associated with the study of heterogeneity. The coverage is thus broad and rich, and the contributing authors have been encouraged to range widely in discussions and reflections.

The chapters are grouped into themes. The first group focuses on the conceptual foundations (Chapters 1–5). These papers examine the meaning of the term, historical developments, and relations to scale. The second theme is modeling population and interspecific interactions in heterogeneous environments (Chapters 6 and 7). The third group is a series of practical problems ranging from sampling methodology to management in a heterogeneous world (Chapters 8–10). Finally, the fourth theme is a collection of examples representing various areas of ecology with their richness of observations, lessons, speculations, and conclusions about heterogeneity (Chapters 11–14). The examples cover various habitats and foci. Desert heterogeneity, large-scale oceanic physical heterogeneity, plant community phenomena, and resource heterogeneity are examined or reviewed.

Can heterogeneity become a successful competitor to the problems of scale? We believe it cannot. Heterogeneity emerges and disappears with alteration of scale. Scale is the window, heterogeneity is a characteristic of the view in it. Depending on the weather, however, the view may change independently. The size of the window, the landscape outside, and the weather may together create many combinations. This book attempts to sample these combinations.

Hamilton and Millbrook

Contents

Contributors

Allen, T.F.H.

Department of Botany
University of Wisconsin
Madison, WI 53706 USA

Armesto, J.J.

Laboratorio de Sistemática &
  Ecología Vegetal
Faculdad de Ciencias
Universidad de Chile
Santiago, Chile

Barry, J.P.

A-001 Scripps Institution of
  Oceanography
La Jolla, CA 92093 USA

Brand, S.

Mitrani Center for Desert Ecology
Blaustein Institute for Desert Research
Ben Gurion University of the Negev
Sede Boqer Campus, Israel 84993

Caswell, H.

Biology Department
Woods Hole Oceanographic Institution
Woods Hole, MA 02543 USA

Chesson, P.   Ecosystem Dynamics Group
                                     Research School of Biological Sciences
                                     Australian National University
                                     Canberra, A.C.T. 2601
                                     Australia

Cohen, J.E.   The Rockefeller University
                                     New York, NY 10021 USA

Colwell, R.K.   Ecology and Evolutionary Biology
                                     University of Connecticut U-42
                                     Storres, CT 06269-3042 USA

Dayton, P.K.   A-001 Scripps Institution of
                                         Oceanography
                                     La Jolla, CA 92093 USA

Downing, J.A.   Départment de Sciences biologiques
                                     Université de Montréal
                                     Montréal, Québec, Canada H3C 3J7

Gardner, R.H.   Environmental Sciences Division
                                     Oak Ridge National Laboratory
                                     Oak Ridge, TN 37831 USA

Hoekstra, T.W.   Rocky Mountain Forest and Range
                                         Experiment Station
                                     240 W. Prospect Street
                                     Fort Collins, CO 80526 USA

Jackson, B.   Computer and Telecommunications
                                         Division
                                     Martin Marietta Energy Systems, Inc.
                                     Oak Ridge, TN 37831 USA

Keddy, P.A.   Department of Biology
                                     University of Ottawa
                                     Ottawa, Ontario, Canada K1N 6N5

Loehle, C.   Environmental Sciences
                                         Section, 773-42A
                                     Savannah River Laboratory
                                     Westinghouse Savannah River
                                         Company
                                     Aiken, SC 29808-0001 USA

McDonnell, M.J.   Institute of Ecosystem Studies
                                   The New York Botanical Garden
                                   Mary Flagler Cary Arboretum
                                   Millbrook, NY 12545 USA

McIntosh, R.P.   Department of Biological Sciences
                                   University of Notre Dame
                                   Notre Dame, IN 46556 USA

Milne, B.T.   Department of Biology
                                   University of New Mexico
                                   Albuquerque, NM 87131 USA

Naeem, S.   Department of Biology/Museum of
                                       Zoology
                                   University of Michigan
                                   Ann Arbor, MI 48109-1048 USA

O'Neill, R.V.   Environmental Sciences Division
                                   Oak Ridge National Laboratory
                                   Oak Ridge, TN 37831 USA

Rollo, C.D.   Department of Biology
                                   McMaster University
                                   Hamilton, Ontario, Canada, L8S 4K1

Shachak, M.   Mitrani Center for Desert Ecology
                                   Blaustein Institute for Desert Research
                                   Ben Gurion University of the Negev
                                   Sede Boqer Campus, Israel 84993

Turner, M.G.   Environmental Sciences Division
                                   Oak Ridge National Laboratory
                                   Oak Ridge, TN 37831 USA

1. Introduction: The Heterogeneity of Heterogeneity: A Glossary

Jurek Kolasa and C. David Rollo

Despite its ubiquitous relevance to most ecological and evolutionary processes, a comprehensive description of the structural and dynamic aspects of heterogeneity has never been constructed. Intuitively, the concept of heterogeneity is clear, but as we scrutinize it our initial impression fractures into complexity. The term appears rather simple when contrasted with homogeneity, the absence of variation. However, one can view heterogeneity from a variety of perspectives, some of which are well known and explored, but many of which are not. Some perspectives have important consequences for ecology and evolution, whereas others appear, at least initially, to be logical curiosities. Furthermore, different perspectives may be inclusive, exclusive, complementary, or overlapping, a problem that hints at the nature of the phenomenon itself.

Here we document various perspectives and their consequences in the form of a glossary, with comments and examples. The purpose, however, is not to provide a set of static definitions. Rather, we compare, contrast, and interpret the landscapes and dynamics revealed with various lenses and relate their implications to ecological and evolutionary paradigms. We begin with the differentiation of spatial and temporal aspects—dimensions conventionally used for classifying heterogeneity. Subsequently, we conduct the discussion along two other major dividing lines that have been relatively unexplored. One involves distinguishing between *measured heterogeneity*, a product of the observer's arbitrary perspective, and *func-*

tional heterogeneity that which ecological entities actually perceive, relate to, and respond to. The former of these frameworks corresponds to *epiphenomenological* and the latter to *organizational* (or systemic) aspects of heterogeneity.

Basic Terms

Heterogeneity and Variability

The classical view (see Chapter 11) distinguishes between heterogeneity (composition of parts of different kinds) and variability (different values of a variable of one kind). As we see later, this distinction is not as sharp as it may appear at first glance. Furthermore, its meaning changes depending on the choice of perspective or approach. Examples are provided by Naeem and Colwell (Chapter 12) and Shachak and Brand (Chapter 11).

Spatial Heterogeneity

Smith (1972) defined heterogeneity in the spatial sense only. He viewed environment as heterogeneous if the rate of a process varies over space in relation to structural variations of the environment. By contrast, the environment is homogeneous if a process has a uniform rate across space. It is interesting to note that Smith's definition required a temporal component to become operational. In general, however, the notion of spatial heterogeneity does not require this component. In fact, a static descriptor suffices. For example, we may say that environment is heterogeneous if a chosen quantitative or qualitative descriptor such as vegetation cover or air temperature assumes different values at different locations. Smith's interpretation is an example of *functional heterogeneity*, whereas the latter interpretation is one case of what we label *measured heterogeneity* (see below). Ecologists often demand that variables be measured independently of the system of interest. We suggest later that although it may be a reasonable formal demand for making testable statements, this approach to heterogeneity, in its extreme interpretation, is usually inappropriate and may prevent key insights.

Point/summary: *Spatial heterogeneity may be viewed from dynamic or static and observer oriented or ecological entity oriented perspectives.*

Temporal Heterogeneity

Formally, temporal heterogeneity is similar to spatial heterogeneity, except that it refers to one point in space and many points in time. There is an intricate relation between these two dimensions. Whenever two adjacent locations or sites differ in their temporal variation (temporal heterogeneity), they represent spatial heterogeneity. This point is true even if their temporal patterns are identical but asynchronous. This relation is not symmetrical, however. Whenever two sites are different at any given moment in time, they may be either homogeneous or heterogeneous temporally. If they are homogeneous, they must have constant but different values of the same descriptor over the specified time interval. Naeem and Colwell (see Chapter 12) provide an interesting example of the interactions between the two dimensions applied to resource levels.

Point summary: *Temporal heterogeneity is not equivalent to spatial heterogeneity.*

Triumvirate of Heterogeneity

Deterministic Approach

Our strategy when preparing the glossary was to first extrapolate all of the consequences and implications of heterogeneity and then narrow the focus onto those aspects most relevant to conventional ecological concerns. The linkages are truly surprising. For example, heterogeneity is a phenomenon associated with discontinuities or modulations that may assume many forms and combinations. There is a triumvirate of discontinuities that can contribute to heterogeneity in both spatial and temporal dimensions (Fig. 1.1). This triumvirate consists of deterministic, chaotic, and random behaviors. Information theory is concerned mainly with identifying predictable elements of temporal signals or spatial configurations. Noise is considered mainly because it often represents unavoidable interference

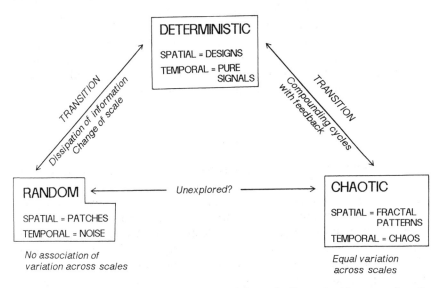

Figure 1.1. Triumvirate of heterogeneity. Where do discontinuities come from?

(Guiasu, 1977). If predictable elements are deterministic, the simplest complete temporal representation would be a pure sine wave. Many cycles can be superimposed to create complex signals. Such complexity may still be deterministic, however, if each signal can be hierarchically decomposed, e.g., time series analysis (Box and Jenkins, 1976).

Ecologists tend to be concerned mainly with spatial patchiness, and yet temporal dynamics, as we bring out here, are equally relevant to heterogeneity. Information theory, having its foundations in the analysis of signal transmission (Shannon, 1948), is largely concerned with extracting complex temporal signals from noise and decomposing them into various components. Spatial aspects are rarely emphasized. Clearly, unification of these areas would be immensely profitable. Probably the closest linkage between spatial and temporal variability has emerged from recent advances in the realm of chaos.

We also identified the important role of perceptual systems for functional heterogeneity. The resolution (perception) and processing (cognition) of environmental signals is the cornerstone of information theory (Guiasu, 1977). Information theory is concerned with the discrimination of signals from noise, and heterogeneity in its classic ecological sense is a major aspect of spatiotemporal noise. The relative nature of deterministic heterogeneity becomes apparent because complex signals that are "absolutely" deterministic seem "relatively" random to organisms that have cognitive systems too simple to decode them. The geometrical equivalent of deterministic heterogeneity is design. Thus a pattern of equally sized, equally spaced dots or stripes is a design. A checkerboard is a design. Designs, as in signals, may become highly complex. As in the case of complex temporal signals, even an absolutely precise complex design may appear as a relatively random pattern to an organism incapable of perceiving or processing the information. In this regard, relative scale is important. A statistical analysis of the frequency of various letters in the pages of a book or the squares occupied by chess pieces over various tournaments yields a random pattern. The information contained at one level may appear as noise at finer scales.

Random

Biologists interested in heterogeneity have largely focused on random spatial elements (e.g., Caswell and Cohen, Chapter 6). Even in an ecological sense, however, heterogeneity may be associated with information. Optimal foragers, for example, utilize intra- and interpatch information when deciding when to switch patches (e.g., Kotliar and Wiens, 1991). However, heterogeneity in an ecological sense is usually viewed as a filter of noise that imposes interference or constraints. It does not mean heterogeneity is necessarily disadvantageous. High heterogeneity interferes with competition and consequently allows higher species diversity (Harris, 1986), reduces the impact of predation and increases population stability (Huffaker et al., 1963), and helps maintain intraspecific genetic polymorphism. In such cases, the consequences of heterogeneity are independent of associated potential information. Geometrically, *randomness* is represented as patchiness. Various statistics texts thoroughly explore the classification of patch dispersion (i.e., aggregation and overdispersion). Temporally, randomness is noise, i.e., perturbations with no deterministic relation with respect to period, amplitude, or phase (Fig. 1.1).

Ecologists have traditionally neglected the temporal aspects of heterogeneity, but clearly they are of equal importance. A unification of ideas from information theorists and ecologists could prove immensely profitable. In ecological systems, one organism's noise may well be another's communication. In this glossary, however, we emphasize noninformational aspects.

Chaotic

A third major category has been recognized whose exploration is just beginning. *Chaos* has the superficial appearance of *randomness* but is generated *deterministically*. *Chaos* is a complex phenomenon, and we do not have enough space to explain it fully here. Readers may consult Gleick (1987) and Glass and Mackey (1988). Temporally, *chaos* emerges from cyclic systems that contain feedback loops, when the process has a feedback-dependent rate that exceeds the capacity of the feedback system to maintain stability. Extremely small variations in the state of a system are compounded, so it is unlikely that any particular path would be repeated. In population dynamics chaos was first recognized in the discrete versions of the Lotka-Volterra equations. At low growth rates populations approach equilibrium smoothly. At faster rates converging or stable cycles emerge. As the system is pushed beyond this point by greater rates of increase or greater density dependence, unstable and unpredictable oscillations (*chaos*) may emerge (May, 1974, 1976; May and Oster, 1976). *Chaos* differs from noise in that it shows the same degree of variation recursively at whatever scale is examined. Interestingly, this aspect was first noted by Berger and Mandelbrot (1963) working on transmission of information signals, again reinforcing the complementary perspective of ecologists and information theorists.

This property of repeated variation at finer and finer scales has a geometrical analogue known as *fractals*. Fractal geometry often generates complex patterns that can be dissected into infinitely small scales. At each level, the pattern differs but always shows the same relative variability. There are a number of texts available illuminating fractals for their aesthetic value. For biologists, strategies based on bifurcating branching processes (e.g., nervous systems, circulatory systems, lungs, tree branches, compound leaves, phylogenetic diversification) may have an underlying fractal geometry.

The importance of *chaos* to studies of heterogeneity can be appreciated if we realize that the bifurcating branching structures of many plants (and consequently forests) may be of fractal nature. Thus the environment of organisms such as caterpillars, monkeys, and birds may be dominated by this realm. Moreover, climate was one of the earliest phenomena recognized as chaotic (Lorenz, 1963), so forest species may be dominated temporally by this realm as well.

Point/summary: *Temporal heterogeneity may involve deterministic random, and chaotic variations.*

Emergent Properties

Because scale is a crucial aspect of heterogeneity, changes in scale are frequently associated with new or emergent properties. For example, a hypothetical green bug, traversing an environment consisting of equal numbers of yellow, green, and blue patches would lose its camouflage when it is in two-thirds of the patches. An organism viewing the world from a higher perspective, however, would see only green. It would be true even if the environment consisted only of equal numbers of yellow and blue patches, as blue and yellow emerge as green as the grain coalesces. At this resolution, the patchiness of the environment fuses into homogeneity. Thus homogeneous green emerges from the heterogeneous blue and yellow environment. Some consequences of changing either the grain or the extent in an observation set have been already identified. Allen (presented at the Annual Meeting of the Ecological Society of America, Minneapolis, 1984) demonstrated how increasing the extent leads to averaging of variation and increasing of the grain transforms a variable into a constant. In fact, in both cases perceptual homogenization takes place.

The importance of the perceptual system can be appreciated by realizing that a color-blind organism might not see the green bug on any of the three colored backgrounds if they were all of equal shade. In this case, we might expect the perceptual system to rely on movement detectors for small moving bugs, as in the case of toads (Lettvin et al., 1959). If static spatial resolution is poor, dynamics may enhance discrimination. During World War II, color-blind men were employed to detect enemy installations: They were not fooled by camouflage nets, as they discriminate using patterns and shades rather than colors. This example is interesting because it illustrates how a major adaptation evolved to help differentiate the world (i.e., color vision) can be turned into a liability by increasing the heterogeneity (noise) in that modality. Naeem and Colwell (Chapter 12, lesson 5) give more examples of the role of dynamics (temporal variation) in determining the performance of species.

Point/summary: *Shifts in scale may produce more than averaqes or constants: They may make homogeneity out of heterogeneity and vice versa.*

Rates/Frequencies

At the base of temporal heterogeneity is the concept of rate. For example, a frog sitting in absolute stillness is invisible to a snake that passes only inches away because the snake requires movement to elicit a strike. At the other extreme, fast organisms are difficult to catch. Possibly one of the best examples of this point concerns orb web spiders. In response to the peck of a bird or the hum of a spider-wasp's wings, the spider pulls on its web and sets the whole tangle resonating at a high frequency dependent on the weight of the spider and the elasticity of the silk. In addition to making it difficult for the predator to aim, some of these spiders nearly disappear, reminiscent of the spokes on a spinning bicycle wheel. If the flicker fusion frequency of any of these spider's predators is much slower than our own, we may in fact have here an example of literal invisibility as a predator avoidance mechanism.

The use of movement in detector systems is widespread. The toad's perceptual system responds to small, dark moving objects (Lettvin et al., 1959) because the visual background is too complex to process otherwise. Toads may have less success if the wind moves leaves so the visual background is also in motion. An adaptation to such dynamic heterogeneity can be seen in walking sticks, which sway deliberately back and forth when walking. If it is breezy, their deliberate movements make them appear like a twig swaying in the wind. The protective effect breaks down when there is no wind. Flocks and herds may also increase a species heterogeneity relative to predators, as there are more moving parts.

Point/summary: *The frequency at which an ecological system operates modifies its interaction with heterogeneity.*

Generated Dynamic Heterogeneity

Temporal heterogeneity may take many forms, some aspects of which are generated by the dynamics of the organisms themselves. Because heterogeneity is relative to the observer, movement by the organism, rather then its environment, also has important consequences. Kotliar and Wiens (1991) pointed out that for any organism the two related aspects of heterogeneity are the resolution of grain and the perceptual extent (see Grain and Extent, below). They defined extent as the home range size of the individual. We suggest a refinement to this concept. Grain and extent may be defined as the degree of acuity of a stationary organism with respect to short- and long-range perceptual ability. Thus *grain* is the finest component of the environment that can be differentiated up close, and *extent* is the range at which a relevant object can be distinguished from a fixed vantage point. This framework has the advantage of allowing a fuller exploration of

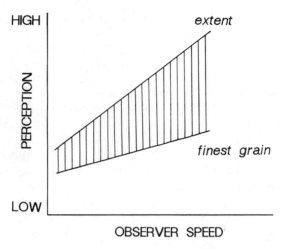

Figure 1.2. Hypothetical relation between the perceptive resolution of grain and extent as a function of speed of movement through the environment. Note the general shift toward a coarser range of heterogeneity.

the consequence of variable rates and longer time scales. For example, if one looks at roadside stones from a parked car, they are discrete objects. At 100 km/hour these same stones become an indistinct blur. At the same time, those larger features on the horizon are sampled at a rate that approaches and then surpasses that obtainable by stationary scanning. The closer objects are, the faster they are sampled by a moving observer. The overall effect of observer motion is to reduce the lower levels of heterogeneity and decrease perception of grain while expanding the higher levels of heterogeneity and increasing extent (Fig. 1.2). Thus an animal searching for patches of berry bushes moves at rates that are appropriate to sampling higher levels of heterogeneity but must slow down and scan more slowly in order to locate berries at lower levels of heterogeneity (higher resolution).

This framework applies to a simple exploratory foray. The operation at longer time scales (e.g., daily activity patterns, seasonal migratory behavior, lifetime dispersal movements, and geographical population expansions) is associated with increasing scales of heterogeneity (expanding extent of environmental variation) appropriate for resource acquisition, habitat selection, life history evolution, and speciation. At finer levels of resolution the information system is largely behavioral, whereas it shifts to physiology and then genetics at coarser levels and longer time scales.

Population ecologists have long been concerned with the population regulation (equilibrium) of organism numbers. Time lags and predator–prey systems generate cycles of various stability, and population studies were one of the founding themes identifying the general phenomenon of *chaos* (May, 1974, 1976). It has already been suggested that adaptations may ensue based on these temporal aspects. For example, the periodicity of cicadas to prime numbers has been suggested as an adaptation to prevent predator tracking. Conceivably, some high-fecundity prey species have population dynamics adapted to generate chaos, as efficient tracking of such species would be impossible for predators.

Another potential source of dynamic heterogeneity may emerge from the cognitive systems in animals. Just as such systems must be designed to deal with external uncertainty in most stages of perception, cognition, and processing, it is likely that they are also selected to generate uncertainty, e.g., chemical defenses in grasshoppers (Jones et al., 1986). Any organism that behaves in a completely predictable manner is vulnerable to exploitation by other organisms that detect that predictability. This point is probably best exemplified in Dawkins' (1982) *The Extended Phenotype.* Consequently, we might well expect nervous systems to generate uncertain behavior. For example, will the rabbit deek or dodge, reverse or jump? Squirrels appear to suffer enormous road kills because they use a wonderfully unpredictable reversal of direction when escaping from predators such as dogs and cats. Consequently, when running in front of cars, they often turn unexpectedly and run back in front of the car with undesired results for both squirrel and motorist. The presence of chaotic dynamics has been demonstrated in squid neurons (Aihara and Matsumoto, 1986), and nervous systems appear based on a fractal geometry. It is likely that nervous systems can generate adaptive chaotic outputs in a range of situations.

Point/summary: *Heterogeneity changes for ecological entities depending on their rate of movement relative to the environment.*

Grain

Grain determines the fineness of the distinctions that can be made in an observation set (Allen et al. 1984; see also Chapter 3). Relative to the organism, however, grain has two components, the size of the components and their distance away. Thus grains of sand may be seen when held in the hand but coalesce into dunes at 100 m. Leaves on a bush may be distinguished for greater distances but eventually become amorphous green on branches. Trees cannot be seen as an entity from too short a distance but remain as distinct units for long distances.

Thus grain is a complex character relative to the resolution of the observer. Snowy owls have almost telescopic farsightedness but may have some difficulty visualizing the mouths of their young without leaning back far enough to focus. The usual meaning of grain is the smallest resolution possible at the shortest distances. The above discussion points out, however, that what we formally defined as extent is the minimum size of resolution at the maximum range of the observer's perception. In this regard,

grain varies continuously as the size and distance of objects covary relative
to the observer. More information can be perceived by telescoping sensory
focus to closer and distant horizons, which is important to experiments.
Depending on the grain of the observation, smaller or larger variations in
space, time, or system organization emerge. For comparative purposes,
therefore, fixing the grain or scaling as appropriate is necessary; otherwise
results differ even for systems of the same heterogeneity. Allen and Hoekstra discuss this point and related topics in Chapter 3.

   Point summary: *An ecological system may experience more than one size
of grain simultaneously depending on the scope of perception.*

Other Terms

Measured Heterogeneity

When we examine attributes of the natural world, we often use standard
and arbitrary yardsticks. These yardsticks may give a misleading impression that an "objective" approach has been employed. Examples include
per capita income, distance in meters or light-years, Shannon's index to
measure species diversity, carbon dioxide assimilation to measure primary
production. Some of the measures used in ecology can be considered inadequate in principle, and this problem applies to heterogeneity as well.
The measures are inadequate when they severely distort ecological relations of interest. For example, distance in meters between prey and predator is an "objective" measure of predation likelihood because it is defined outside the system. By contrast, the frequency of encounters between
predator and prey is a measure defined by the system. The latter measure is
more relevant but requires calibration each time it is applied it to a new
system. Such a measure cannot be viewed as standard or absolute. General
measures of heterogeneity are few and not widely known (see Chapter 8),
but they fall clearly into the category of absolute measures. Variance
associated with an estimate of the mean, for example, is one of such
measures. O'Neill et al. (Chapter 5) employ it to examine heterogeneity
changes across scales. With this particular application they use variance to
test predictions about the hierarchical structures of landscapes.

   A general characteristic of these measures is their unidimensionality or
near-unidimensionality. By this statement we mean that one evaluates
heterogeneity from some minimum value to some maximum value along
one axis. It is immaterial that such an axis may be composite. Also, the
variable measured (e.g., length of the patch border, number of different
patches, diversity index) usually represents one or a small number of
aspects of the system in question. This approach is most accessible and
convenient, although critics point out its low biological relevance (Addicott et al., 1987). Species' perspective (e.g., Addicott et al., 1987; Kolasa,

1989; see also Chapter 4) offers an alternative that is biologically much richer. We label this approach *functional heterogeneity*.

The measurement of heterogeneity requires methods of simplification and summarization. A measurement may or may not allow reconstruction of the actual system. Measured heterogeneity can be summarized without loss of structure if the system is partially deterministic, so a descriptive model or equation can be derived that can regenerate the same system (e.g., an equation for a circle). For random components, statistical models may summarize the system but cannot regenerate the same structure. Chaotic systems have underlying deterministic dynamics, but the models generate novel outputs each time.

Point/summary: *Arbitrary measures of heterogeneity are tempting and popular, but their ability to reflect the relevant properties of the system of interest is unclear and questionable.*

Functional Heterogeneity

Functional heterogeneity is the heterogeneity an ecological entity (individual, population, species, or multispecies) perceives *and* responds to. It is the heterogeneity that is most and directly relevant to the ecological entity of interest. For any given habitat there are many heterogeneities. For a single individual, perceived heterogeneity depends on the temporal and spatial scale at which the individual operates, and this heterogeneity differs from that of a population. The scaling of organisms with respect to numerous features of metabolism, physiology, behavior, and morphology follows well documented allometric curves, e.g., the mouse to elephant curve for respiration (Peters, 1983; Calder, 1984; Schmidt-Nielsen, 1984). The physiological time scales of organisms become faster with decreasing size. This fact must be integrated with constant environmental cycles such as the 24-hour day (Calder, 1984, p. 226). This integration is largely accomplished by biological clocks that govern much of the circadian and seasonal functions of organisms (Beck, 1968; Saunders, 1976). Thus many aspects of design of organism are tradeoffs and compromises between the internal scale of ecological entities and scales at which environmental characteristics come to play. At least some sensory modalities are also scale-dependent. Thus the hearing of elephants has a typical range but is shifted an octave lower than in humans. Similarly, mice are shifted higher. Their voices scale appropriately (Calder, 1984, pp. 236–243). In addition to simple scaling factors, organism size is also correlated with degree of sensory/cognitive sophistication (e.g., light-sensitive eye spots and nerve nets versus binocular vision and encephalization) (Bonner, 1988). Thus for any individual of a given species, age, or size, functional heterogeneity is likely to vary for different environmental variables and the corresponding sensory modality. For example, air temperature may appear variable to a thermoregulating lizard on an hourly basis, whereas food may appear variable at weekly scales. Predation intensity has no meaning for an individual that is not eaten even if the intensity varies at the same temporal scale as temperature. Chesson (Chapter 7) addresses the question of perceptual shift at different spatial, temporal, and organizational (individuals, populations) scales.

A further complication superimposed on such variation is asymmetry in the scale of perception and response by individuals or populations. It may be a thoroughly different matter simply to register change than to respond effectively. Furthermore, an ecological entity may filter the same variable in different ways depending on past experience. The formation of search images by predators and perceptual/cognitive sets for puzzle-solving are good examples. Phenotypic expression of spines and protective helmets by *Daphnia* in response to the presence of *Chaoborus* predators (Hebert and Grewe, 1985) is an example of the same phenomenon at the population level.

Thus not only different relations may hold for different variables, but these relations may change as a function of interaction with the variables. For example, we may stand and appreciate the beauty of a tropical beach or an alpine slope, but we could, if we were so inclined, turn our attention to the form and structure of sand grains or snow flakes, respectively. We discuss this problem in some depth later in the chapter.

Summarizing, a general insight is that functional heterogeneity is necessarily multidimensional. Whereas *measured heterogeneity* applies across infinite spatiotemporal scales and may be arbitrarily bounded, *functional heterogeneity* arises from the interaction between scales relevant to the ecological entity and to its environment. Thus functional measures as such may provide better description of relations. Johnson (1970) pointed out that one reason information theory has largely failed to make any major contribution to biology is that it measures absolute quantities of information (= measured information) but ignores that of qualitative relevance to biology (= functional information). Johnson (1970) pointed out that two rabbits in a box contain about the same amount of information regardless of their sex. The biological future of the system is different, however, if the sexes are different than if they are the same.

Whether the two classes of heterogeneity are interchangeable, translatable, or even comparable remains an open question. Johnson (1970) suggested that functional measures could be translated from the larger class of measurable information by the use of an "intensity" modifier. Such a modifier could define the biological relevance of information.

Investigations of interactions among various dimensions of heterogeneity are highly promising. Caswell and Cohen (Chapter 6) explore the behavior of a model with only two dimensions of heterogeneity: patchiness and disturbance. Even this simple system reveals the complexity and ample range of possible responses.

Point/summary: *Functional heterogeneity is heterogeneity from the per-*

spective of participating ecological entities. It has many dimensions and many potentially important interactions among them. There are many functional heterogeneities in a system as simple as local population and many more in a system as complex as an ecosystem.

Interaction Between the Measured and Functional Concepts of Heterogeneity

The distinction between measured and functional heterogeneity sometimes appears indistinct because the distinction depends on the resolution of the study. The logical distinction, however, is clear. In cases where the choice of heterogeneity measures is often strongly influenced by prior knowledge of the organisms involved, the *functional* and *measured* heterogeneities begin to converge. Thus researcher makes indices of heterogeneity more and more functional while interpreting them as if they were entity-independent, or "objective." If the scale, extent, and resolution of the study are narrow, measured heterogeneity may approach congruence with functional heterogeneity. For example, Plowright and Galen (1985) focused on the effects of habitat landmarks such as large plants or vegetation boundaries on foraging routes of bumblebees. Even though they did not start the analysis from the bumblebee perspective, there is little question that the study addressed one relevant aspect of the bumblebee habitat heterogeneity. Thus their arbitrary measure of habitat heterogeneity may be close to some aspect of functional heterogeneity.

Point/summary: *Measured heterogeneity evolves and converges toward functional heterogeneity as the knowledge of the system increases.*

Continuous Heterogeneity

When a variable changes gradually and continuously in space or time, the habitat can be considered to be continuously heterogeneous. Examples include light attenuation in water, change in partial oxygen pressure with altitude, change in mean temperature with latitude, or concentration of molecules or individuals diffusing from a central point. Continuous heterogeneity, however, is a close relative of the "red herring." Although our instruments can find it, it often disappears when we change the scale of the measurement (e.g., the earth appears round or flat depending on the resolution and scope of the instrument). Furthermore, because of the pervading nonlinearity of perceptual filtering of biological entities, continuous heterogeneity is likely to disappear when functional heterogeneity is the methodological tool adopted. We view continuous heterogeneity as a conceptual consequence of the measured view identified earlier. Yet we suggest that it carries little meaning for ecological entities.

Indeed, not all heterogeneous variables must be perceived as abruptly heterogeneous. There is a suite of biological mechanisms that amplify minor discontinuities to ecologically meaningful ones (e.g., color vision in primates). Should these mechanisms be ignored, the variable will likely appear homogeneous, even if it is not. A classic example is the response of the human breathing reflex to blood concentrations of carbon dioxide and oxygen. Both concentrations change along rather smooth curves, but some particular values stimulate a breath-in reflex. Thus to an organism the carbon dioxide continuum is broken into two discrete phases: acceptable level and unacceptable level. An ecological/genetic example might be the switch points in frequency-dependent predators in response to densities of various clones of polymorphic prey. Naeem and Colwell (Chapter 12) supply further examples how interaction of species resolution and resource variation produces qualitatively new situations, particularly when they discuss Levins' (1979) "clock control."

Point/summary: *Continuous heterogeneity is an artifact of the indiscriminate use of an arbitrary approach to heterogeneity.*

Patchy Heterogeneity

The simplest case of patchy heterogeneity occurs when a variable shows two discrete states at one level of resolution. Day and night, loud and quiet, and water and land are examples of such heterogeneity. Dynamics of species interactions in such landscapes are explored by Caswell and Cohen (Chapter 6). Any natural system is likely to be more complex than that. First, there may be more than two states. Second, each of the states may be patchily heterogeneous in its own right at higher levels of resolution. These obvious possibilities strongly suggest that nested patchiness is a common feature of natural habitats. Community consequences of such a hierarchy of patches are many and sometimes surprising (see Kolasa, 1989, for more details). One of the as yet unexplored properties of hierarchical patchiness is *threshold heterogeneity*. By threshold heterogeneity we mean the level of resolution at which the grain (patch size) becomes so fine that an individual or a species stops responding to it (e.g., how far apart do two pinpricks have to be before you know there are more than one?). Threshold heterogeneity is, in other words, the grain resolution at which the species stops changing its distribution, energy transfer, reproductive output, or other features of interest to ecologists. At this level of resolution the world becomes functionally homogeneous. Thus the notion of threshold heterogeneity has a strict link to functional heterogeneity. The notion of threshold heterogeneity applies only within the methodology of functional heterogeneity, and it is incongruent with the measured heterogeneity. In fact, threshold heterogeneity may be considered as one of the many dimensions of functional heterogeneity. As a result, it is likely to differ among species, particularly between habitat specialists and habitat generalists.

In addition to a lower threshold heterogeneity, there can be an upper threshold heterogeneity level. This threshold is determined by the interaction between the scope (Kotliar and Wiens, 1990) and habitat heterogeneity. When the size of a patch perceived as homogeneous exceeds the scope of the ecological entity (individual, population, metapopulation), the habitat becomes virtually homogeneous for this entity. Note that for a single species several thresholds may be identified depending on the organizational level. A habitat may appear homogeneous for an individual wolf, heterogeneous for a pack, and homogeneous again with respect to the population (i.e., if the whole population lives in the same habitat type). In the last case, Hutchinsons's term "homogeneously diverse" might also apply.

Point/summary: *For any ecological entity there is be an upper and a lower level of resolution at which it stops to respond to heterogeneity.*

Habitat Complexity

Some authors distinguish between horizontal and vertical patchy heterogeneity (August, 1983). The latter then acquires a special label: *complexity*. Although the two types of heterogeneity may differ in their effects on local communities, they do not warrant the special distinction accorded. They are just two of many dimensions patchy heterogeneity can have in space–time. Attaching a special label to the vertical habitat heterogeneity and contrasting it to any other type of heterogeneity offers little or no terminological advantages. That the distinction is logically weak is further demonstrated by consideration of fish habitat. For example, fish ecologists view fish thermal habitat as three-dimensional. Nevertheless, the vertical component has qualities that are different from those of the horizontal component, and a comparison between them may be fruitful. Yet the term "habitat complexity" would be a misnomer in this case because "complexity" has other, more useful meanings in ecology.

Point/summary: *Habitat complexity may be an important component of heterogeneity for some ecological entities but does not constitute a class of heterogeneity of its own.*

Homogeneous Heterogeneity

Hutchinson (1957, 1967) used the term *homogeneously diverse* to indicate the relativity of species perceptions. He defined the habitat as being homogeneously diverse when the mosaic of habitat patches contained small patches relative to the "normal" ranges of individuals of a species. When these patches are large, the habitat is heterogeneously diverse. The view of functional heterogeneity presented earlier is broader because it goes beyond the consideration of mere range of movement. The notion of threshold heterogeneity can also be interpreted in the context of Hutchinson's terminology. It means a transition between a habitat that is heterogeneously diverse to one that is homogeneously diverse.

Point/summary: *Hutchinson's earlier terms can be replaced by a broader perspective of functional heterogeneity and threshold heterogeneity.*

Disturbance versus Heterogeneity

A common assumption is that there is some interaction between disturbance and heterogeneity. Disturbance creates heterogeneity according to some authors (Christensen, 1985; Denslow, 1985). This interaction is, however, far from clear. First, the nature of disturbance requires better understanding. The notion of disturbance proposed by Pickett et al. (1989) demands that disturbance be defined by relating it to the system structure. Second, the various faces of heterogeneity leave much room for interpretation and misinterpretation. A more common but less helpful view of disturbance is that of the damage to biomass. Despite the basic difference between these two views, a common relation appears to hold.

Consider a single spatial scale only. At this scale the system comprises one or several more or less homogeneous areas. For simplicity consider just one such area. A disturbance that wipes out a portion of this homogeneous area leaves at least two types of habitat: disturbed and undisturbed. The originally homogeneous system becomes heterogeneous. This heterogeneity continues to increase until 50% of the area is disturbed (Fig. 1.3A). When the extent of disturbance exceeds 50% of the area in question, the disturbance begins to homogenize the patch. The result of this simple process coincides with the predictions of the intermediate disturbance hypothesis effects on diversity. The heterogeneity patterns are significantly different, however, for habitat patches involving two scales of resolution. In particular, when the extent of disturbance spans several patches, it homogenizes the system at this particular scale almost from the beginning of the process (Fig. 1.3B). The last specification is needed because those several patches might have looked homogeneous at a larger scale. If we now consider this new, larger scale instead, a large disturbance may increase heterogeneity again while simultaneously decreasing it at a lower scale (higher resolution).

An analogy occurs in population genetics. A new allele that is favored by natural selection increases in a homozygous population, increasing heterogeneity until its frequency reaches 50%. After this point, heterogeneity declines and reaches homogeneity again as the new allele is fixed

---▷

Figure 1.3. Habitat heterogeneity as a function of the spatial scale of disturbance. Heterogeneity—a combination of disturbed and undisturbed areas—is measured by the Shannon diversity index H, ($\log_{10}$). The shaded areas symbolize stages of the simulated spatial progression of disturbance (expressed as area). (A) The square represents a single, initially homogeneous habitat. Habitat size (200 units) is arbitrary. Disturbance is allowed to increase to the size of the habitat. (B) The square represents a heterogeneous habitat composed of four habitat types: A, B, C, and D. Habitat size (400 units) is arbitrary. Disturbance is allowed to increase to the size of the habitat.

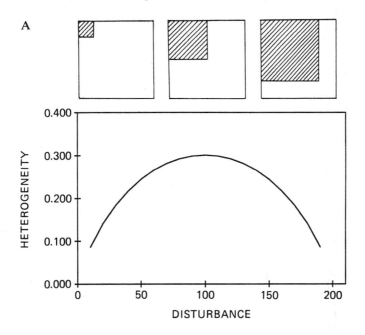

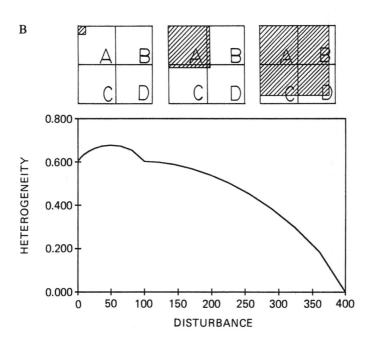

(Lewontin, 1974). It must be pointed out that the homogeneity at either end of such a process must, by definition, be different in nature; otherwise no disturbance has occurred. Thus the long-term temporal view retains the imprint of heterogeneity, even though the beginning and end products are homogeneous.

Point/summary: *The relation between disturbance and heterogeneity is nonlinear. At a single level of resolution intermediate disturbance create the highest heterogeneity, although this result may differ when several levels are considered.*

Dual View

Most of the above discussion focused on the heterogeneity of the habitat, or environmental template on which the organisms or populations operate. This view is incomplete, however. A population of ecological entities may itself be heterogeneous or homogeneous in relation to the environment. Differential performance of individuals and species on the same environment is a staple of population ecology, evolutionary ecology, and community ecology (e.g., see Chapter 7). It is a two-way street, however, and no accident that heterogeneity in gene pools is thought to be largely controlled by environmental heterogeneity. Genomes might be expected to move to fixation if habitats are relatively stable or homogeneous over time. Some explicitly argue that it is the heterogeneity among individuals of a population where sources and causes of ecosystem stability should be sought (Lomnicki, 1982). It might be useful to distinguish the effects of homogeneous environment on heterogeneous population from the effects of heterogeneous environment on homogeneous populations (e.g., see models by Caswell and Cohen, Chapter 6). An interesting, albeit brief, attempt to interpret the interaction of the two heterogeneities was provided by Janzen (1986). He discussed differences between "invasion" of crop plants and wild plants and the interaction with the species pool of potential insect colonizers. A clear view of these two aspects of heterogeneity may greatly enhance the understanding of its impact and interaction. A complete view of the performance of any ecological system requires a joint consideration of both aspects.

The dual view of heterogeneity in ecological systems is probably easy to accept and widely practiced, even though it is not always explicitly stated. However, it is not the only view possible. One can think of organizational heterogeneity where the number and composition of integral system components varies at different levels of organization. This alternative is discussed in the next section.

Point/summary: *Heterogeneity of the environment can be contrasted to heterogeneity of the ecological entity under study. Performance of individuals or populations may be influenced by both.*

Integral View

To explain the integral view we use an example: A population of lions may be heterogeneous in terms of size, intelligence, age, and other characteristics of individuals. This heterogeneity represents one of the two parts of the dual view. Yet the population can also be heterogeneous at levels higher than the individual, i.e., involving structures such as families, hunting packs, and herds. Individual lions are members of prides, and within prides individuals form groups of different relatedness, function, or rank. Aggregation of individuals into higher level entities, such as hunting teams and prides, adds an important aspect to heterogeneity of the lion population. Performance of an individual with respect to its environment may undergo deep modification as a result of this organizational heterogeneity. For example, weak individuals may survive in times of plenty owing to more successful hunting by other members of the group, but these same weak individuals (e.g., cubs) are the first to starve in times of scarcity. Another example involves heterogeneity among the sexes.

The general effects of functional heterogeneities at different levels of ecological organization have never been explored in an explicit way. This lack is not surprising because a general framework for describing ecological organization is a relatively recent proposition (Kolasa and Pickett, 1989). An important question is whether heterogeneity of the system as such can be assessed. Indeed, as for the case with diversity (Kolasa and Biesiadka, 1984), heterogeneity of the whole system is infinite. Because such heterogeneity is hierarchical, it is only through application of a rigid constraint that heterogeneity can be measured. This constraint is the necessary specification of the level of organization and consequently of the entities involved. Once there is no ambiguity as to which level is involved, one can proceed to determine the number of different components and spatial or temporal relations among them. A new question emerges regarding whether heterogeneities of various levels are additive. The answer is "no, they are not." Adding heterogeneities of entities at lower levels to those of higher levels forfeits the information obtained at any given level. For example, if a pack of wild dogs has 12 members of six ranks, the heterogeneity can, by some convention, be said to be six. The heterogeneity of interest at the next lower level may be skin patterns. Whether individual dogs have uniform or patchy skins has nothing to do with our previous determination that heterogeneity of pack composition is six. Adding the two values—the heterogeneity of ranks and heterogeneity of skin patterns—results in a higher value. Unfortunately, we will never know the value of what. The relation among the whole and its parts versus the external heterogeneity is briefly discussed below under Filters.

Other attempts to examine the role of heterogeneity in biological organization have been rooted in the theory of information. The preliminary results are not radically different from the expectations presented above. For example, Atlan (1985) theorized about the role of increased heterogeneity in the reduction of redundancy at a given level of integration. Possible outcomes involve either an increase in disorder and malfunction or an increase in the complexity of a system (complexity here means the number of components, connectivity, and interaction strength).

Point/summary: *A budding framework for the study of functional heterogeneity is available. This framework is based on recognition of hierarchical structure of ecological systems. Analyses of heterogeneity and the resulting generalizations may lead to new and important insights about the various roles heterogeneity can play at different levels of organization.*

Scales

Although the notion of scale has already appeared several times, it is useful to state explicitly the relations among scales and other concepts. The relation is complex, and this book devotes a considerable amount of space to it (e.g., Chapters 3, 7, and 14). Therefore we limit ourselves here to a few general observations.

First, most of the discussion, generalizations, or suggestions make sense either in an entirely general sense or at a specified scale. For example, contrast properties of a heterogeneous habitat template in general to heterogeneity of a template defined at a particular scale. Whereas the general notion is scale-independent, heterogeneity of a specific template is scale-dependent and changes with altered scale. One must be aware of this distinction to avoid methodological traps.

From a functional viewpoint, scale can create heterogeneity from homogeneity even though only the size of the habitat patch changes. For example, if a homogeneous habitat is large, some individuals fail to find mates, whereas they would have if the habitat were small. Diseases that either kill their hosts or create permanent immunity require large habitats, so by the time the infection revisits a site new hosts are available.

Another aspect of scale appears when we try to sample systems whose components operate at several scales. Chesson (Chapter 7) reviews the effects that between-patch and within-patch heterogeneity may have on ecological systems. Barry and Dayton (Chapter 14) point out the importance of the unresolved issue of scale coupling between biotic and abiotic components of the system. Without recognizing the scale differences, quantitative and interpretational errors abound (see Chapter 4). Many community ecology studies are in fact burdened by this problem. The so-called community patterns may be a product of simultaneous sampling of organisms perceiving and responding to the environment at several scales. This factor, when unrecognized, presents a serious limitation and even invalidates many studies. On the other hand, recognition of the factor may lead to new insights into community organization (e.g., Kolasa, 1989).

Point/summary: *The importance of scale is pervading and commonly recognized, but some nuances such as unintended lumping of data sets spanning more than one scale need more attention.*

Filters

We have already stressed the importance of the system definition, i.e., its structure in terms of lower level entities and its integrity in terms of completeness. It is important to realize that environmental heterogeneity experienced by the whole system may be different from that experienced by the parts (Allen and Starr, 1982; Allen et al., 1984, see also earlier in this Chapter). It is because the external influences are filtered out by the higher level entity. A *filter* is thus always associated with the surface of the higher level entity. A paradoxical situation arises in that a lower level ecological entity may experience a coarser grain of environmental heterogeneity. For example, a fledgling may be cushioned against variation in food availability by a more intense search effort of its parents. However, both the parents and the fledgling, when viewed as a higher level entity, experience this variation directly. There is no filter. The degree to which this paradox becomes manifest depends on the integration of the system. Alternatively, the greater the joint effect of parts on modifying the signal of an environmental variable, the coarser is its perception for an individual part. For example, variations in temperature more affect individual snakes than snakes aggregated for warmth. On average, snakes in aggregation experience less variation than the aggregation as a whole.

Point/summary: *Effectiveness of filters is not fixed. It varies depending on the system's integration.*

Conclusions

We have reflected on a number of aspects of heterogeneity. That number turns out to be much higher than one book can cover. Not all of the issues can be clearly presented or related to others at present. It becomes obvious, however, that the richness of the existing and potential conceptual views may yield a comparable richness of empirical discoveries. Although our understanding of the concept may still be rudimentary, its complexity and heterogeneity is unquestionable. We believe that further progress in all areas of ecology will involve and perhaps even be dominated by considerations of heterogeneity.

Acknowledgments. This research was supported by NSERC Operating Grants to J.K. and C.D.R.

References

Addicott JF, Aho JM, Antolin MF, Padilla DK, Richardson JS, Soluk DA (1987) Ecological neighborhoods: scaling environmental patterns. Oikos 49:340–346

Aihara K, Matsumoto G (1986) Chaotic oscillations and bifurcations in squid giant axons. In Holden AV (ed) *Chaos*. Princeton University Press, Princeton, pp 257–269

Allen TFH, Starr TB (1982) *Hierarchy: Perspectives for Ecological Complexity*. University of Chicago Press, Chicago

Allen TFH, O'Neill RV, Hoekstra TW (1984) Interlevel relations in ecological research and management: some working principles from hierarchy theory. USDA Forest Service General Technical Report RM-110, pp 1–11

Atlan H (1985) Information theory and self-organization in ecosystems. Can Bull Fish Aquat Sci 213:187–199

August PV (1983) The role of habitat complexity and heterogeneity in structuring tropical mammal communities. Ecology 64:1495–1507

Beck SD (1968) *Insect Photoperiodism*. Academic Press, Orlando

Berger M, Mandelbrot B (1963) A new model for the clustering of errors on telephone circuits. IBM J Res Develop 7:224–236

Bonner JT (1988) *The Evolution of Complexity*. Princeton University Press, Princeton

Box GEP, Jenkins GM (1976) *Time Series Analysis: Forecasting and Control*. Holden-Day Publishers, San Francisco

Calder WA (1984) *Size, Function and Life History*. Harvard University Press, Cambridge, MA

Christensen NL (1985) Shrubland fire regimes and their evolutionary consequences. In Pickett STA, White PS (eds) *The Ecology of Natural Disturbance and Patch Dynamics*. Academic Press, Orlando, pp 85–100

Dawkins R (1982) *The Extended Phenotype*. Oxford University Press, Oxford

Denslow JS (1985) Disturbance-mediated coexistence of species. In Pickett STA, White PS (eds) *The Ecology of Natural Disturbance and Patch Dynamics*. Academic Press, Orlando, pp 307–323

Glass L, Mackey MC (1988) *From Clocks to Chaos: The Rhythms of Life*. Princeton University Press, Princeton

Gleick J (1987) *Chaos: Making a New Science*. Penguin Books, New York

Guiasu S (1977) *Information Theory with Applications*. McGraw-Hill, New York

Harris GP (1986) *Phytoplankton Ecology: Structure Function and Fluctuation*. Chapman & Hall, New York

Hebert PDN, Grewe PM (1985) *Chaoborus*-induced shifts in the morphology of *Daphnia ambigua*. Limnol Oceanogr 30:1291–1297

Huffaker CB, Shea KP, Herman SG (1963) Experimental studies on predation: complex dispersion and levels of food in an acarine predator-prey interaction. Hilgardia 34:305–330

Hutchinson EG (1957) Concluding remarks. Cold Spring Harbor Symp Quant Biol 22:415–427

Hutchinson EG (1967) *A Treatise of Limnology. Vol 2. Introduction to Lake Biology and the Limnoplankton*. Wiley, New York

Janzen DH (1986) Does a hectare of cropland equal a hectare of wild host plant? Am Nat 128:147–149

Johnson HA (1970) Information theory in biology after 18 years. Science 168:1545–1550

Jones CG, Hess TA, Whitman DW, Silk PJ, Blum MS (1986) Idiosyncratic variation in chemical defenses among individual generalist grasshoppers. J Chem Ecol 12:749–761

Kolasa J (1989) Ecological systems in hierarchical perspective: breaks in the community structure and other consequences. Ecology 70:36–47

Kolasa J, Biesiadka E (1984) Diversity concept in ecology. Acta Biotheor (Leiden) 33:145–162

Kolasa J, Pickett STA (1989) Ecological systems and the concept of biological organization. Proc Natl Acad Sci USA 86:8837–8841

Kotliar NB, Wiens JA (1990) Multiple scales of patchiness and patch structure: a hierarchical framework for the study of heterogeneity. Oikos 59:253–260

Lettvin JY, Maturana HR, McCulloch WS, Pitts WH (1959) What the frog's eye tells the frog's brain. Proc Inst Radio Eng 47:1940–1951

Levins R (1979) Coexistence in a variable environment . Am Nat 114:765–783

Lewontin RC (1974) *The Genetic Basis of Evolutionary Change.* Columbia University Press, New York

Lomnicki A (1982) Individual heterogeneity and population regulation In *King's College Sociobiology Group: Current Problems in Sociobiology.* Cambridge University Press, Cambridge, pp 153–167

Lorenz EN (1963) Deterministic nonperiodic flow. J Atm Sci 20:282–293

May RM (1974) Biological populations with nonoverlapping generations: stable points, stable cycles, and chaos. Science 186:645–647

May RM (1976) Simple mathematical models with very complicated dynamics. Nature 261:459–467

May RM, Oster GF (1976) Bifurcations and dynamic complexity in simple ecological models. Am Nat 110:573–599

Peters RH (1983) *The Ecological Implications of Body Size.* Cambridge University Press, Cambridge

Pickett STA, Kolasa J, Armesto JJ, Collins S (1989) The ecological concept of disturbance and its expression at various hierarchical levels. Oikos 54:129–136

Plowright RC, Galen C (1985) Landmarks or obstacles: the effect of spatial heterogeneity on bumble bee foraging behavior. Oikos 44:459–464

Saunders DS (1976) *Insect Clocks.* Pergamon Press, New York

Schmidt-Nielsen K (1984) *Scaling: Why Is Animal Size So Important.* Cambridge University Press, Cambridge

Shannon CE (1948) A mathematical theory of communication. Bell Syst Techn J 27:379–423, 623–656

Smith FE (1972) Spatial heterogeneity, stability, and diversity in ecosystems. Trans Conn Acad Arts Sci 44:309–335

2. Concept and Terminology of Homogeneity and Heterogeneity in Ecology

Robert P. McIntosh

The terms of an ideal scientific language and the concepts they express are clear and precise. Ambiguity is a barrier to progress. Ecologists have long felt the lack of such a terminology appropriate to a rigorous science. Historical evidence of the problem is readily seen in the vagaries of the meanings of words such as association, community, competition, niche, and a host of others.

The Ecological Society of America (ESA) established a Committee on Nomenclature during the 1930s to "make ecological terms more efficient tools of thought." The Committee published several lists of definitions between 1933 and 1952, but formal efforts to codify ecological terminology by committee were abandoned. Haskell (1940) called for "mathematical systematization" of key terms of ecology, but only Hutchinson (1957) heeded him. Although ecological glossaries or dictionaries were published by Carpenter (1956), Hanson (1962), Lewis (1977), and Lincoln et al. (1984), none seems to have stabilized the terminology of ecology; and problems of proliferation, vagueness, and confounding of terms noted by Greig-Smith (1961), among others, persist. Review of current journals, textbooks, and symposium volumes does not suggest convergence on meanings of significant terms, especially those that relate to complex or conceptually and theoretically significant topics.

In recent decades philosophers and historians of science have considered scientific language, the merit of scientific terms as metaphor, and even the usefulness of ambiguity, as suggested by Hardin (1960) in his evaluation of the phrase "the competitive exclusion principle." Niven (1982), a philosopher, offered to formalize basic concepts of ecology. Loehle (1988), an ecologist, argued that "philosophical tools," especially of terminology and linguistics, can assist in clarification of terminology and produce better ecological theory. The productivity for ecology of philosophy at large or of particular approaches to philosophy has been in dispute (McIntosh, 1987; Wiegleb, 1989; Shrader-Frechette and McCoy, 1990; Loehle, 1990).

Among the earliest and most persistent concepts in ecology, under various pseudonyms, were homogeneity and its antonym heterogeneity. The significance attached to these concepts is evident in the concern of some prominent ecologists that investigators embrace a comfortable homogeneity and avoid the difficulties posed by heterogeneity. MacArthur (1972), in his own earnest search for patterns "discovered and described by theory," deplored the tendency of many ecologists to study organisms in homogenized media and urged them to become more concerned with the structure of the environment and the effects of spatial and temporal patterns. Colwell (1984) criticized "cooking pot ecology" for treating ecological communities as homogeneous mixtures. He listed four assumptions for what he called the homogeneity of "old traditional" theoretical ecology, which was predicated on the following.

1. Homogeneity of component organisms
2. Continuity or unchangeability of component organisms
3. Homogeneity of interaction probabilities
4. Continuity of interactions assuming equilibrium

Colwell asserted that there is a growing, presumably recent, body of evidence that the spatial and temporal distributions of organisms, environments, and interactions are not homogeneous and that biological reality violates the assumptions of most ecological theory. Kareiva (1986) agreed and emphasized the importance of "patchiness" in contrast to homogeneity and allowed that spatial heterogeneity, in the rubric of patchiness, had "become a common theme in ecology only after modelers had formalized its significance."

The tendency of naturalists and their ecologist successors to seek order in nature has commonly led to assumptions concerning uniformity, homogeneity, and constancy. Traditional approaches to description and classification of the enormous variation recognized as biologists explored the complexities of nature expanded as concepts and methods became more diverse and as qualitative judgments were supplemented by quantitative description and analysis. I have undertaken here to examine ways in which protoecologists and their ecologist successors recognized and addressed problems of homogeneity or heterogeneity of ecological systems and to consider some of the terminology that developed in the process.

Eighteenth and Nineteenth Century Concepts of Heterogeneity in Biogeography and Protoecology

The great tradition of balance of nature, going back to antiquity, imputed to nature homogeneity, constancy, or equilibrium and abhorred thoughts of extinction and randomness. Order and coherence were commonly believed in Christian tradition to be characteristic of Divine providence. Such ideas die hard; and elements of them, although not usually claiming Divine sanction, persist in some views of nature. The transformation during the eighteenth century of ideas of taxonomy and biogeography from the relative homogeneity of the post-Ark world was traced by Browne (1983). A notable contributor was J.R. Forester, naturalist on the second voyage of Captain Cook (1772–1775). Forester recognized geographical entities, relating organisms to combinations of physical environment and noting that floras changed, becoming less similar as distance increased. Following earlier ideas and anticipating later concepts, he related distribution of organisms to "heat." The eighteenth century naturalist Peter Kalm recognized the effect of smaller-scale change in North American forests caused by tree windfall and described its effects in some detail, anticipating recent observations (McIntosh, 1961).

Ecology developed as a science in the context of nineteenth century biogeography, which was formalizing concepts and criteria for describing patterns of distributions of organisms. Large-scale faunal and floral regions, or realms, were the stuff of traditional biogeography; but a significant departure was transfer of attention from taxonomic criteria to another type of natural object recognized by aggregations of organisms defined by nontaxonomic criteria. The pioneer of this transformation was Alexander von Humboldt, who early in the nineteenth century described and measured, where possible, patterns or "zones" of vegetation and coincident climate. He categorized them by physiognomy of dominant plants. Humboldt was perhaps the first to use the term "association"—which was to become a key concept of ecology—to describe coincident occurrence of organisms. He also opened Pandora's box by using "botanical arithmetic," a numerical method for comparing floristic similarity of two areas. This method was a precursor to measures of similarity introduced early in the twentieth century (Browne, 1983; Nicolson, 1983).

Humboldt influenced the thinking of the major nineteenth century biologists who were the ancestors and founders of ecology. Two parallel sequences may be recognized—in terrestrial and aquatic biogeography. Humboldt's coarse segregation of recognizable units of terrestrial vegetation was pursued by numerous continental, particularly German, biologists, among them J.F. Meyer, J.F. Schouw, and A. Grisebach (Nicolson, 1983). They expanded Humboldt's ideas about vegetational units—their physiognomy, composition, and nomenclature. By the midnineteenth century, vegetation was considered as segregated into distinct coummunities or formations that were "ever-recurring elements" in the landscape and that "never transgressed the orderly and correct conposition of their kind" (Kerner, 1863). Such ideas permeated European biogeography and are the background of the proximate developers of plant ecology at the turn of the century: O. Drude, A.F.W. Schimper, and E. Warming. Warming pioneered the description of communities based on growth form rather than taxonomic criteria. He recognized horizontal differences, vertical strata, and temporal change such as periodicity and succession. All of these ideas were to become major components of heterogeneity in twentieth century concepts of ecology.

Edward Forbes, a British naturalist of the midnineteenth century, extended his studies of terrestrial molluscs and plants to the marine benthos, where by extensive dredging he recognized "provinces" and "zones" of organisms on the sea floor. Because they were known only from the dredge samples, he necessarily distinguished them by taxa and counts of individuals. Forbes's provinces were, he said, areas derived from "special manifestations of the Creative Power." His "zones" or "facies" were more mundane assemblages of organisms related to mineral composition or depth of the bottom sediments. Forbes recognized the ideas of association of organisms, of change in space in number of species and quantities of each, and of the effect of the organism on the environment (Rehbock, 1983). Studies of the marine benthos provided additional ideas about spatial distribution and communities of organisms and, in 1877, the earliest name for a distinctive community—biocoenosis—by the German zoologist Mobius (McIntosh, 1985).

Forbes and Ernst Haeckel, who in 1866 coined the word "ecology," also studied the distribution of organisms floating in the water column; but the refinements of method and the name "plankton" were not established until the 1880s. Victor Hensen developed sampling methods and early statistical descriptions of plankton and used them to estimate numbers and biomass of plankton in a much larger volume of seawater (Lussenhop, 1974). This work set the stage for a strenuous and perhaps first explicit dispute about homogeneity in ecology. Hensen's calculations of biomass of plankton, as in his earlier studies of earthworm distribution, assumed "uniform" (regular or equidistant) distribution in space. Given that assumption, beyond some threshold of sample size the mean number of organisms in an area of a given size was constant, and the number in a large area could be calculated from a sample of the water column under 1 m^2.

Haeckel vigorously attacked Hensen on the ground that his own experience showed that plankton was irregularly distributed, occurring in clumps; thus extrapolation from a sample was improper and, he said, "utterly worthless." The dispute persisted into the early decades of the twentieth century. Hensen argued that, in the sea, the species and number of individuals remain "to a certain extent" constant everywhere. The phrase, to a certain extent, is reminiscent of the ubiquitous "more or less," later commonly inserted in descriptions of homogeneity or constancy of populations or communities. Although Hensen claimed uniformity of distribution of plankton in the sea, he said it was not true of organisms on land because the environment varied over short distances and a small area could not be a true sample of a large area. Johnstone (1908), in a book on marine biology, noted the annual periodicity of plankton. He reviewed the dispute over uniform spatial distribution and questioned the validity of extrapolating from a sample if distribution is not uniform.

Early limnologists, like marine biologists, confronted the problem of homogeneity in lakes. A.J. Forel, the founder of limnology, implicitly recognized heterogeneity of a lake by starting his work in the profundal zone. E.A. Birge, during the 1890s and with his colleague C. Juday during the early 1900s, recognized vertical layering of water in lakes and heterogeneous distribution of plankton in both vertical and horizontal dimensions (Beckel, 1987). Beckel commented that Juday wrote in a letter an extended criticism of G.E. Hutchinson's later use of mathematics because it was based on the assumption of uniformity when nonuniform factors were involved. Forbes (1887) described the lake as a "microcosm" operating to produce a "just" equilibrium. As early as 1880, Forbes had developed a verbal model of predator—prey population cycle, such as appeared later in theoretical mathematical models, recognizing regular fluctuations in populations. Forbes was cognizant of heterogeneity within a lake but subordinated it to the understanding of the whole and emphasized regularity of plants and animals maintaining an "economical balance of supply and demand." As data on lakes accumulated, limnologists shared with terrestrial and marine ecologists the problem of heterogeneity in space and time. Where Murray and Pullar (1910) had presumed one sample per lake reasonable for bathymetrical survey, later limnologists faced the problem of increasing variation as more lakes were studied in detail showing heterogeneity. Like plant and marine ecologists, limnologists adopted the device of classification to manage heterogeneity. A number of methods based variously on benthic organisms, plankton, and trophic status were developed; and, for some, classification became an end in itself (Rodhe, 1975). Classes carry the assumption of internal homogeneity, an assumption deplored by Brinkhurst (1974).

Studies of streams lagged behind studies of lakes, perhaps because streams are clearly among the least homogeneous of natural entities. Kofoid (1897) noted the difficulty of classifying zones of macrophytes. He was among the first to examine the assumption of homogeneity in streams or, as he put it, "the assumption that a collection at a single point is representative of a large area," a much disputed issue among marine ecologists as well. Kofoid considered longitudinal, transverse, and chronological variation of streams; he introduced in the last of these variations the concept of "pulse," which he noted varied in duration and magnitude and added to the problems of heterogeneity facing ecologists.

Biogeographers and protoecologists, before there was a formal discipline of ecology, addressed problems of spatial distribution, or heterogeneity, qualitatively at a regional or landscape level. Particularly in aquatic systems, they were concerned about spatial heterogeneity or clumping and considered quantitative approaches to dealing with it. The crucial significance of homogeneity or its converse heterogeneity in space had already led to disputes among the forerunners of ecology during the late nineteenth century as it would during the twentieth century.

Heterogeneity in Early Twentieth Century Ecology

Aquatic ecologists addressed heterogeneity before 1900 in terms of its qualitative as well as its quantitative aspects. Terrestrial plant ecologists had addressed qualitative heterogeneity through much of the nineteenth century, but it was not until 1898 that F.E. Clements and his early associate Roscoe Pound discovered the quadrat and elaborated on quantitative heterogeneity of terrestrial vegetation. Pound and Clements (1900), in their study on the vegetation of Nebraska, addressed "abundance" of species in terms of number of individuals and mode of distribution, recognizing that some species occurred as solitary individuals and others "in patches of greater or less extent scattered about the floral covering." They developed a crude numerical formulation and a series of code names for degrees of patchiness or aggregation in space. Clements (1905), in pioneering work, spelled out the use of quadrats and simple quantitative methods in the study of communities. He addressed the issue of homogeneity, stating that in a "homogeneous" community the quadrat is representative in the same way a type specimen represents a species. In a less "uniform" (which he used as a synonym for homogeneous) community, the quadrat sample is less representative, and additional quadrats must be located in markedly different "zones, consocies, or patches" to sample its heterogeneity.

Clements, although sometimes advocating a view of the community as a homogeneous superorganism at least at the climax stage, was an early expositor of spatial and temporal heterogeneity of environment and organisms at various scales. He described *alternation*, or occurrence of formations at different places, as a response to heterogeneity of the surface of the earth. He created an arcane, and now largely abandoned, system of classifying heterogeneity of vegetation in space and time. He elaborated the process of temporal change, or *succession*. On a longer time scale, he was one of the earliest to consider the use of tree rings to examine climatic and vegetational change. He coined the term paleoecology and recognized that lake sediments were the result of long-term climatic and vegetational changes and stored in their depths the record of those changes. Given that Clements was one of the earliest to propose ideas of the terrestrial community in a quantitative sense, although he rarely used quantitative methods, it is not surprising that a certain ambiguity exists. He wrote, "In consequence, association in its largest expression, vegetation, is essentially heterogeneous, while in those areas which possess physical or biological definiteness, habitats and vegetation centers, it is relatively homogeneous."

During Clements' era multiple schools of phytosociology arose that recognized "homogeneity" or "uniformity" of vegetation by different criteria and on different scales. The early years of the twentieth century were the heyday of the "unit community theory" (Whittaker, 1962). A landmark statement of this idea was given at the Third International Botanical Congress in 1905: "An association is a community of definite composition, possessing a uniform physiognomy and growing in uniform habitat conditions." This statement corresponds with one definition of biotope later offered by the Committee on Nomenclature of the ESA. The key problem of specifying the meanings of "definite" and "uniform" exercised ecologists for decades, calling for much use of the familiar modifier "more or less."

Thus during the formative years of ecology the concept, recognition, and terminology of homogeneity and heterogeneity were primary concerns of ecologists. Words relating to the concept were often used without explicit definition. They sometimes referred to membership in a class or a qualitative or logical difference, or they were sometimes concerned with quantitative differences. "Homogeneous" was often used interchangeably with "uniform" and commonly implied uniform or regular spacing between individuals. It also connoted having shared characteristics, usually species or at least dominant species. Uniformity also implied lack of change (constancy or, at least, similarity) temporally or spatially.

Sampling, Quantitative Methods, and Homogeneity

The subjective ideal of uniformity or homogeneity of communities persisted during the early decades of the twentieth century. "Uniform" and "homogeneous" were used to describe different attributes of populations or communities. Uniform was sometimes used to mean equidistant spacing of individual organisms and the constancy of the sampling results. The Committee on Nomenclature of the ESA defined homogeneity in this sense of uniform or regular spacing of individual organisms but also defined it as a property of similar composition of a community. Plant ecologists had long used homogeneity (sensu similarity) of sets of samples as a criterion of community. The Committee on Nomenclature described a measure of homogeneity based on a combination of constancy of occurrence of species in several stands and frequency of species in individual stands. "Homoteneity" was later proposed to define the similarity of a group of separate stands (Dahl, 1960), leaving "homogeneity" to describe the similarity of samples within a single stand.

Proliferation of quantitative methods or sampling and analysis in aquatic biology and terrestrial ecology during the twentieth century exacerbated the problems of assessing homogeneity. Many early ecologists, notably Continental plant ecologists, introduced quadrat samples (one sample per stand) into a tradition of expert eyeballing of homogeneous stands based on long experience. Assurances of homogeneity (relative or more or less) of a study area are still commonly encountered in current publications often with little or no documentation other than the intuition or expert eyeball of a new generation of ecologists.

An early approach to assessing homogeneity or, as it sometimes came to be called, "similarity," among stands or samples of a community was the introduction of a numerical coefficient by the plant ecologist Jaccard in 1902 followed by others through the 1930s, reaching a crescendo during the 1950s and 1960s (Greig-Smith, 1964). The burden of these methods, as an elaboration of the plant geographers' "botanical arithmetic," is to compare samples to determine if they share the same kinds (and sometimes quantities) of species or other attributes. If the samples share the same species, or quantities thereof, they are presumably drawn from a homogeneous community.

A variant on the quantitative approach to homogeneity of distribution was the introduction in 1907 of a coefficient of association by the entomologist-limnologist S.A. Forbes. Forbes' coefficient was an index of the frequency of joint association of species. Forbes' coefficient was criticized by Michael (1921) on statistical grounds, but Michael gave Forbes credit for recognizing the quantitative nature of the problem of association of species. Used in numerous subsequent indices, a measure of association between species, positive or negative, indicates departure from randomness and suggests patterns in the spatial and temporal distribution of organisms (Greig-Smith, 1964; Janson and Vegelius, 1981). Absence of association (or correlation) is a criterion of homogeneity, or lack of pattern, in later uses of statistical ecology.

Numerical methods of assessing heterogeneity, or pattern, as aspects of it came to be called, were based on data derived from samples usually of a given area and number. Perhaps the earliest detailed examination of sampling and the effect of quadrat size on sample data and the problem of assessing homogeneity was that of C. Raunkiaer, a Danish plant ecologist. Beginning during the early 1900s, Raunkiaer studied the distribution of frequency of occurrence of species in samples. In Raunkiaer's usage, a formation is homogeneous if its composition, as determined from the samples, is everywhere the same. As a result of his studies, Raunkiaer formulated a "law" based on the distribution of species frequencies for five 20% class intervals. This method, he said, produced a reversed, J-shaped curve, increasing in the highest frequency class. Raunkiaer's "law" was much discussed and criticized but persisted as an indication of homogeneity of a community (McIntosh, 1962). Reference to Raunkiaer's early work has reappeared in studies of species distributions that assert an underlying pattern (Hanski, 1982; Gotelli and Simberloff, 1987).

Community Concept and Homogeneity

Comparison of different stands of a presumably homogeneous community, as a test of homogeneity in the sense of similarity, using numerical methods was further developed during the 1930s and 1940s. Perhaps the best known and most widely used of the indices proposed is that devised by Sorenson (1948). Sorenson's method compared samples or stands on the basis of total species content, avoiding the subjective judgments commonly used when comparing stands. Similar methods were advocated for insect synecology (Renkonen, 1949). Various such coefficients have been proposed and commonly used, often in conjunction with multivariate analysis (Campbell, 1978).

The underlying idea of some uses of homogeneity was the ubiquitous concept of the community unit. According to this concept the landscape consisted of representative stands, homogeneous in some respect within the community unit but heterogenous as between different units and thus subject to classification into a limited number of types. The degree of heterogeneity depended on the underlying concept of community. The concepts of constant species, fixed minimal area of community, and asymptotic species–area curves were associated with the uniform community. Into this world of recognizable, relatively homogeneous entities during the 1920s appeared the so-called individualistic concept of H.A. Gleason. Gleason's idea was heterogeneity rampant. Its essence was that the environment varied continuously, each species had individualistic properties, and the community was an individualistic concatenation of environment and coincident aggregation of species with a large element of chance. It is not surprising that Gleason's idea went largely unsung and unaccepted until the 1950s (McIntosh, 1967, 1975, 1985).

The concept of a community continuously changing with the environment also developed in studies of marine benthic ecology. Stephen (1933) described gradual transitions without separation into unit communities. Subsequently, Sanders (1960) wrote that the benthic fauna of Buzzards Bay constituted a continuum similar to that which was by then commonly described for terrestrial vegetation. Issues of heterogeneity and conmmunity concept are perpetuated in current animal ecology. Colwell (1984) and Price (1984) described the resurgence of interest in heterogeneity in ecology and resurrected the individualistic idea of Gleason. Haila (1952) wrote that the existence of natural communities assumed in some theories was in doubt; James and Boecklen (1984) wrote that a bird community, like

Gleason's association, is merely a coincidence. Wiens (1981) questioned the validity of community patterns recognized on the basis of single samples, and Brown (1987) reported enormous variation in assemblages of desert rodents on all spatial and temporal scales. The long unresolved question of the community as in some sense homogeneous in composition and structure and differentiable from other communities reappeared in more recent symposia. Schoener (1986) wondered if each community was unique or if there are types. Diamond and Case (1986) asked if communities could be partitioned among a modest number of types. Gleason's individualistic concept is commonly thought to be the ancestor of the continuum, or gradient, approaches to community in which the community is viewed as changing gradually in relation to spatial or environmental gradients (McIntosh, 1967; Whittaker, 1967). It is not surprising to find Gleason's concept continuing in plant ecology and surfacing in the literature of animal community ecology as questions about heterogeneity persist.

The idea of homogeneity of a community in the sense of spatial distribution had been questioned early by perceptive ecologists. Cooper in 1913 and Watt in 1924 had studied patterns of spatial distribution in forests and agreed that the forest was composed of a "mosaic" of patches of different ages following small-scale disturbances at different intervals (McIntosh, 1985) . The earliest examination of heterogeneity of distribution in space using probability statistics was during the 1920s. Plant ecologist H.A. Gleason and marine biologist T. Svedberg (Clark and Evans, 1954) used results of samples and compared them to those expected by chance to show that most organisms are nonrandomly distributed and are most often aggregated or clumped. This point focused attention on the distribution of organisms in space, and numerous studies conducted between the 1930s and 1960s explored statistical properties of samples and their relation to heterogeneity, or pattern, of distribution (Goodall, 1952, 1962). A.R. Clapham, during the 1930s, used measures of variance as an index of spatial heterogeneity under the unclear rubrics of over- and underdispersion (Greig-Smith, 1964).

With the switch to probability statistics, various tests of the statistical properties of sampling methods, and the exposition of pattern in distribution of organisms, ecologists pursued the problems of spatial heterogeneity using colored disks, nuts and bolts, artificial populations of diverse sorts, and finally computer-constructed populations. The importance of spatial heterogeneity and the problems of its measurement in ecology became increasingly apparent during the 1940s and 1950s. Cole (1949) elaborated on earlier measures of interspecific association. Dice (1952) and Clark and Evans (1954) switched from quadrat sampling to measurement of spacing between organisms about the same time that the plant ecologists John Curtis and Grant Cottam were turning to distance measurements (Greig-Smith, 1964). Hutchinson (1953) provided his typically insightful summary of the concept of pattern in ecology, adopting the terms "superdispersion" for clumped distributions and "infradispersion" for even distributions. Dispersion became a common synonym for distribution (Pielou, 1977). Greig-Smith was probably the first to make a virtue of using scale to study spatial heterogeneity, or "pattern" (Greig-Smith, 1952, 1964). He anticipated the urging of modern students concerning the importance of scale in ecology: "It is indeed the varying behavior of non-random distributions with different sample sizes that enables information to be obtained on the scale at which non-randomness is operating."

The switch to the use of probability statistics transformed the dual meanings of spacing to a triumvirate.

1. Uniform, regular, and over- or infradispersed describing equal spacing among individuals
2. Random distribution of individuals independently of each other (sometimes called homogeneous)
3. Clumped, aggregated, or superdispersed distribution

It soon became apparent that measures of distribution were influenced by quadrat size. Elaborations of the idea of pattern led to additional concepts and terminology. *Intensity* specified the degree of change in density from place to place in a community; *grain* described the scale of patch size (Pielou, 1977). Hutchinson (1978) advanced the idea of spatial patterns as ranging from "homogeneously diverse" to "heterogeneously diverse." This discription added to the character of the habitat the scale of movement of the organism using it. *Homogeneously diverse* is a pattern of elements in space that is small relative to the normal range of movement of an organism. *Heterogeneously diverse* is a pattern in which the elements are large relative to the range of movements of an organism. Thus the same pattern may be scaled as either, depending on the scale of movement of the percipient organism. Hutchinson provided a definition of *biotope* as an area that is horizontally homogeneously diverse relative to the larger mobile organisms within it.

By the 1950s it was abundantly clear that most organisms were distributed in space in clumps or aggregates and that this form of spatial heterogeneity had profound effects on understanding species populations, their interactions with other species, and consequent community properties and functions. Standard expressions of mean density were criticized because of the underlying assumption that each individual occupied the same area and that they were equally spaced. Extensive studies of distribution of organisms, recognition of aggregation, size and spacing of such aggregations, and concern about distance of an organism to its neighbors appeared. MacArthur (1957) began his distinguished career with a study of distribution of warblers vertically and horizontally in space and time. Pimentel (1961) studied the influence of pattern of plant distribution on insects. MacArthur and Pianka (1966) examined optimal use by animals of a patchy environment. Lloyd (1967) introduced his concept of "mean crowding," modifying the misleading mean density per area by a factor measuring the effect of aggregation. Marine littoral ecologists emphasized the importance of disturbance and patch dynamics to community pattern and regeneration (Levin and Paine, 1974). Taylor elaborated his "power law," which he claimed relates degree of clustering to density (Taylor, 1987). Rosenzweig et al. (1975) studied patterns of species diversity with respect to habitat complexity as an aggregate measure of habitat variables. Spatial heterogeneity, or pattern, has the concomitant of scale; and in addition to recognizing heterogeneity, it is essential to assess its scale. The scale of pattern, patch size, or grain is addressed by pattern analysis. Greig-Smith (1952) introduced a technique for detecting pattern and its scale by examining variance in a series of sample quadrats increasing in size. Marine biologists introduced spectral analysis to characterize scale in ecosystems in space and time (Platt and Denman, 1975). Problems with the measurement and interpretation of scale of pattern persist, and not all methods are equally successful in detecting the elements—patch and gap—of a pattern (Carpenter and Cheney, 1983).

Clustering, Gradient Analysis, and Ordination

Recognition that species and environments are enormously heterogeneous in space and time, and the difficulty dealing with it, provided the rationalization for intense effort to order heterogeneity by means of numerical methods of clustering, gradient analysis, or any of a large battery of methods of multivariate analysis (McIntosh, 1985). Beginning during the mid-1950s with the work of J.T. Curtis, D. Goodall, and R.H. Whittaker, patterns of curves, or clouds of points in one-, two-, or three-dimensional arrays were constructed and analyzed to organize heterogeneity into interpretable patterns. The familiar idea of interspecific association was the basis of an early method of clustering—association analysis (Williams and Lambert, 1959). The essence of such methods is to reduce heterogeneous groups to more homogeneous groups in which there are no significant associations between species. Some argued that creating such homogeneous groups was a desirable preface to methods of ordination.

Simple plots of species quantities on environmental gradients (direct ordinations) or axes derived from measures of species quantities (indirect ordinations) initiated the technique of ordination, which was used by plant ecologists to buttress a concept of continuously varying communities (Whittaker, 1967). More sophisticated mathematical methods of multivariate analysis were introduced into ecology in the hope of assisting ecologists to address heterogeneity. For three decades ecological patterns were discerned by association analysis, principal component analysis, detrended correspondence analysis, and cluster analysis, alone or together.

Heterogeneity of ecological material was amplified by heterogeneity of data input, data transformation, multiple similarity coefficients, and multivariate methods in search of "robust" interpretations (Gauch, 1982). Heterogeneity of simple two- or three-dimensional geometry was complicated by the geometry of many dimensions, euclidean and otherwise. Ventures into many dimensions and the return to a few have complicated the ecological scene (Hutchinson, 1957; Gauch, 1982). During the present era heterogeneity is taken for granted; and pattern, patchiness, and scale are key words of ecology, with "hierarchy," "fractals," and even "chaos" waiting in the wings as new means of dealing with heterogeneity.

Mathematical Population Ecology and Community Theory

During the 1920s formal mathematical population theory emerged in the work of Raymond Pearl, A.J. Lotka, and Vito Volterra (Kingsland, 1985). In the earliest models the parameters did not vary in relation to spatial or temporal dimensions. The models are described as "homogeneous" in respect to space and time (Smith, 1972). They were, in effect, the epitome of the homogeneity described by Colwell (1984). Lotka, in fact, developed his ideas on the basis of an analogy between biological populations and chemical systems that were homogeneous in contrast to biological systems. Lotka's ideas were seized on by a Russian biologist, G.F. Gause, who extended the mathematical theory into experimental manipulations of mixed populations of organisms. A critical aspect of some of Gause's experiments, as in later experiments, was the homogeneity or heterogeneity of the experimental environment. The classic population theory and the community theory based on it were equilibrium theories (Chesson and Case, 1986), and homogeneity of environment and random distribution in space were essential assumptions that, in the view of many ecologists deeply immersed in the heterogeneity of ecological systems, greatly limited their utility (Wiens, 1989).

Theoretical population ecology developed during the 1920s and 1930s from a "law" of population growth under equilibrium conditions with only three functional parameters and assuming constant and homogeneous individuals and environments (Kingsland, 1985). Although frequently criticized on the grounds of these homogenizing assumptions, this "law" was expanded into an elaborate body of theory and experiment extending to two species interactions, especially competition and predation (May, 1976). Competition theory and experiment, supplemented by extended inference, provided a law, or principle, of competitive exclusion that became the basis of a new stage in homogeneity of theoretical community ecology during the 1960s and 1970s. The term constant was reintroduced to describe the interaction coefficients between pairs of species; and equilibrium was the condition in which the interactions occurred leading to stable endpoints. Theoretical constructs assumed random distribution and interactions. The key word frequently encountered was "pattern," and effective science was equated with a search for pattern. Communities were assembled by rules as widespread patterns in nature were recognized. Kingsland (1985) noted that Robert MacArthur, the major figure in this school of community ecology, asserted that groups to be studied must be large enough that history played a minimal role but small enough that the patterns were clear, thus minimizing heterogeneity in time.

Ecological Terminology of Homogeneity and Heterogeneity

Despite hopeful assumptions about balance, uniformity, homogeneity, constants, equilibrium, and in more refined circles *ceteris paribus*, protoecologists and subsequently self-conscious ecologists faced the problems of heterogeneity in a variety of ways. Left out of this account herein is the increasing recognition of genetic variation and the mutual recriminations sometimes encountered between ecologists and geneticists for failing to recognize genotypic heterogeneity and ecological heterogeneity, respectively. The inherent heterogeneity of ecological systems is evident in the myriad terms developed by the often maligned descriptive ecology. It is more formally evident in the substantial number of terms used by ecologists that connote homogeneity or heterogeneity, spatially, temporally, or abstractly.

A crude index of the interest of ecologists in questions of homogeneity and heterogeneity is the lists of selected terms and the frequency with which they are indexed in ecology journals (Table 2.1) and several symposium volumes (Table 2.2). Obviously, the interpretations are restricted because the numbers represent varying spans of time, volumes of publication, and the idiosyncracies of the makers of indices. Some of the terms have rather heterogeneous meanings, meanings that have changed over the decades or entered the ecologist's lexicon for only a short time, later to be forgotten or replaced by a new term. Grain, for example, as a term referring to spatial heterogeneity is of relatively recent vintage and would not be expected in journals before the 1960s. Although it was introduced in theoretical mathematical ecology during the mid-1960s, it does not appear in indices of recent ecological journals surveyed or of the recent symposium volumes.

Most of the terms used refer to heterogeneity because homogeneity by its very nature requires little description. Terms related to homogeneity (e.g., constant, equilibrium, stability, uniformity) were indexed less frequently in journals than those suggesting heterogeneity. Stability was most frequently indexed, particularly after 1970, although its meaning, like that of equilibrium, was variable. Terms that are general in meaning (e.g., distribution and variation) are indexed frequently. "Variation" is the leading

Table 2.1. Terms Indexed in Journals

Term	J Ecol 1–20 (1913–1932)	J Ecol 21–50 (1913–1962)	Ecol Monogr 1–20 (1931–1950)	Ecol Monogr 21–40 (1951–1970)	Ecology 1–30 (1920–1949)	Ecology 31–50 (1950–1969)	Ecology 51–60 (1970–1979)	Ecology 61–67 (1980–1987)	Total
Area	0	0	1	1	0	1	5	1	9
Constant(ce,cy)	3	3	1	0	0	0	3	1	11
Dispersion	0	3	0	1	1	4	5	7	21
Distribution	7	22	28	27	79	95	71	8	317
Diversity	0	2	0	13	0	17	49	29	100
Equilibrium	0	0	1	2	1	5	5	8	22
Gradient	0	2	0	16	0	13	18	9	58
Grain	0	0	0	0	0	0	0	0	0
Heterogeneity(homo)	0	1	0	0	5	6	5	8	25
Patch(iness)	0	1	0	0	0	0	5	43	49
Pattern	0	24	1	5	0	11	42	6	89
Scale	1	0	0	1	0	0	1	2	5
Similar(ity)	0	1	0	5	0	2	6	5	19
Stable(ity)	0	2	1	1	1	6	23	20	54
Strata(ification) or layer(ing)	3	7	8	1	3	4	6	0	32
Structure(al)	1	2	2	7	5	14	52	4	87
Uniform(ity)	0	1	0	0	0	2	0	1	4
Variation(ability)	1	5	1	5	9	12	11	11	55
Zone(ation)	10	1	0	14	7	6	6	3	47
Total	26	77	44	99	111	182	299	176	

Table 2.2. Terms Indexed in Recent Books

	Cooley & Golley (1984)	Price et al. (1984)	Strong et al. (1984)	Diamond & Case (1986)	Kikkawa & Anderson (1986)	Total
Area	1	1	0	0	1	3
Constant(ce,cy)	2	0	1	0	1	4
Dispersion	0	0	4	4	0	8
Distribution	22	7	0	0	0	29
Diversity	36	11	1	88	19	155
Equilibrium	3	15	22	60	6	106
Gradient	19	0	0	20	1	40
Grain	0	0	0	0	0	0
Heterogeneity(homo)	0	1	16	28	0	45
Patch(iness)	2	15	20	50	0	87
Pattern	6	0	0	0	2	8
Scale	0	18	0	35	0	53
Similar(ity)	0	0	1	3	1	5
Stable(ity)	19	3	10	29	8	69
Strata(ification) or layer(ing)	0	1	1	0	0	2
Structure(al)	0	1	0	0	1	2
Variation(ability)	28	41	0	110	0	179
Zone(ation)	0	0	2	11	0	13
Total	138	114	80	438	40	

term indexed in recent symposium volumes (Table 2.2), and "distribution" is the leading term in journals (Table 2.1). "Structure," which is similarly general in meaning, is relatively frequently (fourth) cited in journals but rarely (next to last) in the symposium volumes. Oddly, more than one-half (n = 52) of the total journal entries for "structure" were in *Ecology* 1970–1979, but *Ecology* 1980–1987 indexed "structure" only four times (Table 2.1). "Structure" has varied meanings and is a catchall term to describe abstract patterns of organization or spatial (especially vertical) distribution. A number of terms were introduced with new developments in ecology. "Gradient" became frequent after 1950 with the resurgence of the ideas of Gleason and the introduction of gradient analysis by Whittaker. "Pattern," a term much favored by theoretical community ecologists of the 1960s and 1970s, was frequently indexed during that period as renewed interest developed in heterogeneity. Although "pattern" commonly connotes spatial distribution, it may be used to describe an abstract relation such as trophic pattern (= trophic structure). "Patchiness," a more specific term relating to spatial distribution when in clumps or aggregates, increased in *Ecology* after 1980, coincident with a decrease in the use of "pattern". "Diversity" ranked second in both journal and symposium indexes and was frequently indexed only after 1960. Diversity continues as a phenomenon of major interest to ecologists, but the flurry of interest in diversity (or evenness) indices prevalent during the 1960s has abated.

The most striking thing about the terms indexed in the several symposia during the mid-1980s is the difference, generally and specifically, among the symposium volumes. The volume edited by Diamond and Case (1986) indexed these terms more frequently than the other four volumes together. The heterogeneity of ecological verbiage is evident in that the term "variations," which is indexed most frequently in the symposia collectively, does not appear at all in two of the volumes. "Diversity," the second term in total frequency of occurrence, appears only once in one of the volumes. "Equilibrium" and "stability" are both frequently indexed but with different incidences from volume to volume. "Patch" is much favored over "pattern," which was infrequently indexed, the reverse of the case in journal indices. Perhaps the most striking comparison is between the publications of Diamond and Case (1986) and Strong (1984). Their volumes are roughly the same size, have indexes similar in length, are multiauthored, include diverse taxa, and, as their titles suggest, deal with the same topic. The former indexes terms relating to homogeneity and heterogeneity five times more frequently than the latter; its two leading terms ("variation" and "equilibrium") rarely appear in the other volume. The Diamond and Case book indexes "scale" relatively frequently (fifth), whereas the book of Strong does not index it at all. The Diamond and Case book indexes "diversity" 88 times and that of Strong only once. Ecologists have apparently not concurred on terms relevant to the general concepts of homogeneity and heterogeneity. To adapt Hutchinson's phrase, the ecological lexicon is heterogeneously diverse.

In Quest of Homogeneity

A traditional belief extending from natural history to early ecology was that nature is balanced and orderly, and it was commonly assumed that nature, or at least substantial aspects of it, is homogeneous. As biologists and ecologists came to recognize how intrinsically heterogeneous, changeable, and complex nature is, hope shifted from an intrinsically ordered nature to one that at least could be made to appear orderly by sufficiently experienced and astute individuals. Among the earlier beliefs was that an environmental control led the distribution and aggregations of organisms to form homogeneous communities. Then it was thought that biotic activities, particularly of widespread organisms called dominants, would create relatively homogeneous conditions. The earliest method of determining relatively homogeneous areas was classification, which was widely adopted for terrestrial and aquatic systems. Numerous systems of classification were devised over many decades to bring order into the heterogeneity of the earth's environment and its organisms. In varying degrees, these systems succeeded for certain places, times, and taxa. None proved generally satisfying in large part because the task at hand was so complex and the bases of classification so divergent.

Among early ecologists, the phytosociologists, marine benthic ecologists, and limnologists converged on classifying units on the basis of some criteria of homogeneity or similarity. Phytosociologists asserted the reality of the "concrete" community comprised of representative examples with an internal homogeneity defined by diverse attributes of the species present. Prominent among them was the American F.E. Clements. Although Clements astutely recognized heterogeneity in time and space on several scales, as noted earlier his theoretical framework was based on large-scale units of vegetation as natural areas with similar or identical climatic characteristics, including a vegetation "formation" with characteristic dominant species and development that he described as an organism. Although Clements' homogeneous formations, by the standards of more recent ecologists, included an enormous amount of heterogeneity, the overall concept provided a reassuring order and homogeneity that was widely adopted by American ecologists beyond the bounds of phytosociology. Among the terms frequently encountered during the early decades of the twentieth century were "constant" or "constance," referring to occurrence of a species in stands of a community; "homogeneity" and "similarity," describing like species composition or habitat characteristics; "stability," implying constancy in time; and "uniformity," meaning likeness in spacing.

Widespread acceptance of the idea of relatively homogeneous communities and underlying assumptions of stability or equilibrium brought forth substantial criticism at critical periods in the history of ecology. During the early decades of the twentieth century, Forrest Shreve, H.C. Cowles, W.S. Cooper, and A.S. Watt questioned the Clementsian assumptions of homogeneous unit communities. Gleason propounded his nominalist ideas of individualistic species, environments, and communities. Gleason's ideas were ignored but emerged during the 1950s in the guise of continuum or gradient in the work of Curtis and Whittaker. These concepts were associated with expansion of quantitative measures of spatial and temporal heterogeneity, and ideas of homogeneity shifted largely to the scale of heterogeneity (McIntosh, 1985).

Hutchinson (1957) addressed problems of heterogeneity in aquatic communities. He developed the idea of "fugitive" species that seized on patches of disturbance in the manner of pioneering or colonizing species. He reviewed the concept of pattern in ecology (Hutchinson, 1953). Hutchinson and others began an effort to incorporate heterogeneity in the homogeneous theoretical models urged on ecologists. Smith (1972) analyzed classic theoretical population models and asserted that it was unlikely that homogeneous models were much help and that heterogeneity was the basis of stability and diversity. Goodall (1962) wrote: "In most ecosystems, spatial heterogeneity is intrinsic and inseparable from their mode of functioning." By the mid-1970s animal ecologists were revolting against the hegemony of theoretical community ecology predicated on homogeneous classic theoretical models (Colwell, 1984; McIntosh, 1987; Wiens, 1989) and urging ecologists to attend to the problems of heterogeneity in all its guises and at all scales—microhabitat to landscape.

The history of ecology has been marked by hopes of finding or devising homogeneity to limit the frustrating heterogeneity of nature. Cowles (1904) described ecology as chaos in 1903, and the mathematics of chaos appealed to some ecologists of the 1980s. Stanley Cain, during the 1940s, expressed the fear that ecological entities may be too complex to comprehend. L.B. Slobodkin, during the 1980s, suggested that the problem of heterogeneity was frustrating ecologists, leading to polemics. Numerous ecologists of the 1980s reacted to the efforts of previous decades, based on mathematical population theory, to homogenize ecological communities by urging "pluralism" in recognition of heterogeneous approaches to the intrinsic heterogeneity of ecological phenomena, and some asserted the relevance of the individualistic ideas of Gleason (McIntosh, 1987). Homogeneity of ecologists has been no less difficult to achieve than homogeneity of populations, communities, ecosystems, or models thereof and probably is undesirable in any case. A putative unified ecology based on a paradigm of predictive or theoretical power of a single approach, such as has been urged on ecology, does not seem in prospect although candidate theories are still being developed. Pluralism and proliferation of theories and approaches may be the appropriate state for ecology in contrast to the dictum that it should "mature" into a "hard," mathematical, unified theoretical science. Recognition of the desirability of pluralistic approaches to ecology should not preclude efforts to clarify terminology and the concepts embodied by it. If ecologists differ, they should differ on reasonably clear grounds. It helps during an argument to be arguing about the same thing. If ecologists include heterogeneous groups, these groups should be recognized as different from, not better than, each other.

Acknowledgments. The author is indebted to his colleagues Dr. Stephen Carpenter and Dr. David Lodge for their patience in reading the manuscript and their helpful comments.

References

Beckel A (1987) Breaking new waters: a century of limnology at the University of Wisconsin. Trans Wisc Acad Sci Arts Lett (special issue) 122p.
Brinkhurst RO (1974) *The Benthos of Lakes.* St. Martin's Press, New York

Browne J (1983) *The Secular Ark*. Yale University Press, New Haven

Brown JH (1987) Spatial and temporal variation in desert rodent assemblages. Bull Ecol Soc Am 68:271

Campbell BM (1978) Similarity coefficients for classifying releves. Vegetatio 37:101–109

Carpenter JR (1956) *An Ecological Glossary*. Haefner, New York

Carpenter SR, Cheney J (1983) Scale of spatial pattern: four methods compared. Vegetatio 53:153–160

Chesson PL, Case TJ (1986) Overview: nonequilibrium community theories: chance, variability, history, and coexistence. In Diamond J, Case TJ (eds) *Community Ecology*. Harper & Row, New York, pp 229–239

Clark PJ, Evans FC (1954) Distance to nearest neighbour as a measure of spatial relationships in populations. Ecology 35:445–453

Clements FE (1905) *Research Methods in Ecology*. University Publishing Company, Lincoln, Nebraska

Cole LC (1949) The measurement of interspecific association. Ecology 30:411–424

Colwell RK (1984) What's new? Community ecology discovers ecology. In Price PW, Slobodchidoff DN, Gaud WS (eds) *A New Ecology: Novel Approaches to Interactive Systems*. Wiley, New York, pp 387–396

Cooley JH, Golley FH (eds) (1984) *Trends In Ecological Research for the 1980s*. Plenum, New York

Cooper WS (1913) The climax forest of Isle Royale, Lake Superior, and its development. Bot Gaz 55:1–44, 115–140, 189–235

Cowles HC (1904) The work of the year 1903 in ecology. Science 19:879–885

Dahl E (1960) Some measures of uniformity in vegetation analysis. Ecology 41:805–808

Diamond J, Case TJ (eds) (1986) *Community Ecology*. Harper & Row, New York

Dice LR (1952) Measure of the spacing between individuals within a population. Contr Lab Vert Biol Univ Mich 55:1–23

Forbes SA (1880) On some interactions of organisms. Bull Illinois State Lab Nat Hist 1:3–17

Forbes SA (1887) The lake as a microcosm. Bull Sci Assoc Peoria Illinois 1887:77–87

Gauch H Jr (1982) *Multivariate Analysis in Community Ecology*. Cambridge University Press, Cambridge

Goodall DW (1952) Quantitative aspects of plant distribution. Biol Rev 27:194–245

Goodall DW (1962) Bibliography of statistical plant sociology. Excerpta Botanica Section B, B and 4, pp 16–322

Gotelli NJ, Simberloff D (1987) The distribution and abundance of tallgrass prairie plants: a test of the core-satellite hypothesis. Am Nat 130:18–35

Greig-Smith (1952) The use of random and contiguous quadrats in the study of structure of plant communities. Ann Bot Lond NS 16:293–316

Greig-Smith P (1961) Ecological terminology. In Gray P (ed) *Encyclopedia of Biological Science*. Reinhold, New York

Greig-Smith P (1964) *Quantitative Plant Ecology*. Butterworth, London

Haila Y (1952) Hypothetico-deductivism and the competition controversy. Ann Zool Fenn 19:255–263

Hanski I (1982) Dynamics of regional distribution: the core and satellite hypothesis. Oikos 38:210–221

Hanson HC (1962) *Dictionary of Ecology*. Philosophical Library, New York

Hardin G (1960) The competitive exclusion principle. Science 131:1292–1297

Haskell EF (1940) Mathematical systematization of "environment," "organism," and "habitat." Ecology 21:1–16.

Hutchinson GE (1953) The concept of pattern in ecology. Proc Acad Natl Sci Phila 105:1–12

Hutchinson GE (1957) Concluding remarks. Cold Spring Harbor Symp Quant Biol 22:415–427

Hutchinson GE (1978) *An Introduction to Population Ecology.* Yale University Press, New Haven

James FC, Boecklen WJ (1984) Interspecific morphological relationships and the densities of birds. In Strong DR Jr, Simberloff D, Abele LG, Thistle AB (eds) *Ecological Communities: Conceptual Issues and the Evidence.* Princeton University Press, Princeton, pp 458–477

Janson S, Vegelius J (1981) Measures of ecological association. Oecologia 49:371–376

Johnstone J (1908) *Conditions of Life in the Sea: A Short Account of Quantitative Marine Biological Research.* Cambridge University Press, Cambridge

Kareiva P (1986) Patchiness, dispersal, and species interactions: consequences for communities of herbivorous insects. In Diamond J, Case TJ (eds) *Community Ecology.* Harper & Row, New York, pp 192–206

Kerner vMA (1863) Das Pflanzen der Donaulander. Wagner, Innsbruck. Translated by Conard HS (1950) *The Background of Plant Ecology.* Iowa State College Press, Ames

Kikkawa J, Anderson DJ (eds) (1986) *Community Ecology: Pattern and Process.* Blackwell, Melbourne

Kingsland S (1985) *Modeling Nature: Episodes in the History of Population Ecology.* University of Chicago Press, Chicago

Kofoid C (1897) On some important sources of error in the plankton method. Science 6:829–832

Levin SA, Paine RT (1974) Disturbance, patch formation, and community structure. Proc Natl Acad Sci USA 71:2744–2747

Lewis WH (1977) *Ecological Field Glossary: A Naturalist's Vocabulary.* Greenwood Press, Westport, CT

Lincoln RJ, Boxshall GA, Clark PF (eds) (1984) *A Dictionary of Ecology, Evolution and Systemics.* Cambridge University Press, Cambridge

Lloyd M (1967) Mean crowding. J Anim Ecol 36:1–30

Loehle C (1988) Philosophical tools: potential contributions to ecology. Oikos 51:97–104

Loehle C (1990) Philosophical Tools reply to Shrader-Frechette and McCoy. Oikos 58:115–119

Lussenhop J (1974) Victor Hensen and the development of sampling methods in ecology. J Hist Biol 7:319–337

MacArthur RH (1957) On the relative abundance of bird species. Proc Natl Acad Sci USA 43:293–295

MacArthur R (1972) Coexistence of species. In Behnke JA (ed) *Challenging Biological Problems: Directions Toward Their Solutions.* Oxford University Press, New York, pp 253–259

MacArthur RH, Pianka ER (1966) On optimal use of a patchy environment. Am Nat 100:603–609

May R (1976) *Theoretical Ecology.* Saunders, Philadelphia

McIntosh RP (1961) Windfall in forest ecology. Ecology 42:834

McIntosh RP (1962) Raunkiaer's "law of frequency." Ecology 43:533–535

McIntosh RP (1967) The continuum concept of vegetation. Bot Rev 33:130–187

McIntosh RP (1975) HA Gleason "individualistic ecologist" (1882–1975): his contributions to ecological theory. Bull Torrey Bot Club 102:253–273

McIntosh RP (1985) *The Background of Ecology: Concept and Theory.* Cambridge University Press, Cambridge

McIntosh RP (1987) *Pluralism in Ecology*. Annu Rev Ecol Syst 18:321–341

Michael EL (1921) Marine ecology and the coefficient of association: a plan in behalf of quantitative biology. J Ecol 9:54–59

Murray J, Pullar L (1910) *Bathymetrical Survey of the Scottish Freshwater Lochs.* Challenger Office, Edinburgh

Nicolson M (1983) *The Development of Plant Ecology*. PhD dissertation, University of Edinburgh, Edinburgh

Niven BS (1982) Formalization of the basic concepts of animal ecology. Erkenntnis 17:307–320

Pielou EC (1977) *Mathematical Ecology*. Wiley-Interscience, New York

Pimentel D (1961) The influence of plant spatial patterns on insect populations. Ann Entomol Soc Am 54:61–69

Platt T, Denman KL (1975) Spectral analysis in ecology. Annu Rev Ecol Syst 6:189–210

Pound R, Clements FE (1900) *The Phytogeography of Nebraska*. Published by the Botanical Seminar, Lincoln, Nebraska

Price PW (1984) Alternative paradigms in community ecology. In Price PW, Slobodchidoff DN, Gaud WS (eds) *A New Ecology: Novel Approaches to Interactive Systems*. Wiley, New York, pp 353–383

Price PW, Slobodchikoff, Gaud WS (eds) (1984) *A New Ecology: Novel Approaches to Interactive Systems*. Wiley, New York

Rehbock PF (1983) *The Philosophical Naturalists: Themes in Early Nineteenth Century British Biology*. University of Wisconsin Press, Madison

Renkonen O (1949) Discussion on the ways of insect synecology. Oikos 1:122–126

Rodhe W (1975) The SIL foundation and our fundament. Verh Int Verein Limnol 19:16–25

Rosenzweig ML, Smigel B, Kraft A (1975) Patterns of food, space and diversity. In Prakash I, Ghosh PK (eds) *Rodents in Desert Environments*. Junk, The Hague, pp 241–268

Sanders HL (1960) Benthic studies in Buzzards Bay. III. The structure of the soft bottom community. Limnol Oceanogr 5:138–153

Schoener TW (1986) Kinds of ecological communities—ecology becomes pluralistic. In Diamond J, Case TJ (eds) *Community Ecology*. Harper & Row, New York, pp 467–479

Shrader-Frechette KS, McCoy E (1990) Theory reduction and explanation in ecology. Oikos 58:109–114.

Smith FE (1972) Spatial heterogeneity, stability, and diversity in ecosystems. Trans Conn Acad Arts Sci 44:309–335

Sorenson T (1948) A method of establishing groups of equal amplitude in plant sociology based on similarity of species content. Det Konge Danske Videns Selskab 5:1–34

Stephen AC (1933) Studies on the Scottish marine fauna; the natural faunistic divisions illustrated by the quantitative distribution of the molluscs. Trans R Soc Edinb 57:391

Strong DR Jr, Simberloff D, Abele LG, Thistle AB (eds) (1984) *Ecological Communities: Conceptual Issues and the Evidence*. Princeton University Press, Princeton

Taylor LR (1987) Synoptic dynamics, migration and the Rothamated Insect Survey. J Anim Ecol 55:1–38

Watt AS (1924) On the ecology of British beechwoods with special reference to their regeneration. II. The development and structure of beech communities on the Sussex Downs. J Ecol 12:145–204

Whittaker RH (1962) Classification of natural communities. Bot Rev 28:1–239

Whittaker RH (1967) Gradient analysis of vegetation. Biol Rev 42:207–264

Wiegleb G (1989) Explanation and prediction in vegetation science. Vegetatio
    83:17–34
Wiens J (1981) Single-sample surveys of communities: are the revealed patterns
    real? Am Nat 117:90–98
Wiens, JA (1989) *The Ecology of Bird Communities*. Vol 1. *Foundations and Pat-
    terns*. Cambridge University Press, Cambridge
Williams WT, Lambert JM (1959) Multivariate methods in plant ecology. I.
    Association-analysis in plant communities. J Ecol 47:83–101

3. Role of Heterogeneity in Scaling of Ecological Systems Under Analysis

Timothy F.H. Allen and Thomas W. Hoekstra

This chapter addresses heterogeneity in the context of scale. Scale is emerging as one of the critical problems that must be adequately considered if different ecological studies are to be either compared in a corroboration or contrasted in a refutation. Some argument in the ecological literature is misdirected because the contentions are differently scaled and so are not competitive (e.g., Belsky, 1986, 1987 versus McNaughton, 1985, 1986, as discussed in Brown and Allen, 1989). Disparately scaled ecological situations cannot be compared in any simple way, even if superficially it appears that it is the same community or site that is being addressed.

The effect of heterogeneity on scaling has received relatively little attention and so is one of the places where scale-based misunderstandings can arise. Usually the scale that is being used in a given study is obvious once it is explicitly stated. However, heterogeneity in a data collection is one of the scaling considerations that is less obvious. Scaling errors or mismatches associated with heterogeneity can therefore be easily overlooked and creep into studies insidiously. Being vigilant in remembering to include heterogeneity in scaling is important, but it does little good if we are not aware of the complexities in the way heterogeneity influences the scale of what we see. We can offer only a relatively clumsy solution; however, the problem of scaling and heterogeneity does not go away by itself, and so any procedure for dealing with it is much better than nothing. This chapter is about the intricacies of the relationship between the material world, the procedures of observation and analysis, and the models of the material system that come from our interpretation of the observations.

The Observer in the System

The entities that emerge in a data set are scaled by virtue of the observation protocol and the filters applied to the data set during analysis (Allen et al., 1987). Heterogeneity in the material system influences the effect of the filtering operation of analysis, and in this way it exerts a scaling effect. Between the material system and any understanding of it we may achieve, there are sets of data that are variously heterogeneous. This heterogeneity depends on the sampling regime, as it addresses the material world, and is treated here separately from heterogeneity in the material system itself. Heterogeneity in the material system arises from the interaction of various ecological processes: dispersal, ecesis, competition, resource capture, resources depletion, resource retention by organisms and whole ecosystems, various disruptive processes, and so on. In this chapter we link the material processes that underlie heterogeneity and their expression in an ecological description.

We do not wish to be heard as saying that processes in nature have a scale in and of themselves, and that it is our job as scientists to find that scale. Scale is not a property of nature alone but, rather, is something associated with observation and analysis (Allen and Starr, 1982). We are speaking here of observed pattern, not nature. There are many ways to detect an ecological flux that relates to, for example, nitrogen availability. Each way of detecting the flux identifies the process operating at a different scale. The scale of a process is fixed only once the actors in the system are specified by the observer (Fig. 3.1). Only then are the processes related to nitrogen cycling given a scale specified by the decision of the observer, not by nature. We use the scaling operation of data analysis as a vehicle for part of this discussion. The recognition of a significant cluster or order in an ordination result is one of the ways to name an ecological actor. Interpretation of results is therefore part of scaling the system. We presume that there is a material system underlying ecological conceptions, but it is a mistake to confuse our specifications and conceptions of that system with the system itself. At this stage in the development of ecology, scale is a heuristic factor. Reifying ecological scales appears only to confuse matters at this juncture.

Reifying heterogeneity is also a mistake, as it is not a simple attribute of the material world independent of observation. One of the stumbling blocks to the intellectual conception of ecology is the willingness of ecologists to suppress the observer in the system and pretend that we have direct access to the material world. The heterogeneity with which we deal has always come through a series of filters before it is ours to consider. Heter-

Actors	Small Scale

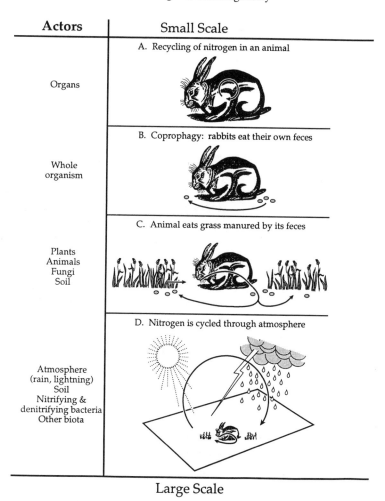

A. Recycling of nitrogen in an animal

Organs

B. Coprophagy: rabbits eat their own feces

Whole
organism

C. Animal eats grass manured by its feces

Plants
Animals
Fungi
Soil

D. Nitrogen is cycled through atmosphere

Atmosphere
(rain, lightning)
Soil
Nitrifying &
denitrifying bacteria
Other biota

Large Scale

Figure 3.1. The process of nitrogen recycling does not have a single scale in nature. It is not scaled until we identify the actors in the cycle, as the particular actors have a scale, which fixes the scale of the cycle in which they are involved. Nitrogen cycling occurs at all scales, from cycling within an organism to biogeochemical cycling at a global scale.

ogeneity has a subjective component by virtue of the observer's taxonomic decision to recognize certain distinctions. Community heterogeneity need not be of species occurrences and could involve instead guilds, genera, life forms, or whatever. Even after we have chosen species as the vehicle to express heterogeneity, a forester might decide to include trees but not microbes or herbs in the data matrix. When considering heterogeneity, it is important to avoid confusing what follows from subjective decisions of the observer with what is so because of the nature of the material system.

General Properties of a Scale

Scale is a matter that pertains to all types of ecological data. General aspects of scaling in ecological data have been considered elsewhere (Allen and Starr, 1982; Allen et al., 1987) but must be at least mentioned here so the scaling effects of heterogeneity in communities can be put in context.

Grain

Grain relates to the level of resolution. The grain of an observation is the finest distinction that is made between isolated datum values. It determines the smallest entities that can be seen in the study. If quadrats are the sampling method, the size of quadrat then determines some aspects of grain. If two phases of environment or vegetation are so small that they regularly fit together inside the quadrat, any heterogeneity ascribable to differences between the two phases is obviously lost because the grain is too coarse. Any two entities so small as to fit inside one period in time or space within the measuring system cannot be distinguished and so is not detected as separate in the analysis. Small, fast entities belong at a low level of organization, so the grain determines the lowest level of organization that can be ascertained (Allen et al., 1987). This point applies not just to the grain of the raw data but also to the finest distinctions that are left in the data after aggregation or data transformation.

Extent

In contrast to the grain, the extent determines the largest entities that can be detected in the data. The extent is the span of all the measurements in a given study. In community analysis the extent is related to the size of the data matrix and the largest degree of difference that exists between columns or between rows. However, extent should not be mistaken for some simple matrix algebraic parameter that can be numerically defined. It is clearly related to mechanical things such as the number of samples; but it is more than that, for extent circumscribes the entire biology and environmental variation that can fit into universe of discourse. The choice of a certain community type as the object of study contributes to limiting the extent in time and space. Anything that takes longer to happen than the full period over which measurements are made cannot be detected. For example, the community structure embodied in the annual cycle of a phytoplankton community cannot be detected in a month of sampling. Any variations of form or function that fall outside the geographic area of study also cannot be detected (O'Neill et al., 1986; Allen et al., 1987).

The scale of the study is an interaction of grain and extent. If the extent is large, the sampling protocol will be prohibitively expensive unless the grain is relatively coarse. Conversely, a study involving fine grain must of necessity have a narrow extent.

Strategy of Data Reduction

Scaling in the Process of Analysis

There are likely to be signals in the data set that survived the filter of the grain imposed by the sampling procedure, only to be missed as the analyst inspects the results. This situation occurs because there are usually too many bits of information of that size for the ecologist to be cognizant of them all. Furthermore, large patterns that do fit within the limits of the extent of the study can still be missed because the observer is distracted by the weight of strong finer grain structure. In general, the patterns that fit between grain and extent compete for attention in the mind of the ecologist (O'Neill et al., 1986). The very strategy of data reduction through ordination is to remove all but a few patterns so a clear view can be achieved. By relegating some signals, even those with biological meaning, to the status of noise, the analyst allows insight into what survives.

The axes of an ordination are a summary of the correlation or covariance structure in the data or something approximating that structure. The correlation structure can be sensitive to even a small number of changes in the data entering the analysis. Single outlier samples are well known to have dramatic effects, and they are removed by the judicious inclusion of only samples that fall within a certain range of heterogeneity. Removing outlier stands from an analysis is in fact a rescaling operation. Note that it is a modification of the extent, not just in terms of giving the data matrix a smaller rank but by removing a sample that contributed dramatically to the heterogeneity of the universe of discourse. The outlier introduced by itself a whole new set of biological and environmental considerations.

The outlier problem is an aberration in sampling that can be ameliorated, but the sensitivity of the correlation structure, through which the outlier problem manifests, is always present. That sensitivity of correlation structure plays a role molding the form of the output in all analyses of vegetation. Even a single correlation has an implied scale by virtue of the size of the grand total of samples across which there is a departure from random expectation. Change the size of the grand total, and the correlation changes. Occurrences that are positively correlated in a large-scale universe can become negatively correlated in a smaller universe. Changes in correlation structure often imply a change in scale (Beals, 1973) (Fig. 3.2).

The correlation structure of a data set is a reflection of the cohesive effects of the critical ecological factors. A given constellation of related ecological processes exhibit robust phenomena, and these phenomena are what particularly interest the ecologist. They are manifested in data reductions as certain axes in ordinations or as major clusters in classifications. Large-scale considerations associated with slow, unwavering processes are taken as unchanging context by fast local processes that fluctuate over

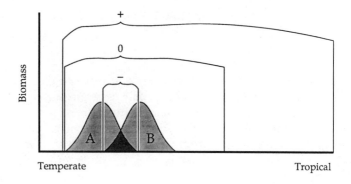

Figure 3.2. On a gradient from temperate Wisconsin to the tropics, the correlation between species A and B changes from negative to positive, depending on the scale. Within Wisconsin there is a negative correlation of species that prevail north of the tension zone to those that predominate to the south of the tension zone. Over a region of the whole northern United States there is no correlation, as the species occur together in the proportions that each is present across the entire region. Across the entire gradient all the way to the tropics, the temperate species are distinctly northern and are positively correlated.

short time periods. Furthermore, the fast local processes affect only the slow processes through some sort of average value that ignores short-term variability (Fig. 3.3). Thus large-scale processes are isolated from the behavior of small-scale processes and vice versa. That is why it is possible to distinguish coherent ecologies from each other by assigning each to its own ordination axis.

The whole strategy of data reduction is to separate the effects of the important constellations of processes. The distinction is usually performed on the basis of scale differences, even if that is not made explicit in the method of analysis. For example, the extraction of the first axis or the first hierarchical cluster essentially changes the correlation structure from that in the original correlation matrix into that of the residual matrix. Remember that a change in correlation structure is often a change in scale. When

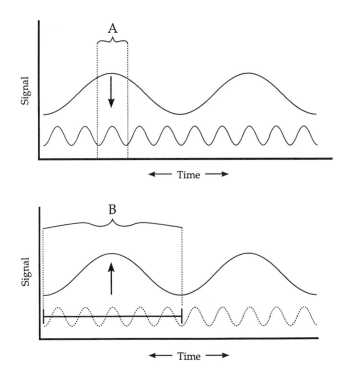

Figure 3.3. Over the period of one cycle for a local fast process, the regional process appears as a constant context. Conversely, averaged over the entire cycle period of the regional process, the local process appears as a constant flat signal.

the strongest correlations are given weight on the first eigenvector, an ecology of a certain scale is being expressed. Those scaled effects are removed in the first axis. The new correlation structure in the residual matrix that is used to extract subsequent axes reflects a differently scaled set of interactions that are now free to be expressed. The system is "nearly decomposable" into processes operating at different scales (Simon, 1962).

Although multivariate techniques can do a remarkable job of distinguishing different levels of organization, the system is only "nearly decomposable." There is a leakage of influence across scales: Note in Figure 3.1 that it is the same nitrogen moving around differently scaled cycles; therefore this fact does link levels to a degree. Similarly, changes in an ordination axis are only "nearly decomposable" from later axes. When one changes a datum value in a matrix, column and row totals change, which affects the grand total. Thus there is a slight shift in orientation of the first eigenvector or its equivalent in other ordination methods. Thus an environmental factor that substantially affects the performance of a major species on the ground has an influence on the first eigenvector. All other axes are held in the context of the first axis extracted, and that is how a

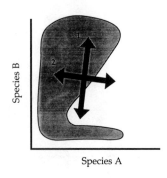

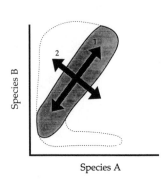

Figure 3.4. In the simple condition of two species, the first case shows species B predominant; and description of its variation consumes the first summary axis, leaving species A the second summary axis. In the second case, some factor has come to suppress each species when it is growing alone. Accordingly, the first summary axis is reoriented to reflect a positive association between species A and B. Note how it forces the second axis, independent as it may be, to express a competitive process between the species rather than a description of the performance of species A. Thus the expression of the processes that underlie the performance of species A is compromised. This situation is a caricature example of a process that takes over the early summary axes and so affects the expression of variation on later axes.

change in the first axis influences all later axes to a degree. Thus by influencing the data structure so as to have an effect on the early eigenvectors, a given ecological process can influence the expression in the ordination of the effects of all processes. This reorganization works even on processes that are not related in any direct fashion in nature (Fig. 3.4). Processes that are in fact related to each other in nature, e.g., pH and carbon availability in a water column, should have a particularly dramatic effect on each other's expression under analysis because processes related in nature are likely to be correlated in their effects in the analysis. Moreover, high correlations often play a large role in orienting the first few axes of the analysis.

Any ordination axis showing clearly one biologically significant pattern of necessity suppresses other, also meaningful patterns that were present in the raw data. With multivariate analyses, the various meaningful signals compete for control of data reduction; that is, they compete for high loadings on the dominant axes. Ordination axes are a reflection of the effects of underlying gangs of processes that work together to give coherent patterns in the field and so in the data. Different patterns are reflections of different sets of processes that compete for the earlier axes. With stress-reducing ordination techniques, local minima in stress are overwhelmed by the global minimum stress.

Although the later axes reflect processes whose influence in the data was overwhelmed by the pattern of the early axes, the underlying ecology of the later axes is often still expressed with a surprising coherence. For example, Allen and Shugart (1983) were able to recover the pattern of the first two axes of a covariance analysis on the fourth and fifth axes of a corresponding correlation analysis. Their success was fairly remarkable because in most analyses the fourth and fifth axes have often been compromised by the earlier axes. The trick is to transform the data matrix so that patterns formerly relegated to later axes can move to the front of the transformed queue and so express their associated ecology with clarity (Allen et al., 1984).

Changing the Filter

Despite the great effectiveness in signal retrieval of these powerful techniques of data reduction, there is a critical problem: The order of ordination axes is not fixed by the ecology of the situation. Rather, it is determined by the happenstance of the details of a particular analysis of a particular data set and how the variance is partitioned between the sampling regime and the analysis. There is much potential for changing the order of the expression of differently scaled ecologies held within in a given set of raw data. Small changes in either the data matrix or the distance measures used to build the correlation matrix or its equivalent can reverse the order of the early axes (Allen and Shugart, 1983). The switch occurs when the variability in the transformed data associated with a losing set of processes becomes infinitesimally larger than that of a formerly winning set of processes. The switch occurs whenever those matrix algebraic criteria are met, but there is no general prescription from the ecology as to when exactly that will happen.

Axes very late in the sequence are so deeply set in the context of the early axes that they become distorted and often give uninterpretable signals when their turn arrives. When one performs some strong data transformation that reapportions the total variability between phenomena, signals so deeply embedded before the transformation that they were uninterpretable can suddenly come to the front of the queue and express an apparently new and meaningful pattern on the first couple of ordination axes. Strong data transformations allow the expression of completely new sets of processes.

Between the grain and the extent of most data sets there is room for several levels of organization. Each level is occupied by a gang of commensurate interacting ecological processes. Together the processes display a phenomenon. All the levels of organization play a role in determining the details in the raw data, but the influence of each in the raw data is lost in a cacophony of all the others. That is why analysis is necessary. The patterns associated with each of the levels are there to be found by the right filter. Imposing that filter is indeed the very strategy of data reduction in the first place.

This style of multivariate analysis is analogous to using a radio receiver to isolate only certain frequencies from the full spectrum of radio waves. All the differently scaled ecological phenomena are candidates to secure the first axes of the analysis, depending on the happenstance of the details of the mode of analysis. Data transformation is like turning the radio dial. We do not know which pattern will prevail during the analysis of a new data set. There is no published guide to the ecological equivalent of radio bands. Worse than that, we do not yet know how to read the radio dial, for we also have little idea as to how far we have to change the procedure of the first analysis to achieve a significantly different result. The assignment of rank to the axes may seem capricious. Even so, the patterns are ordered in such a way that we can say something about the change in pattern should it occur. That knowledge comes from the fact that both the gangs of processes in the material system that generate the patterns, as well as the data transformations that can reveal the pattern, are scale-ordered. Scale is the common factor, and we can use it to make predictions. We may have difficulty reading our ecological analogue of a radio dial, but at least scale lets us ascertain in which direction we are turning the knob.

Movement up-scale refers to a coarsening of the grain or widening of extent, thereby causing an observation to be of a larger-scale ecological circumstance. Movement down-scale refers to a narrowing of the extent or deployment of finer grain, thereby causing an observation to be of a smaller-scale ecological circumstance (Fig. 3.5). Some transformations and changes in analytical procedures move the analysis up-scale, whereas others move it down-scale, and we do know which is which. Both widening the extent by including more material heterogeneity and coarsening the grain to remove fine distinctions move the analysis up-scale. Conversely, both narrowing the extent to focus on a less heterogeneous material system and making the grain finer reveal patterns of a smaller-scale ecology. Although we cannot say which scale will prevail in the first analysis, transformations do change the result in some generally predictable way relative to prior analyses. We can keep on pressing the data set with transformations that progressively exert influence up-scale by widening the extent or

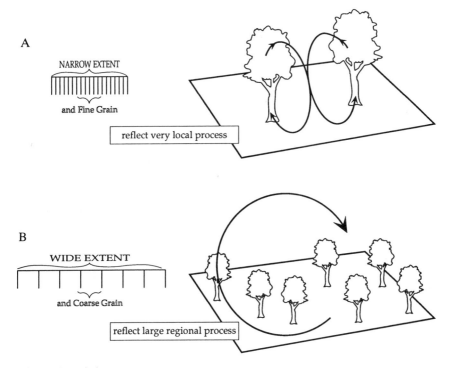

Figure 3.5. (A) Narrow extent and fine grain reflect relatively local ephemeral processes. (B) Wide extent and coarse grain reflect larger-scale regional processes.

coarsening the grain. Eventually a new pattern comes to the fore. We can predict that this pattern will pertain to a set of processes that operate at some larger scale in time and space and will be contextual to a wider range of special ecological circumstances (Allen, 1987).

Example Transformations at Work

To exemplify transformations we have discussed so far in the abstract, we now introduce a contrast between two particular transformations and the way they bring out their respective sets of ecological processes. There are some processes and factors that relate to the vigorous growth of the dominant plants at a site. However, these factors can become important only after the establishment and survival of the plants generate the contestants in the race to grow large. Vigor is predicated on establishment in the first place, and therefore vigor pertains to a lower level of organization than establishment. A quantitatively based pattern coming from cover data reflects factors that allow certain individuals to grow large while others remain small. Accordingly, both sets of influence—those for establishment and those for growth—are found in data collected on the plants from the site. Suitable data collections and transformation can bring each set of factors to the fore in turn.

In a binary matrix (where all values not equal to 0 become 1), the processes that allow survival underlie the principal pattern. In that case, the fine distinctions between amounts of vegetational cover are suppressed. The pattern that emerges is one that is contextual for, not reflective of, the vigor of dominant plants. Accordingly, all the fine grain distinctions between the performance of the largest plants at microsites has a background of arrival at the site and factors for ecesis. That background is what is brought out in the binary matrix.

Now consider a cover matrix derived from the same site. The variance in cover values in a quantitative matrix usually swamps the variance of the distinctions between low cover and complete absence. Low cover versus absence is highlighted in presence data, but it is swamped here in the cover data. Thus the influence of processes of ecesis and last-ditch survival is relegated to some later axis of the analysis. Chances are the pattern pertaining to establishment processes will be unrecognizable. Thus both signals of environmental influence exist in relations in the data as a whole, but only one set can be displayed with clarity in any given analysis. Binary transformation is coarser grained, and so we should not be surprised that it is reflective of ecological processes that are of larger scale in both time and space. Larger-scale processes of the material system emerge predictably with larger-scale data transformations that use wide extent and coarse grain (Allen and Wileyto, 1983).

Data transformations determine the grain that is allowed into the process of analysis. By contrast, taking subsets of the total data matrix so as to change the rank of the matrix is clearly related to a shift in the extent. The larger the sheer size of the data matrix, the coarser and the more contextual is the pattern that emerges. In general, the larger the data matrix, the higher is the level of organization that appears and the larger scale is the controlling process. Conversely, the finer the grain of the distinction between datum values that is allowed by the data transformation, the more the variance associated with local fast processes undercuts the influence of the larger context in the analysis. Fine grain exhibits low levels of organization.

Heterogeneity and Research Protocol

Having defined scale in terms of grain and extent, we are now in a position to see how scaling relates to heterogeneity. There are two considerations here. The first is concerned with differences in heterogeneity of the data collection protocol and analytical procedure. The second is the scaling effect of differences in the heterogeneity of the material system from which data are collected.

Manipulating Heterogeneity Through Sampling

Heterogeneity is certainly related to the extent. In fact, it is through increases in heterogeneity across the data that a wider extent allows a study to move up-scale. With increases in the temporal and spatial context of the study, the sampling regime encounters more variability. This fact can be related to spatial extent in species–area curves. Because various organisms become conspicuous at different times of the year or over a succession of years, increasing the temporal extent of the study increases heterogeneity. It is a truism that larger areas can contain processes that are more widespread. Sampling over a wider spatial extent often encounters larger-scale ecological systems. By sampling over a longer time span it is possible to identify slower processes, another larger-scale consideration. Widening the extent relates simply to a move up-scale.

However, the situation becomes more complicated when one considers grain. A simple increase in spatial extent is not the only way to increase heterogeneity. More thorough and detailed sampling of a fixed area often reveals more types of organisms and so increases the heterogeneity across the sample universe. The extra information puts more subsystems into the whole, thereby moving it up-scale. Thus merely intensifying sampling without sampling for a longer time or over a wider area adds heterogeneity and so moves the study up-scale. One might expect a finer grain in the observation protocol to decrease the scale, but here it inserts more differences and so increases the scale of what is found by increasing heterogeneity.

The resolution to the dilemma is that the increase in heterogeneity comes from a veiled increase in extent. When one samples with greater resolving power to obtain a more detailed account of an area, incidentally there are more individual samples to compensate for their being smaller and making finer distinctions. Otherwise the finer-grained samples would in aggregate sample a smaller part of the universe, and perceived heterogeneity would be reduced. More ephemeral and more local processes would come to the fore in the patterns revealed under analysis. Sampling a system defined by a certain area and time span so as to obtain a more detailed account of everything present must involve an increase in the number of samples in the universe, which puts more into the entire data set. This maneuver often causes a move up-scale because more of the material world has been captured in the data set.

More detail here does not just mean finer distinctions, although it does involve that aspect. Rather, it means a greater mass of detail in aggregate, something that just finer distinctions do not necessarily give if the entire universe sampled is smaller. Heterogeneity does not involve just grain-defined differences between data points. It is the aggregate of all differences between all data points, which necessarily brings in extent as well as grain. We stated earlier that the scale is fixed by the relation between grain and extent, and we have just explained that heterogeneity is a measure of that relationship. Thus relative heterogeneity becomes the benchmark for a shift in scale rather than the grain and extent themselves.

There is a price to be paid for trying to sample in greater detail so as to obtain a more thorough account of what is present in an area. Because it involves a larger number of finer-grained data points, the grain and extent are varied, respectively, down-scale and up-scale in opposite directions at the same time. There is no simple formula for saying a priori whether it is the finess of the grain or the wideness of the extent that wins in the tug of war to rescale the system of observations. If indeed a fuller account of the details of the system is achieved, heterogeneity increases and the move is up-scale. However, the search for a more detailed account may lead to a description of a homogeneous subsystem, not the whole system. Then the move is down-scale. The situation is an invitation for confusion.

The larger-scale pattern relates to the context of what is found in a smaller-scale pattern. A smaller-scale pattern might indicate a mechanism for the factors determining the larger-scale pattern. The scaling effect of wider extent and narrower grain depends on unknown factors regarding thresholds of the spatiotemporal interaction of critical processes in the material system. The final outcome depends on how these thresholds in nature interact with the sampling thresholds embodied in the happenstance of the grain of the sampling units. Without knowledge of whether the new analysis is a move up- or down-scale, it is not possible to argue causal relations. Causality implies a set of subsystems that interact to cause behavior of the whole. Clearly, if one does not know which pattern represents the larger system, all ability to cross-reference so as to invoke a causal relation between levels is lost.

There is much to be said for comparing studies only when the relative scale across the comparison is unambiguous. This condition can be achieved by changing only the grain while maintaining the extent as defined by number of samples in a spatially and temporally defined universe. It can also be achieved by changing only the extent while holding the grain constant. If both grain and extent are to be changed at the same time, it is advisable that they be changed in the same direction. The scale change is unambiguously up-scale if the grain is coarsened and the extent is widened. It is down-scale if the extent is narrowed and the grain is made finer. In any of these scenarios, the scale being known, it is possible to infer what are contextual relations up-scale and what are candidates for mechanistic explanation down-scale.

Heterogeneity and Analytical Methods

We cannot conclude the discussion of the role of heterogeneity in the research protocol without reference to the method of analysis. The choice of the distance measure (Austin and Greig Smith, 1968; Noy-Meir et al., 1975) plays a large role determining how the pattern belonging to a smaller- or larger-scale set of processes would be ranked in the reduced space of the ordination. The way that a particular analytical technique handles and weights variability also determines which scale comes to the fore in the results of the data analysis. The differences between many techniques depends on how the grain-defined individual samples are related to the extent-defined whole set (Allen, 1987).

Consider, for example, the techniques that give straight gradients as a rule, e.g., step-across (Williamson, 1978), detrended correspondence analysis (Hill and Gauch, 1980), and multidimensional scaling (Kruskal, 1964). They all emphasize the global space, and they all place an implicit quasi-noneuclidean transformation on the variability of the local spaces. Local parts of the global space are rescaled so that the long gradient across the entire space is straight. Accordingly, such techniques give precedence to the global space as defined by the entire data matrix rather than the values of particular stands. Thus gradient straightening techniques weight extent over grain when determining which pattern emerges on the first few ordination axes.

Each in their own way, various techniques of analysis place weight on heterogeneity from alpha diversity as opposed to heterogeneity from beta diversity (Pielou, 1975). Gradient straightening techniques give more weight to beta-diversity-based heterogeneity in that the entire space takes precedence when ranking the patterns that emerge in the ordination. Other techniques are more sensitive to the signal in individual columns or rows in the data matrix, and so these techniques give more weight to the effect of grain on heterogeneity, an alpha diversity consideration. The ability of various techniques to change the rank of differently scaled ecological signals has not been tested in an organized way. It would make a worthy dissertation topic. Our intuition is that the techniques that are particularly subject to curvature, such as principal component analysis, will prove more flexible and revealing. However, we know little of how it would work.

Heterogeneity in the Material System

Up to this point we have considered how data collection and transformation work together to give certain levels of heterogeneity in the data. We have shown how this apparent heterogeneity changes the scale of what emerges under analysis. It is now time to turn to heterogeneity on the ground in the material system and see how it feeds into what we see in the results of analysis.

Material Heterogeneity and Results of Analysis

Greater heterogeneity is encountered if there is greater variability in microsites. An area with fine grain variability in habitat involves more heterogeneity if the sampling regime is comparable to that used in a more homogeneous site. Variability in habitat allows losers in competition in one habitat to persist in one of the other types of microsite in the system. Variability in habitats within a site holds the competitive regime away from the final solution, the equilibrium of competitive exclusion. A rule of thumb in hierarchy theory is that higher levels hold lower levels away from equilibrium. Higher levels are populated by larger-scale entities. The bottom line is that variability of habitat makes for a larger-scale ecology.

The proportions of the various microhabitats at a study site would have an effect on what emerges under analysis. If only one type of microhabitat predominated in the area sampled, the signal that would emerge from analysis would reflect a small-scale ecology. The other uncommon types of microhabitat would be incidental background to the system that emerged. A larger-scale ecological system would be apparent in the data if the various microhabitats were represented equally in the field or at least in the sampling regime. The latter situation is analogous to high diversity through evenness (Pielou, 1975). What was the predominant system in the low diversity uneven site or data collection would be contained as a working part of a larger-scale, more diverse, and equitable ecological circumstance. Thus increased heterogeneity in the material system moves the system observed up-scale in the same manner as did the increased heterogeneity resulting from manipulating extent in the data-processing protocol.

Because heterogeneity is a determinant of scale, and scale determines what the observer encounters, heterogeneity is a principal player in establishing what an ecologist finds. It is well known that the first eigenvector in a heterogeneous data set is almost always an abundance or richness axis. The same is true for the main stem, or the sequence in the chain of classification techniques. Sometimes gradients of abundance or richness are of interest because important ecological processes are responsible for the pattern. In that case the community analysis is often particularly successful.

Such was the situation in the work of Day et al. (1988). They encountered gradients in riverine marsh vegetation for the effects of flooding and fertility. Their first axis was high biomass and low diversity sites to low biomass and species-rich sites. Undisturbed sites had high biomass and litter; competitive exclusion reduced species richness. Kimmerer and Allen (1982) found a similar effect in riverine bryophyte communities. The homogeneous stands with low diversity and high biomass were at one end of the ordination, whereas the species-rich stands held away from competitive equilibrium stands were at the other end. Often successful community analysis interprets a gradient of heterogeneity of habitat and vegetation. Researchers can take advantage of meaningful differences in heterogeneity. They reliably emerge in the analysis because differences in heterogeneity are an overriding factor when orienting the early axes.

Essentially no experience and little understanding of the techniques of data reduction are needed to reveal in almost any data set the factors that determine heterogeneity. First time users picking canned programs at random get that far so long as there is something ecologically interesting about the differences in heterogeneity. In the work of Day et al. (1988) experienced researchers laid out a much needed model for riverine vegetation, but often the factors that underlie differences in heterogeneity are obvious and trivial. This fact could explain the prevalent criticism of ordination—that it does not tell you anything you did not already know. Certainly the literature is full of analyses reporting the obvious.

The difficult part of multivariate community analysis is getting at other factors that do not influence heterogeneity. Placing their vegetational effects in front of heterogeneity in the queue for the principal axes is difficult to achieve. Often we are not interested in heterogeneity per se because it is either obvious or does not matter in the material ecological system. Nevertheless, its effects, meaningful or not, dominate most analyses. On the first axis, at least we expect heterogeneity to be a big determinant and can recognize it for what it is, but not all the influences of changes in heterogeneity are so transparent. Differences in heterogeneity resulting from changes in either research protocol or the material system affect stand totals, proportionalities, and grand totals. All three effects, in turn, affect what appears in the results of multivariate vegetation analyses. Heterogeneity is such a powerful influence on so many analyses that it is wise to assume it is significant in determining what we see in situations where it fails to take over the first axis in an obvious way. The problem is that we cannot expect to recognize the more subtle influences of heterogeneity in scaling analyses, even though they are commonplace.

Material Heterogeneity and Mismatches of Scale

We can expect heterogeneity to influence greatly the scale of any analysis of vegetation in all of the above ways. Anything that influences the scale of perception is of overwhelming importance in determining what appears to happen. When a situation appears suddenly and unexpectedly, and is radically different, the usual explanation is that there has been a change in the scale of perception. The radical change is attributable to a differently scaled set of processes coming to the fore. The ecological processes that determined what was seen before the change are still at hand, but their expression is muted by their losing control of the channels of the observer's perception. In vegetation analysis they have lost control of the stream of the analysis. The object of study is basically the same. In such a study little in material nature is different, and the apparently new situation is mainly a matter of changes unwittingly made in the scale of observation.

This artifact of scaling is responsible for much contention in the literature. Well meaning, fair-minded scholars achieve results that are apparently at odds with published accounts. With the realization of the importance of scale, some workers have tested various community models and found that they work at one scale but not at others. Moore and Keddy (1989) were able to show how Grime's (1973, 1979) intermediate disturbance/ stress model applies to communities at large scales but does not seem consistent with data pertaining to lower levels of organization involving a narrower scope. In their work they changed grain and extent in the manner recommended here, so their study is a good example of how many of the abstractions laid out here are only generalizations of explicitly data-based investigations already in the literature. Competition is a contentious issue, perhaps because of scale questions changing the significance of results. With so many places and so many ways for heterogeneity to change results, and with ecologists having so little predictive capability as to the scale that will emerge in a given isolated analysis, the situation is an invitation for misunderstanding. Even publishing the exact protocol of data collection and comparing the results only to those of investigations that used exactly the same protocol does not solve the problem. It is because incidental differences in heterogeneity in the material system can have as large an effect on the outcome of analysis as changes in the protocol for data collection, transformation, and analysis.

As an example, consider two prairies that have almost the same processes at work. In this hypothetical case, the differences between them are trivial and are not of interest with regard to how nature works, given the question to be investigated. The problem is that there is a difference in heterogeneity; one prairie has different microsites represented in roughly even numbers, whereas the other is predominantly of one type of microsite, although the other type of microsite is well represented and should be considered a part of the whole system. Even with a standard sampling regime, the two prairies could appear far more different than they are because under analysis their differences in heterogeneity emerge as scale differences that reorder the appearance of the important controlling factors. Data sets from each do indeed contain signal from the influence of all the differently scaled critical processes, but the patterns assigned to early axes in one analysis appear on the late axes of the other. The situation appears different, but it is not.

The sorts of differences that can appear are large. Changes in scale basically change what is labeled as inside the system and what is asserted to be outside the system. Allen and Wileyto (1983) showed how rescaling prairie data turned fire from a disturbance in a fine-grained analysis to a working part of the system in a coarse-grained analysis. Furthermore, that system respecification emphasized a different temporal pattern. With fire as a disturbance coming from outside the system, the vegetation appeared to have a memory of only one growing season. Winter erased memory of fire every year. On the other hand, with fire as part of the system, the vegetation appeared to be responding to events that happened at least 5 years earlier.

If such differences can emerge within a single data set under transformation, comparing different prairies seems a risky business. Given the above discussion, differences emerging across such comparisons are not a safe basis for starting a contentious argument as to disparity of findings. How then are we to compare anything regarding vegetation or ever feel confident about observed differences?

Solution

The solution is to make comparisons relative to some explicit ecological process in the material system whose scale is clearly defined by explicit identification of the actors and their actions. We cannot standardize data-handling procedures by using simple parameters such as a standard-sized data matrix or a particular field method and be assured that we have commensurate scaling on each side of the comparison. There are too many steps between data collection and interpretation of analysis for incidental differences in heterogeneity to work their mischief. Therefore the items to be compared must be rescaled relative to each other while the analysis proceeds. Analysis needs to be iterative. The first pass gives the preliminary scale of each system. There is a rescaling in light of the first pass. The second pass checks that the rescaling has made the parties comparable. If the scales exhibited are found to be commensurate, the result of the second pass itself may be used to compare one ecological system with the other. In the case of prairies, it is important that the results to be compared are scaled so that fire is either part of the system in both patches of vegetation or is a perturbation coming from outside the system in both cases. Note that it may involve analyses that are different in their details of sample number or degree of sample aggregation.

The general method of making the parties to the comparison compatible is to perform the analysis and see how the patterns that emerge relate to some reference phenomenon of ecological significance. The reference phenomenon could be fire, grazing pressure, successional status, or other significant factor. It is unlikely that all analyses of the different sites will relate to the phenomenon at the same scale if for no other reason than incidental differences in heterogeneity on the ground. Remember that although we do know how to rescale relative to a previous analysis of a given data set we do not appear to have the ability to predict the scale that will emerge in a first analysis of a new data set. One decides what is going to be the scale of the systems relative to the phenomenon and then rescales with changes in the analytical protocol of the analyses that did not fit. If fire is the reference and it has been decided that it should be scaled to be inside the fire–vegetation entity as a working part, some analyses must be redone with a coarser grain or wider extent so fire is seen as having become incorporated. Only when the analyses address the same level of organization, i.e., prairie as a fire-maintained not as a fire-disturbed vegetation, can comparisons be made.

It may not seem satisfying to be forced to compare results that come from different analytical procedures. However, the alternative is to use a universal data-collecting method combined with a standard analysis that only by chance gives universally scaled results. The problem is that subtle differences in the patterns of heterogeneity scale the analyses differently, leading to results that are not comparable. The critical similarity that is sought is a scale equivalence; then at least one knows that differences are not artifacts of differences in the scale of perception.

Conclusion

The early data reduction studies in ecology were at pains to perform the analyses with focused and therefore scaled questions in mind. The importance of having an explicit question turns on the difficulty of asking a question or erecting a hypothesis without fixing the scale used to address the system. That is where the power of the null hypothesis resides. Curtis and McIntosh (1951) used forest succession as the unifying view. Dix (1959) made careful comparisons of the effects of grazing. At all his sites the role of grazing was explicit.

One of the problems with modern styles of data analysis is that they are not based on phenomena but, rather, depend on some mathematical or statistical attribute such as variance maximization. Ecological concepts are not tied to a particular scale until the actors in the system and the relationships that bind them together are specified. Variance maximization does not have a scale that applies to nature. Rather, it is scale-malleable, depending on accidents of heterogeneity allowed by data collection, transformation, and analytical techniques and the way they relate to the heterogeneity of the material system.

Consider the paired grazed and ungrazed sites that Dix (1959) used to frame his study. There are many scales at which grazing occurs (Brown and Allen, 1989), but at least if one has a phenomenon such as grazing in mind the scaling can be done relative to some aspect of nature that makes comparison less likely to be bogus. The message is that data reduction in ecology should return to its origins. The paradigm-founding work was much more ecologically focused than the prevailing approaches, which are mathematically but not ecologically defined. It is not to say that we disapprove of the splendid advances made in methodology (Legendre and Legendre, 1987). It is the use of methods without a prescribed focus that deserves reproach. We cannot afford an attitude to community data analysis that waits to see what happens before taking a point of view.

As we said at the outset, our solution to the scaling problem as regards heterogeneity is not elegant. We have not offered a special scale-proof mathematization of the effects of heterogeneity on data analysis. However, we have tried to indicate where the problems might arise and the form they might take. We have offered a general protocol for erecting a frame within which comparison is likely to be valid. What we have done is to standardize after heterogeneity has done its scaling. Because that scaling is likely to be complex, we cannot correct for the effects of heterogeneity before the fact. We let heterogeneity do what it will, and only then do we rescale. It is a standard strategy of the hierarchist to incorporate difficulties into the system of analysis and make corrections for it later. Only in that way can we cut off all the heterogeneous heads of the gorgon at one swipe.

References

Allen TFH (1987) Hierarchical complexity in ecology: a non-euclidean conception of the data space. Vegetatio 69:17–25

Allen TFH, Shugart HH (1983) Ordination of simulated forest succession: the relation of dominance to correlation structure. Vegetatio 51:141–155

Allen TFH, Starr TB (1982) *Hierarchy: Perspectives for Ecological Complexity*. University of Chicago Press, Chicago

Allen TFH, Wileyto EP (1983) A hierarchical model for the complexity of plant communities. J Theor Biol 101:529–540

Allen TFH, Sadowsky DA, Woodhead N (1984) Data transormation as a scaling operation in ordination of plankton. Vegetatio 56:147–160

Allen TFH, O'Neill RV, Hoekstra TW (1987) Interlevel relations in ecological research management: some working principles from hierarchy theory. J Appl Syst Anal 14:63–79

Austin MP, Greig-Smith P (1968) The application of quantitative methods to vegetation survey. II. Some methodological problems of data from rain forest. J Ecol 56:827–844

Beals EW (1973) Ordination: mathematical elegance and ecological naivete. J Ecol 61:23–26

Belsky AJ (1986) Does herbivory benefit plants? A review of the evidence. Am Nat 127:870–892

Belsky AJ (1987) The effects of grazing: confounding of ecosystem, community and organism scales. Am Nat 129:777–783

Brown BJ, Allen TFH (1989) The importance of scale in evaluating herbivory impacts. Oikos 54:189–194

Curtis JT, McIntosh RP (1951) An upland forest continuum of the prairie-forest border region of Wisconsin. Ecology 32:476–496

Day RT, Keddy PA, McNeill J, Carleton T (1988) Fertility and disturbance gradients: a summary model for riverine marsh vegetation. Ecology 69:1044–1054

Dix RL (1959) The influence of grazing on thin soil prairies of Wisconsin. Ecology 49:36–49

Grime P (1973) Competitive exclusion in herbaceous vegetation. Nature 242:344–347

Grime P (1979) *Plant Strategies and Vegetation Processes*. Wiley, Chichester

Hill MO, Gauch HG Jr (1980) Detrended correspondence analysis: an improved ordination technique. Vegetation 42:47–58

Kimmerer RW, Allen TFH (1982) The role of disturbance in the pattern of a riparian bryophyte community. Am Midl Nat 107:370–382

Kruskal JB (1964) Nonmetric multidimensional scaling: a numerical method. Psychometrika 29:28–42

Legendre P, Legendre L (eds) (1987) *Developments in Numerical Ecology*. NATO Advanced Sciences Institutes Series G: Ecological Sciences 14. Springer-Verlag, New York

McNaughton SJ (1985) Ecology of a grazing ecosystem: the Serengeti. Ecol Monogr 55:259–294

McNaughton SJ (1986) On plants and herbivores. Am Nat 128:765–770

Moore D, Keddy PA (1989) The relationship between species richness and standing crop in wetlands: the importance of scale. Vegetatio 79:99–106

Noy-Meir I, Walker D, Williams WT (1975) Data transformations in ecological ordination. II. On the meaning of data standardization. J Ecol 63:779–800

O'Neill RV, DeAngelis DL, Waide JB, Allen TFH (1986) *A Hierarchical Concept of Ecosystems*. (Monographs in Population Biology 23). Princeton University Press, Princeton

Pielou EC (1975) *Ecological Diversity*. Wiley, New York

Simon HA (1962) The architecture of complexity. Proc Am Philos Soc 106:467–482

Williamson MH (1978) The ordination of incidence data. J Ecol 66:911–920

4. Heterogeneity as a Multiscale Characteristic of Landscapes

Bruce T. Milne

The consequences of heterogeneity have been a central theme in ecology at least since Cowles (1899) studied the successional pathways of Great Lakes vegetation. The different abilities of species to tolerate burial, inundation and competition create vegetational gradients that are correlated with proximity to the shore (e.g. Milne and Forman, 1986). Thus each location within the landscape contains a subset of the species pool. The composition of the subset is determined by the differential responses of species to the abiotic and biotic conditions present (Gleason, 1926; Whittaker, 1967; Huston, 1979; Sousa, 1979; Austin, 1985; Tilman, 1988) or to conditions in the past (Marks, 1974; Cole 1985).

In this view communities result from species' responses to abiotic and biotic constraints projected onto the landscape (Whittaker and Levin, 1977; Root, 1988). A similar view explains the assembly of tree communities across North America following the retreat of the Wisconsonian glaciers (Delcourt and Delcourt, 1987). At kilometers-wide scales (Forman and Godron, 1986) spatially heterogeneous communities result from the differential sorting of species over spatially complex landscapes.

Neilson and Wullstein (1983) related species distributions to constraints operating at several scales. *Quercus gambelii* seedling survivorship varies as an interaction between local and regional precipitation. At the regional scale oak mortality is lowest at southerly, dry latitudes. Conversely, at local scales mortality decreases as precipitation increases with elevation. The local response is modified further by the openness of the canopy. The fragmented regional distribution of oaks reflects responses to fine-scale canopy variation overlaid on broad-scale topographic and latitudinal constraints. Thus several levels of selection are evident, and species distributions (e.g., Little, 1971) may be viewed as an interaction between constraints operating at several scales (Senft et al. 1987; Urban et al., 1987).

In this chapter the above observations of species' differential responses and the multiscale distribution of constraints (Allen and Starr, 1982; Allen et al., 1984) motivate an analysis of heterogeneity. Spatial heterogeneity is the complexity that results from interactions between the spatial distribution of environmental constraints and the differential responses of organisms to the constraints. A diffraction grating used to separate light into its component colors provides an analogy. The grating represents the complex spatial patterning of abiotic and biotic constraints. A beam of white light projected onto the grating represents the species pool. Light of various wavelengths interacts differently with the grating, thereby producing a rainbow. The rainbow, or interference pattern, is analogous to the community composition observed within a given landscape (T.F.H. Allen, personal communication). The interaction between the species pool and a given environment results in a community.

As a corollary, spatial patterns in the biotic interactions may exist, just as a diffraction grating exhibits variation in the interference pattern across its surface. The theory of fractal geometry (Mandelbrot, 1983; Peitgen and Saupe, 1988), involving simultaneous measurements at many scales, provides a means for identifying the spatial scales over which an ecological hypothesis is valid. Multiscale analysis also enables the extrapolation of ecosystem characteristics across scales. Multiscale approaches to understanding heterogeneity and its ecological consequences are discussed below following a formal description of heterogeneity.

Heterogeneity Ensuing from Scale-Dependent Structure

Ecologists recognize spatial heterogeneity as a major factor regulating the distribution of species (Wiens, 1976; Risser et al., 1984; Urban et al., 1987). Yet until recently no quantitative theory has explained the origin, dynamics, and consequences of heterogeneity in ways that could increase the accuracy of predictions about ecological processes in complex environments. Facing global climatic changes (Bradley et al., 1987; Cicerone, 1987), ecologists now require tools for relating atmospheric dynamics to the processes that are traditionally studied in 1-m^2 plots. This challenge requires a framework for transforming estimates of biotic and physical conditions across scales ranging several orders of magnitude. The scale-dependent variation in ecological conditions, e.g., diversity (Milne, 1988), is implicit in problems of scale transformation.

A comparison between the characteristics of uniform and heterogeneous sets suggests how the characteristics of heterogeneity may be used to make predictions. First, consider an environment with a constant value for some quantity, e.g., biomass (grams per square meter). Here it is crucial that the quantity be constant even if the environment is examined at the highest resolution possible. This environment is "uniform," and we expect all moments of the statistical distribution of biomass (i.e., mean, variance, skewness, kurtosis) to be equal among subsamples taken from within the environment. Only a euclidean plane composed of points and with a topological dimension $d = 2$ satisfies these criteria (recall that a point has $d = 0$, a curve has $d = 1$, and a plane has $d = 2$). Other sets violate these assumptions primarily by having a scale-dependent quantity that varies with scale for sufficiently coarse samples.

Mandelbrot (1983) recognized the ubiquity of sets that violate these assumptions of uniformity. In heterogeneous sets, estimates of quantities such as biomass vary precisely with the scale at which measurements are made (Burrough 1981, 1983, 1986; Milne, 1988). Changes in biophysical quantities can be predicted precisely, albeit statistically, by relations formulated from observations made at many scales.

For example, consider the density of a block of Swiss cheese. An estimate of cheese density would be zero if the block were viewed from within an air bubble having a 1-mm radius. Expanding the view radius would reveal a positive density as the cheese mass was penetrated. The density estimate would change continuously until a sufficiently long radius (i.e., length scale) was found that incorporated the relatively small variation imparted by air bubbles. Similar scale-dependent "density" occurs in two-dimensional maps (Orbach, 1986).

In fractal sets similar to Swiss cheese, density (i.e., mass/volume) varies as a power of the length scale used to estimate mass; the exponent is the fractal dimension (D). By definition, $D \leq d$, where $d =$ the dimension of the euclidean space in which the fractal is embedded (Mandelbrot, 1983; Orbach, 1986). The fractal dimension is obtained by regression for a rich set of fractal models (Lovejoy and Schertzer, 1985; Milne, 1988, 1991). For density we solve for D in

$$<p(s)> = \frac{B(s)^D}{C(s)^d} \qquad (1)$$

where $s =$ the length scale, B and $C =$ constants, and $d = 2$ or 3 for measurements in the plane and in 3-space, respectively (Orbach, 1986). Density fractal dimensions are estimated using many starting points owing to "local" variation that alters the estimate of D depending on the coordinates of the starting point—hence the "ensemble" mean density $<p(s)>$, as typically represented (Orbach, 1986). The density of a fragmented mass decreases with increasing scale.

The fractal dimension is usually constant over a finite range of scales

(Orbach, 1986; Milne, 1988). For example, mountainous topography appears fractal, but at the scale of the sand grain the topography may be treated as a series of "euclidean" spheres for which volume varies as s^3, i.e., $D = 3$. Landscapes exhibit ranges of length scales over which the fractal dimension of patch perimeters varies little but then suddenly increases (Krummel et al., 1987). Scale-invariant models of nature are limited to a finite range of length scales.

This discussion has described how fragmented patterns deviate in an important way from abstract euclidean sets such as the plane. Furthermore, fractal models allow fragmentation to be described precisely across a range of length scales, thereby facilitating predictions of how estimates of biophysical parameters vary with scale. Some ecological consequences of scale dependence are discussed below.

Ecological Consequences of Scale Dependence

The flow of gases across the landscape is relevant to understanding the mechanisms of global climatic change. In particular, the source–sink relations between soils, microbes, and plants potentially alter gas flux across the landscape (Gosz et al., 1988). Exactly how the spatial complexity of terrestrial environments modifies gas flux at the scale modeled in global circulation models is virtually unknown. A major challenge is to use information about the spatial distribution of sources and sinks to translate ecosystem processing of gases to regional scales.

Ecosystem gases may be divided into two classes. Gases without source–sink relations are described as "additive" (e.g., SF_6). For additive gases, estimates of the mass of gas in a well mixed atmosphere are the sums of the masses obtained from subsets of a more voluminous sample (M.E. Watwood, personal communication).

In contrast, ecologically interactive "nonadditive" gases (e.g., CO_2) are produced by sources and taken up by nearby sinks. The fractal distribution of soil and grass patches implies that source and sink density varies with scale, and that the flux of nonadditive gas is scale-dependent. For example, using a small chamber to measure CO_2 flux over separate patches of bare soil and grass yields estimates different from those obtained with a large chamber encompassing both soil and plants. Within the large chamber the source–sink interaction modifies gas flux relative to measurements taken in small chambers. The fractal geometry of soil and plant cover (Palmer, 1988) provides an opportunity for modeling the behavior of source–sink relations.

For example, satellite imagery of the Sevilleta National Wildlife Refuge (a Long-Term Ecological Research Site) was used to study the fractal geometry of bare soil (Milne, in press). Regression analysis indicated that 86% of the variation in the coverage of bare soil measured in the field was

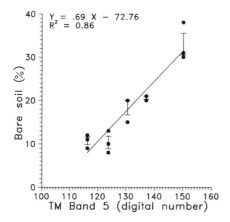

Figure 4.1. Relation between aerial coverage of bare ground and radiance. Radiance was measured using the short wave infrared band of the Landsat Thematic Mapper (TM) sensor (TM band 5). Vertical bars represent 1 SEM.

significantly related to reflectance in the short-wave infrared (Thematic Mapper band 5) (Milne, in review) measured using 30 m wide pixels (Fig. 4.1). The radiance of light measured by remotely sensed imaging was used to map soil coverage across the landscape.

The fractal geometry of bare soil was determined as follows. First, image pixels representing 0 to 9%, 10 to 19%, and 20 to 29% bare soil were separated into three distinct "patch types" on the image. Next, a digital map of each patch type was overlaid with grids of various mesh sizes (i.e., resolution 150 to 1500 m wide cells). Each grid was inspected, and the cells containing any portion of the patch type were tallied. For a uniform plane the number of cells containing the patch type increases as the square of the number of grid cells along one side of the map ($D = 2$) (Mandelbrot, 1983; Morse et al., 1985). In fragmented patch mosaics the number of cells increases exponentially with resolution according to the fractal dimension, where $D \leq d$; low dimensions indicate highly fragmented patch mosaics. This "plane-filling" fractal dimension describes the tendency of a mosaic to fill the plane.

The patch types differed significantly in their fractal dimensions (Fig. 4.2). The 0 to 9% bare soil patches were most plane-filling, with $D = 1.97$. The dimension of each patch type was constant across the range of scales studied (i.e., regression $R^2 > 0.99$), thereby providing a precise parameter describing fragmentation.

The plane-filling properties of the patch types diverged exponentially as resolution was increased (Fig. 4.2). At coarse resolution any two patch types occupied approximately the same number of grid cells (i.e., had similar density). At high resolution patch types with low dimensions occupied exponentially fewer cells than patch types having high fractal dimensions.

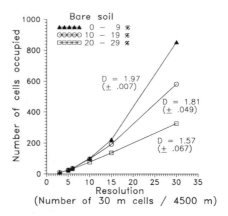

Figure 4.2. Scale-dependent variation in the aerial coverage of three classes of bare soil in the Sevilleta National Wildlife Refuge. Coverage was measured using 30-m pixels spanning a 4500 m wide, square region of grassland. The fractal dimension of each class is shown with 1 standard error of the estimate.

Of what consequence is this finding to studies of gas flux in landscapes? Presumably 1 cm³ of bare soil produces CO_2 at a given rate, and we can estimate the total gas produced per unit area if we know the proportion of the area occupied by bare soil. However, in fractal soil mosaics the proportion may vary with scale owing to the heterogeneous distribution of soil. The density of sinks may also vary with scale according to a different fractal dimension, making the interactions between sources and sinks scale-dependent.

At a given scale (s) the proportion (P_s) of a particular field sample (e.g., s = a 1 m wide chamber) having bare soil present over, for example, 15% of the area is

$$P_s = \frac{Bs^D}{Cs^2} \tag{2}$$

where B and C = numerical constants, Cs^2 = the area of the study region, and D = the dimension of the patch type having 15% bare soil. The numerator is analogous to "mass" in equation (1). Expression (2) is the portion of the study area occupied by the patch type at a given scale.

The proportions of the study area occupied by a given patch type at one scale versus another scale are related by Z, where

$$Z = P_s/P_{s-k} \tag{3}$$

P_{s-k} = the proportion of the study area occupied by the patch type when measured at a finer scale (i.e., higher resolution). Substituting from equa-

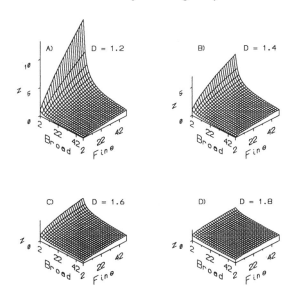

Figure 4.3. Factors (Z) by which fractal patch density at broad scales is related to density at fine scales. Variation in Z is illustrated for fractal dimensions ranging from 1.2 to 1.8. The discrepancy in density at one scale versus that in another varies with the fractal dimension of the pattern; it is highest for low dimensional patch mosaics (i.e., sparse coverage at all scales).

tion (2) indicates that Z varies as the ratio of the scale-dependent proportions.

$$Z = \frac{Bs^D \, C(s-k)^2}{Cs^2 \, B(s-k)^D}$$
$$= s^{D-2} \, (s-k)^{2-D} \qquad\qquad (4)$$

For a uniform plane with $D = 2$, $Z =$ unity, reflecting the lack of scale dependence for a euclidean set. For $D < 2$ there are nonlinear relations between Z, s, and $s - k$ (Fig. 4.3); see equation (4). Density in patch mosaics that nearly fill the plane (e.g., $D = 1.8$) exhibits little discrepancy when examined at disparate scales. However, the discrepancies grow exponentially with decreasing D (Fig. 4.3). Thus the relation in equation (4) prescribes how to transform measurements across scales, given the scale dependence of patch density. For gas flux, the aerial proportions at each scale need to be weighted by the net gas production and gas absorption rates for soil and grass, respectively.

These analytical results suggest that the difference in nonadditivity at one scale versus another is regulated by the fractal dimension of the bare soil and grass mosaic. Furthermore, extrapolating gas flux across scales involves calculating Z for many possible ratios of soil and grass and integrating across the ratios. Figure 4.3 illustrates the consequences of fractal geometry for translating biophysical processes across scales.

Scale-Dependent Relations Among Patch Types

Many problems in landscape ecology are multivariate, including the flux of nonadditive gases. An infinite number of soil classes could be defined, each affecting gas flux. Related problems exist in community ecology, where the associations of species in communities are scale-dependent, and the ecological interpretations are affected by both the algorithm used for analysis (Kenkel and Orloci, 1986) and the scale of the data (Noy-Meir and Anderson, 1973; Gardner et al., 1987; Getis and Franklin, 1987; Turner et al., 1989).

Scale dependence in community composition arises from the differential responses of species to environmental conditions. Some species are unconstrained and therefore omnipresent, whereas others are highly constrained and patchy at all scales. The quadrats used by ecologists often sample the joint spatial distribution of the species in a "cookie cutter" fashion, lumping the species into communities without consideration of the scaling behavior of component species (Milne, in press) or of temporal variation (Wiens, 1981). The scale-dependent association of species or patch types may be assessed by conducting measurements across a range of scales or by proper data transformation (Allen and Shugart, 1983). The origin of scale-dependent correlations among patch types in landscapes using landscape data aggregated at several levels of resolution is addressed next.

Example 1. Grassland Soil Distribution

Using the data represented in Figures 4.1 and 4.2, Spearman rank correlations were made comparing the aerial coverage of patch types within grid cells. Scale dependence of the correlations was clear (Fig. 4.4), and two major scale effects were observed.

First, the statistical significance of correlations varied with scale. Patch types 2 and 3 were not significantly correlated unless examined at a resolution of 30, i.e., using grid cells 900 m wide. The rate at which the correlation approached a significantly non zero value would be unknown without conducting the analysis at a variety of scales.

Second, the rate at which the correlations changed with scale reflected the numerical *difference* in the fractal dimensions of the two patch types considered. In Figure 4.2 the discrepancy between the aerial coverage of any two patch types increased with resolution. The discrepancy grew fastest for the two patch types differing greatly in their fractal dimensions. The fractal geometry of the patches imparts scale dependence to correlations because of the differential change in density predicted by the fractal

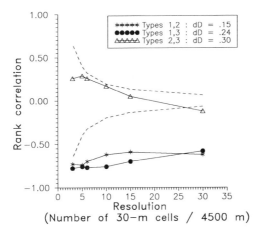

Figure 4.4. Scale-dependent Spearman rank correlations between patches of bare soil in the Sevilleta National Wildlife Refuge. Patch types 1, 2, and 3 represent 0 to 9%, 10 to 19%, and 20 to 29% bare soil, respectively. Each curve describes the correlations between two patch types at several levels of resolution. Curves outside the dashed lines represent correlations that were significantly different from zero ($p < 0.05$). The numerical difference in the fractal dimensions of each pair of patches is indicated as "dD", (for "delta D").

relation (equation 1). Co-occurrence generally is higher at low resolution and decreases with increasing resolution. At extremely high resolution (i.e., at one pixel) the correlation between patches must be zero because only one patch type can occupy a given pixel.

Example 2. New Jersey–Pennsylvania Landscapes

Multivariate analyses of landscape composition are also affected by scale-dependent relations. How the array of landscape elements present (e.g., hedgerows, streams, agricultural fields, woodlots) varied with the aerial extent of the sample in eastern Pennsylvania and New Jersey was examined. Fine-scale samples were predicted to exhibit greater among-sample variation in composition than samples collected over broad areas. In terms of example 1, above, the correlation of landscape elements approached zero as resolution was increased and more "unique" combinations of landscape elements were revealed.

Field data were collected to test the hypothesis that landscape composition is scale-dependent. The presence of 32 landscape elements was recorded for 140 segments of roadway (each segment 1.5 km) located randomly along a 400-km route through New Jersey and eastern Pennsylvania. Observations were limited to within approximately 1 km of the roadway. The 1.5-km data were aggregated in a nested fashion to provide samples at

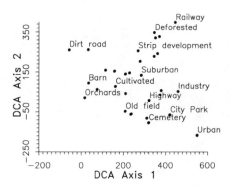

Figure 4.5. Detrended correspondence analysis ordination of landscape elements observed in 1.5 km long samples and aggregations of the samples in New Jersey and Pennsylvania. Each point represents a landscape element. Selected elements were labeled to preserve clarity.

5-, 15-, 30-, 60-, and 120-km scales. The 1.5 km long samples and the aggregated samples were ordinated together using detrended correspondence analysis (DCA) (Hill and Gauch, 1980). Although other techniques may have been equally effective for this purpose, DCA was chosen because of (1) its ability to provide a simultaneous ordination of samples and landscape elements, (2) its relative insensitivity to the number of landscape elements present in each sample, and (3) its nonparametric treatment of abundance distributions along major environmental axes (Peet et al., 1988). The latter characteristic ensured that samples with extreme ordination scores represented extreme positions in the original data space, and that the most central locations on the ordination axes represented the centroid of the data cloud. Thus DCA provided an appropriate framework for examining changes in landscape composition with scale.

The ordination indicated that landscape composition in New Jersey and Pennsylvania varied along a major gradient similar to one described by Forman and Godron (1986). The first ordination axis was readily interpreted as a gradient from undeveloped forested landscapes (dirt roads and conifer forests), to agricultural landscapes (barns and cultivated fields), to suburban areas, and finally to urban and industrial landscapes at the high end of DCA axis 1 (Fig. 4.5). High axis scores on the second axis represented sites exposed to a zinc smelter that deforested rural landscapes in eastern Pennsylvania.

Landscape composition varied with scale. High resolution samples exhibited greater among-sample variation in composition than the low-resolution samples (Fig. 4.6). Clearly, the probability of observing the same array of landscape elements in two samples decreased with sample area (Meentemeyer and Box, 1987). Aggregating samples at coarser scales resulted in a convergence of landscape composition, with the centroid of

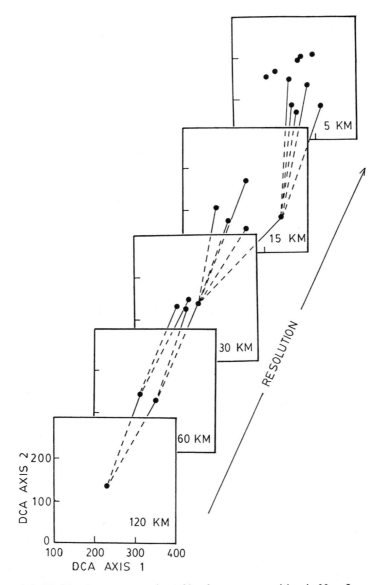

Figure 4.6. Multiscale representation of landscape composition in New Jersey and Pennsylvania. Each panel represents the detrended correspondence analysis (DCA) axes from Figure 4.5. Fine-scale samples (1.5 km) were aggregated at 5-, 15-, 30-, 60-, and 120-km scales and ordinated together. Dashed lines connect samples that were nested at a coarser resolution. To preserve clarity, incomplete nesting is shown for high-resolution samples.

the ordination represented by the 120-km sample, i.e., all landscape elements present.

Multiscale analyses illustrate how the scale dependence of patch density may alter ecological interpretations (Figs. 4.2, 4.4, 4.6). Hypotheses concerning the relative roles of competition, history, or abiotic factors are expected to hold over a limited range of length scales. Length scale limits arise when the scale-dependent juxtaposition of patches alters ecosystem processes or regulates the resources available to species (van Dorp and Opdam, 1987; Fahrig and Paloheimo, 1988). For example, Krummel et al. (1987) observed an increase in the perimeter fractal dimension of forest patches at broad-scales. The change was attributed to broad scale topographic features that increased perimeter/area ratios of forests relative to small, straight-edged patches that were regulated by agricultural practices. Regulatory processes may change with scale.

Conclusions

Species differ in terms of diet, metabolic rate, niche breadth, and the scale at which they encounter the environment (MacArthur, 1972; Brown, 1984; Ricklefs, 1987; Woodward, 1987). Consequently, the distribution of resources in the landscape appears differently to each species (Milne et al., 1989). Predicting the composition of species assemblages within a landscape may require specifying the utility of resources to each species and understanding the effects of scale-dependent variation in resource abundance on each species (Milne et al., in press).

Patch types or resources may exhibit any number of fractal distributions both within and among landscapes (O'Neill et al., 1988). Providing an organism locates a "patch" of resources, the mean density of resources is amplified as the resolution at which the organism samples the environment increases (equation 1). For example, Peters (1983) illustrated relations between body mass, home range area, and ingestion rates. Body mass and metabolic rate determine caloric requirements that are satisfied by foraging for sparse resources in the landscape. Species are constrained to occupy home ranges large enough to provide the necessary resources, as determined by the scaling of caloric requirements with body mass. Thus species assemblages within landscapes are partially regulated by the differential resource requirements of species and the scale dependence of resource density (Senft et al., 1987).

The scale dependence of natural patterns (Burrough, 1981) provides an opportunity to apply the precise calculus of fractal geometry to diverse problems of landscape ecology. Extrapolation of ecological measurements across scales requires a portrait of landscape heterogeneity such that scale-dependent variation in correlations and patch density are clear (Fig. 4.4). Ready applications include analyses of gas flux spanning meter to kilometer scales.

Landscape composition (Fig. 4.5) also varies with the scale at which observations are made. Thus the density of various patches in a landscape may appear differently to species that integrate information at disparate scales. In New Jersey and Pennsylvania, 5 km wide samples of the landscape exhibited tremendous variation that was reduced by aggregating data at coarser scales (Fig. 4.6). Thus wide-ranging species may see a more homogeneous landscape than do fine-scale species.

The interaction between the scale at which species sample the environment and the fractal distribution of a resource complements other explanations of species abundance (May 1988). High-resolution species should perceive exponentially greater resource density (equation 1) (Morse et al., 1985) and therefore should achieve greater abundance than coarse-resolution species. Consider the simplest case, in which just one individual occurs for each spatial unit that contains sufficient resources to support one individual. Clearly, the total abundance is high if individuals obtain their minimal requirements from small spatial units and if there are many such units. The relevant difference between species relates to their ability to detect resources within their home ranges, for example, which may be of different sizes.

The prediction that abundance is proportional to perceptual resolution can be rectified with Kolasa's (1989) apparently contradictory prediction that coarse-resolution species are most abundant. Kolasa's definition of resolution is based on the frequency of occurrence among potential habitat sites; his coarse-resolution species occupy most sampling sites. Equation (1) (and the associated definition of resolution) implies that, for a given total amount of resource, high-resolution species will occupy many sites because they detect exponentially more sites with suitable resource levels by virute of the scale at which they view the landscape. With many suitable sites to occupy, such species will indeed be abundant, and an ecologist applying a fixed sampling area will find high-resolution species in great abundance in most samples. The ecologist's use of a constant measuring scale aggregates the habitat units used by high-resolution species, thereby making the species appear to be indiscriminate in their habitat use. The possibility that species perceive different resource densities suggests that asymmetrical exploitative competition may ensue if one species perceives greater resource abundance than another in a given landscape.

This discussion may be extended to other fractal models, e.g., models sensitive to perimeter lengths (Krummel et al., 1987; Milne, 1988). For example, Brittingham and Temple (1983) implicated forest patch edges in the success of the brown-headed cowbird, a common nest parasite. The bird's approximately 300-m penetration distance into forests is the "scale" at which it perceives forest edges. Other parasites have different penetration distances and thereby "perceive" an exponentially different distribution of patch edges in the landscape, e.g., bark beetles (Rykiel et al., 1988). Likewise, scale-dependent perimeter effects could alter the spread of fire if fuel loads or wind conditions vary with distance from the edge

(Franklin and Forman, 1987). Edge perception is likely to vary as a function of the fractal dimension of the forest patch perimeter.

Ecology is interesting because of the differential responses of species to a given environment. Fractal theory provides a means of integrating the spatial characteristics of complex environments, thereby providing models of heterogeneity in landscapes. A key to understanding heterogeneity is to conduct studies across a range of scales and to extract parameters that are robust to changes in scale.

Acknowledgments. I thank Timothy F.H. Allen and Tom Hoekstra for voicing the interference pattern concept of communities, and Brad Musick for constant criticism and advice concerning remote sensing. Richard T.T. Forman and the 1986 Landscape Ecology class at Harvard University assisted with the New Jersey–Pennsylvania field study. Sterling Grogan and Sandra J. Turner provided helpful comments on the manuscript. Funding was provided by NSF grants BSR-8806435 and BSR-8614981 and DOE grant DE-FGO4-88ER60714. Sevilleta LTER contribution no. 2.

References

Allen TFH, Shugart HH (1983) Ordination of simulated complex forest succession: a new test of ordination methods. Vegetatio 51:141–155

Allen TFH, Starr TB (1982) *Hierarchy: Perspectives for Ecological Complexity.* University of Chicago Press, Chicago

Allen TFH, O'Neill RV, Hoekstra TW (1984) *Interlevel Relations in Ecological Research and Management: Some Working Principles from Hierarchy Theory.* USDA General Technical Report RM-110 US Department of Agriculture Forest Service, Washington, DC

Austin MP (1985) Continuum concept, ordination methods, and niche theory. Annu Rev Ecol Syst 16:39–61

Bradley RS, Diaz HF, Eischeid JK, Jones PD, Kelly PM, Goodess CM (1987) Precipitation fluctuations over northern hemisphere land areas since the mid-19th century. Science 237:171–175

Brittingham MC, Temple SA (1983) Have cowbirds caused forest songbirds to decline? Bioscience 33:31–35

Brown JH (1984) On the relationship between abundance and distribution of species. Am Nat 124:255–279

Burrough PA (1981) Fractal dimensions of landscapes and other environmental data. Nature 294:241–243

Burrough PA (1983) Multiscale sources of spatial variation in soil. I. Application of fractal concepts to nested levels of soil variations. J Soil Sci 34:577–597

Burrough PA (1986) *Principles of Geographical Information Systems for Land Resources Assessment.* Clarendon Press, Oxford

Cicerone RJ (1987) Changes in stratospheric ozone. Science 237:35–42

Cole K (1985) Past rates of change, species richness, and a model of vegetational inertia in the Grand Canyon, Arizona. Am Nat 125:289–303

Cowles HC (1899) The ecological relations of the vegetation of the sand dunes of Lake Michigan. Bot Gaz 27:95–117, 167–202, 281–308, 361–391

Delcourt PA, Delcourt HR (1987) *Long-term Forest Dynamics of the Temperate Zone*. Springer-Verlag, New York

Fahrig L, Paloheimo J (1988) Effect of spatial arrangement of habitat patches on local population size. Ecology 69:468–475

Franklin JF, Forman RTT (1987) Creating landscape patterns by forest cutting: ecological consequences and principles. Landscape Ecol 1:5–18

Forman RTT, Godron M (1986) *Landscape Ecology*. Wiley, New York

Gardner RH, Milne BT, Turner MG, O'Neill RV (1987) Neutral models for the analysis of broad-scale landscape pattern. Landscape Ecol 1:19–28

Getis A, Franklin J (1987) Second-order neighborhood analysis of mapped point patterns. Ecology 68:473–477

Gleason HA (1926) The individualistic concept of the plant association. Bull Torrey Bot Club 53:7–26

Gosz JR, Dahm CN, Risser PG (1988) Long-path FTIR measurement of atmospheric trace gas concentrations. Ecology 69:1326–1330

Hill MO, Gauch HG (1980) Detrended correspondence analysis, an improved ordination technique. Vegetatio 42:47–58

Huston MA (1979) A general hypothesis of species diversity. Am Nat 113:81–101

Kenkel NC, Orloci L (1986) Applying metric and nonmetric multidimensional scaling to ecological studies: some new results. Ecology 67:919–928

Kolasa J (1989) Ecological systems in hierarchical perspective: breaks in the community structure and other consequences. Ecology 70:36–47

Krummel JR, Gardner RH, Sugihara G, O'Neill RV, Coleman PR (1987) Landscape pattern in a disturbed environment. Oikos 48:321–324

Little EL (1971) *Atlas of United States Trees. Vol 1. Conifers and Important Hardwoods* USDA Miscellaneous Publication 1146. US Department of Agriculture Forest Service, Washington, DC

Lovejoy S, Schertzer D (1985) Generalized scale invariance in the atmosphere and fractal models of rain. Water Resources Res 21:1233–1250

MacArthur RH (1972) *Geographical Ecology*. Princeton University Press, Princeton

Mandelbrot B (1983) *The Fractal Geometry of Nature*. WH Freeman, New York.

Marks PL (1974) The role of pin cherry (*Prunus pensylanica* L.) in the maintenance of stability in the northern hardwood ecosystem. Ecol Monogr 44:73–88

May RM (1988) How many species are there on earth? Science 241:1441–1449

Meentemeyer V. Box EO (1987) Scale effects in landscape studies. In Turner MG (ed) *Landscape Heterogeneity and Disturbance*. Springer-Verlag, New York. pp 15–34

Milne BT (1988) Measuring the fractal geometry of landscapes. Appl Math Comput 27:67–79

Milne BT, 1991. (in press) Lessons from applying fractal models to landscape patterns. In Turner MG, Gardner RH (eds) *Quantitative Methods in Landscape Ecology*. Springer-Verlag, New York, pp 199–235

Milne BT (in review) Remote sensing and habitat measurement: a new tool for the Galapagos. In Snell H, Fritts T (eds) *Herpetology of the Galapagos: Research and Management*

Milne BT (in press) Spatial aggregation and neutral models in fractal landscapes. Am Nat

Milne BT, Turner MG, Wiens JA, Johnson AR (in press) Interactions between the fractal geometry of landscapes and allometric herbivory. Theor. Pop. Biol.

Milne BT, Forman RTT (1986) Peninsulas in Maine: woody plant diversity, distance, and environmental patterns Ecology 67:967–974

Milne BT, Johnston K, Forman RTT (1989) Scale-dependent proximity of wildlife habitat in a spatially-neutral bayesian model. Landscape Ecology 2:101–110

Morse DR, Lawton JH, Dodson MM, Williamson MH (1985) Fractal dimension of vegetation and the distribution of arthropod body lengths. Nature 314:731–734

Neilson RP, Wullstein LH (1983) Biogeography of two southwest American oaks in relation to atmospheric dynamics. Biogeogr 10:275–297

Noy-Meir I, Anderson DJ (1973) Multiple pattern analysis, or multiscale ordination: towards a vegetation hologram? In Patil GP, Pielou EC, Waters WE (eds) *Many Species Populations, Ecosystems, and Systems Analysis, Statistical Ecology*. Vol 3. Pennsylvania State University Press, University Park, pp 207–231

O'Neill RV, Krummel JR, Gardner RH, Sugihara G, Jackson B, DeAngelis DL, Milne BT, Turner MG, Zygmunt B, Christensen SW, Dale VH, Grahm RL (1988) Indices of landscape pattern. Landscape Ecol 1:153–162

Orbach R (1986) Dynamics of fractal networks. Science 231:814–819

Palmer MW (1988) Fractal geometry: a tool for describing spatial patterns of plant communities. Vegetatio 75:91–102

Peet RK, Knox RG, Case JS, Allen RB (1988) Putting things in order: the advantages of detrended correspondence analysis. Am Nat 131:924–934

Peitgen H-O, Saupe D (eds) (1988) *The Science of Fractal Images*. Springer-Verlag, New York

Peters RH (1983) *The Ecological Implications of Body Size*. Cambridge University Press, New York

Ricklefs RE (1987) Community diversity: relative roles of local and regional processes. Science 235:167–171

Risser PG, Karr JR, Forman RTT (1984) *Landscape ecology: Directions and approaches*. Illinois Natural History Survey Special Publication 2, Champaign, Il

Root T (1988) Energy constraints on avian distributions and abundances. Ecology 69:330–339

Rykiel EJ Jr, Coulson RN, Sharpe PJ, Allen TFH, Flamm RO (1988) Disturbance propagation by bark beetles as an episodic landscape phenomenon. Landscape Ecol 1:129–139

Senft RL. Coughenour MB, Bailey DW, Rittenhouse LR, Sala OE, Swift DM (1987) Large herbivore foraging and ecological hierarchies. Bioscience 37:789–799

Sousa WP (1979) Experimental investigations of disturbance and ecological succession in a rocky intertidal algal community. Ecol Monogr 49:227–254

Tilman D (1988) *Plant Strategies and the Dynamics and Structure of Plant Communities*. Princeton University Press, Princeton

Turner MG, O'Neill RV, Gardner RH, Milne BT (1989) Effects of changing spatial scale on the analysis of landscape pattern. Landscape Ecology 3:153–162

Urban DL, O'Neill RV, Shugart HH (1987) Landscape ecology. Bioscience 37:119–127

Van Dorp D, Opdam PFM (1987) Effects of patch size, isolation and regional abundance on forest bird communities. Landscape Ecol 1:59–73

Whittaker RH (1967) Gradient analysis of vegetation. Biol Rev 42:207–264

Whittaker RH, Levin SA (1977) The role of mosaic phenomena in natural communities. Theor Pop Biol 12:117–139

Wiens J (1976) Population responses to patchy environments Annu Rev Ecol Syst 7:81–120

Wiens J (1981) Single sample surveys of communities: are the revealed patterns real? Am Nat 117:90–98

Woodward FI (1987) *Climate and Plant Distribution*. Cambridge University Press, New York

5. Heterogeneity and Spatial Hierarchies

Robert V. O'Neill, Robert H. Gardner, Bruce T. Milne,
Monica G. Turner, and Barbara Jackson

To apply the traditional scientific method, ecologists ordinarily focus on the mean or central tendency of a data set. For example, a typical hypothesis test would involve demonstrating that the mean is significantly different from a control measurement. However, ecological systems are heterogeneous, and much information may be lost if the variance of a data set is ignored. This chapter shows that a specific prediction of hierarchy theory can be tested by examining how variance changes as measurements are taken across a range of scales.

A number of authors (e.g., MacMahon et al., 1978; Webster, 1979; Eldredge, 1985; Salthe, 1985) have suggested that hierarchy theory can make significant contributions to the study of ecological systems. The theory views system dynamics as isolated into discrete scales. Interacting entities, such as organisms and populations, operate at a similar spatiotemporal scale and are relatively isolated from dynamics at much large or smaller scales (Allen and Starr, 1982).

The theory does not maintain that every ecological system must necessarily be hierarchical (O'Neill, 1989). Rather, it points out that stable complex systems often take on such a structure. O'Neill et al. (1986) developed the concept that hierarchical structuring permits complex systems to develop by combining relatively stable lower level systems. Though hierarchical structuring is not the only way to achieve stable complex systems, such structure appears to be common in physical, chemical, and biological systems (O'Neill et al., 1986). When a level structure is found, the investigator can take advantage of this fact to simplify the investigation of complex systems (Overton, 1972; Allen et al., 1987).

Although the theory shows considerable promise for organizing the study of ecological systems, it is difficult to test theory predictions. As pointed out below, the simplest test requires estimates of heterogeneity across a broad spectrum of time–space scales, and most ecological investigations focus on one or a few scales. However, as McIntosh (1985) pointed out, much of the current interest in scale and hierarchy builds on a considerable tradition of work on vegetation analysis on landscapes (e.g., Greig-Smith, 1957). Therefore it is not surprising that it is at the landscape scale that test data are available.

In the present study the data were available in the form of digitized maps. A digitized map can be sampled at consecutive scales from the finest resolution to the extent of the total map. It provides one of the few sources of data that is virtually continuous over a significant range of scales.

The purpose of this study was to test predictions about how heterogeneity changes with scale on a landscape. First, theoretical predictions were developed on how heterogeneity should change if the landscape is hierarchically structured. Then the predictions were tested by analyzing the patterns sampled on digitized maps of land cover.

Theoretical Development

The analysis begins with the simple observation that the variance (S^2) associated with an estimate of the mean is inversely proportional to sample size.

$$S^2 \sim 1/n \tag{1}$$

Where $n =$ the number of samples or the size of a sample quadrat. Taking the logarithm of both sides of equation (1) and introducing a proportionality constant (a) we find

$$\ln S^2 = a - \ln n \tag{2}$$

which is a form of the equation for a straight line. Therefore if $\ln S^2$ is plotted as a function of $\ln n$, the result is a straight line with a slope of -1. This result holds whenever the sample population is randomly distributed in space so that each new sample is independent. However, if spatial correlations exist with correlation coefficient (r), the next sample will not be independent and the slope of the log-log plot will lie between -1 ($r = 0$) and 0 ($r = 1$) (Smith, 1938). Wiegert (1962) used this approach to study vegetation on quadrats of different scales. Levin and Buttel (1986) proposed that the deviation of the slope from -1 is a measure of the spatial scale or patchiness of the landscape.

In previous studies it has been assumed that the landscape has no scale structure; i.e., the same structure or pattern occurs at all scales. In contrast, hierarchy theory (Allen and Starr, 1982; O'Neill et al., 1986) predicts that complex dynamic systems such as landscapes often have a hierarchical structure. For a hierarchically structured landscape, we would expect processes that affect pattern to be isolated at discrete scales. At these scales, the slope of the $\ln S^2/\ln n$ relation should be significantly less than -1.0. At intermediate scales, we would expect no pattern, i.e., slopes of -1.0. The expectation, then, is that the log-log plot would appear like a staircase, having discrete scales with shallow slopes (steps) alternating with slopes of -1.0 (risers).

Materials and Methods

To test the hypothesis we used the land use and land cover digital data produced by the US Geological Survey (Fegeas et al., 1983). Six data tapes were made available through the Oak Ridge Geographic Data Systems Group. The tapes are at a resolution of 1:250,000 and contain land use information on a 200-m grid. Each landscape tape contains 525×850 grid points. The tapes and land uses used for this study are shown in Table 5.1.

The sampling design involved a set of radiating transects. Starting near the center of the map, 32 line transects radiated outward, separated by approximately 11.25^0. The shortest transects were 5 grid points (1000 m) in length and the longest was 150 (30,000 m). At each of 30 transect lengths, it was assumed that the 32 transects sampled an area equal to a circle with radius equal to the transect length. The percentage of the grid points along each transect in a specific land use (Table 5.1) was recorded. This method permitted 32 samples at each of 30 scales for estimating mean and variance.

Table 5.1. Landscape Scenes and Land Uses Used for the Analysis of Heterogeneity

Landscape Scene	Abbreviation	Land Use	% Cover
Goodland, Kansas	GD	Grasslands	29
West Palm Beach, Florida	WP	Grasslands	18
Macon, Georgia	MC	Forest	54
Greenville, South Carolina	GV	Forest	62
Natchez, Mississippi	NZ	Forest	55
Knoxville, Tennessee	KX	Forest	76

The abbreviations are those used in Table 5.3. Percent cover refers to the percentage of the grid points on the total landscape scene covered by the land use in the third column.

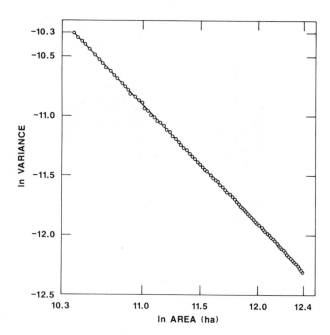

Figure 5.1. Variance in percent barren ground in Goodland, Kansas over a range of sample sizes. Barren ground shows no hierarchical structuring, and variance decreases with a slope of − 1.0.

Heterogeneity in natural vegetation cover, e.g., grassland and forest, should be the best indicator of underlying hierarchical structure. Land uses such as urban development or agriculture are likely to be influenced by human scales of activity and reveal little about hierarchical structuring of natural landscapes. Land uses such as barren ground that are rare (1.77% on the Goodland, Kansas landscape) and randomly scattered are unlikely to be structured by any specific process. Figure 5.1 shows the log-log graph for barren ground on the Goodland, Kansas landscape. The data lie exactly on the − 1.0 line expected for unpatterned land uses (equation 2).

Two methods were used to determine if the expected hierarchical structure was present on the landscape. The first method used regression analysis on a sliding window covering three successive transect lengths, i.e., three successive scales. For example, the first regression of ln variance versus ln area covered transects of length 5, 6, and 7. The second regression covered transects of lengths 6, 7, and 8. To decide that a set of scales represented a hierarchical level, the following criteria had to be satisfied.

1. Slopes had to alternate from more than − 0.5 (ideally approaching − 1.0) to less than − 0.5 (ideally approaching 0) and back to more than − 0.5. The greater the deviation above or below − 0.5, the greater is the confidence that a level had been established.

2. More than a single scale (ideally many consecutive scales) would be found on each step and riser. It is possible that a discrete level or the interval between levels could be represented by a single slope (i.e., three consecutive points). However, confidence is increased if there is a series of points representing each step or riser.

For landscapes that satisfied these criteria, a further analysis was applied in the form of piece-wise linear regression. In essence, the analysis takes the first two points and fits them with a straight line. It then estimates the next point. If the estimate is within a criterion supplied by the user (in our case 0.05), the point is lumped with the first two and a new regression calculated. This process continues until the linear regression fails to predict the next data point ± 0.05. At this point a break is defined, and the process begins anew. In this way the fitted curve is a series of linear segments. According to the hierarchical hypothesis, the linear segments should alternate between line segments with slopes close to -1.0 and line segments with much shallower slopes.

Results

The first task was to determine the range of scales that would accurately reflect any hierarchical structuring on the landscape. We are interested in the pattern over the range where the variance behaves according to equation (2). This expected pattern, in fact, occurs only over intermediate scales on the maps. At the smallest scales, i.e., transect lengths between 5 and 10, variance tends to increase with scale. Each successive sample moves into new pattern elements, and the variance increases until a scale is reached that integrates across the microheterogeneity. In general, variance began to follow equation 2 at transect lengths of 10 and larger, but the general criterion was to begin the analysis when the variance turned downward.

At the opposite extreme, variance would rise at the largest scales sampled. Imagine, for example, that you are sampling an area of mixed agriculture and grassland. At the next larger scale, you suddenly cross a barrier such as a river, estuary, or topographical barrier. For subsequent samples, the variance increases until you reach a scale at which both the old and new patterns are being adequately sampled. For most maps, the variance began to increase at transect lengths of about 45. For purposes of our analysis, we ignored scales beyond 45 that showed increased variance. In general, we considered transects from 10 grid points (1,257 hectares) to 45 grid points (25,447 hectares).

There is a potential bias that results from the radial design. At small scales, the transects are close to each other and may show greater correlation than would be expected. If this bias exists, the smallest scales should show shallow slopes. In fact, the variance increases, as explained above.

Table 5.2. Sliding Window Regressions

Linear Regression (Ln) Range (area)	GD	WP	MC	GV	NZ	KX
7.14–7.50	−0.07		−0.11			
7.33–7.66	−0.09		−0.06	−0.06		
7.50–7.81	−0.03		−0.19	+0.07		
7.66–7.95	−0.15		−0.38	+0.002		
7.81–8.08	−0.27		−0.38	−0.22		
7.95–8.20	−0.04	0	−0.28	−0.28	−0.63*	
8.08–8.31	−0.07	0	−0.56*	−0.17	−1.09*	
8.20–8.42	−0.60*	−0.04	−0.82*	−0.36	−1.04*	
8.31–8.52	−0.93*	−0.33	−1.09*	−0.57*	−1.00*	
8.42–8.62	−0.75*	−0.65*	−1.55*	−0.60*	−1.00*	
8.52–8.71	−0.68*	−0.63*	−1.22*	0.32	−0.84*	
8.62–8.80	−0.68*	−0.55*	−0.94*	0.00	−0.83*	
8.71–8.89	−0.74*	−0.39	−1.00*	−0.11	−0.83*	−0.22
8.80–8.97	−0.58*	−0.29	−0.94*	−0.12	−0.59*	−0.18
8.89–9.05	−0.40	−0.25	−0.44	−0.38	−0.31	−0.25
8.97–9.12	−0.27	−0.14	+0.07	−0.66*	−0.13	−0.40
9.05–9.20	−0.46	0	−0.07	−0.66*	−0.13	−0.26
9.12–9.26	−0.73*	0	0	−0.84*	−0.61*	−0.21
9.20–9.33	−0.86*	−0.08	+0.21	+0.76*	−0.60*	−0.38

Range						
9.26–9.40	− 1.01*	− 0.07	+ 0.14	− 0.93*	− 0.07	− 0.36
9.33–9.46	− 1.05*	+ 0.07	− 0.07	− 1.68*		− 0.31
9.40–9.52	− 0.87*	+ 0.17	− 0.08	− 1.92*		− 0.17
9.46–9.58	− 0.40	+ 0.17	0	− 1.66*		+ 0.08
9.52–9.64	− 0.04	+ 0.08	− 0.08	− 1.33*		− 0.17
9.58–9.70	− 0.27	− 0.33	0	− 0.75*		− 0.50
9.64–9.75	− 0.51*	− 0.64*	+ 0.09			− 0.46
9.70–9.81	− 0.49	− 0.73*	− 0.19			− 0.55*
9.75–9.86	− 0.47	− 0.82*	− 0.39			− 0.72*
9.81–9.91	− 0.83*	− 0.80*	− 0.50			− 0.90*
9.86–9.96	− 10.7	− 0.50				− 1.00*
9.91–10.01	− 0.97	− 0.43				− 0.79*
9.96–10.05		− 0.32				
10.01–10.09		0				

Each entry represents the slope of three successive samples, covering the range indicated in the first column. Slopes greater than − 0.05 are indicated by an asterisk.

Even after the variance begins to decrease, the first slope might be shallow because of this bias. However, on at least one scene, Natchez Table 5.2), the initial slope is steep rather than shallow. Furthermore, for land uses that are distributed randomly, e.g., barren ground at Goodland (Fig. 5.1), the slope of the line is exactly − 1.0, indicating no bias due to the sample design. It seems, therefore, that the potential bias at small scales does not seriously affect the results presented. Nevertheless, some bias is possible on the first slope after the variance starts to decrease, and it may not be circumstantial that most maps show shallow slopes first. Some caution is necessary when interpreting the first shallow slope as indicating a clear hierarchical level.

An additional source of potential bias is the lack of independence of successive samples, a problem shared by all nested quadrat designs. However, this bias tends to make it more difficult to find breaks in the line, as successive points tend to be similar. Therefore this bias works to make the analysis conservative and increases the confidence one can place in changes in slope when they are located.

As a preliminary indication of the hierarchical structuring discovered in this study, Figure 5.2 shows the data for Goodland, Kansas with line segments fitted by piece-wise linear regression. The data appear to break into

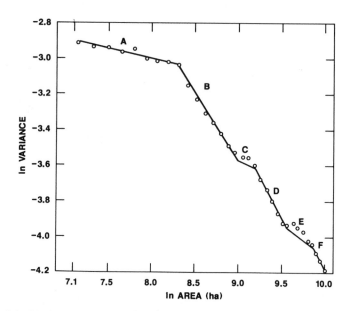

Figure 5.2. Variance in percent grassland at Goodland, Kansas over a range of sample sizes. Hierarchical structuring is indicated by scales with slopes much less than − 1.0 (A, C, E) alternating with scales with slopes approaching − 1.0 (B, D, F).

six line segments with slopes that alternate from shallow to steep. The curve is in marked contrast to the random, unpatterned data in Figure 5.1.

The piece-wise regressions for six landscape scenes are given in Table 5.2. The pattern that most closely matches the expected hierarchical structuring is that of Goodland, Kansas (GD). On this landscape, from 7.14 (1257 ha) to 8.08 (3217 ha) there is a sequence of seven windows showing shallow slopes with the smallest at -0.03, followed by an abrupt change over the range from 8.20 (3631 ha) to 8.80 (6648 ha), with seven windows having large slopes, the largest being -0.93. This grouping followed by three windows with the smallest slope being 0.27, five windows with slopes reaching -1.05, three windows with a smallest slope of -0.04, and six windows with slopes up to $=1.07$. There appears clear evidence here of three levels of scale separated by scales showing random distributions.

A multilevel structure also is evident in the data for Macon, Georgia (MC). Here, a set of shallow slopes is followed by a set of steep slopes, then by another set of shallow slopes. The shallow slopes approach 0 and the steep slopes approach -1.0.

In the case of West Palm Beach (WP) and Greenville (GV), a similar pattern is evident: shallow, steep, shallow, steep. However, in both cases, the intermediate steep slopes do not approach -1.0 (-0.6 to -0.65). Furthermore, the steep slope is found only over two or three windows. Thus the evidence that there are two distinct scales separated by a random interval is not strong. Nevertheless, the evidence is clear that there is at least one set of patterned scales and another distinct region where the data appear to be random.

For Natchez and Knoxville, there is clear evidence only for one level, either preceded or followed by a random section. There is no clear evidence for a multilevel structure, but there is evidence that pattern exists at one set of scales with conditions approaching randomness at other scales.

Table 5.3 shows the results of the piecewise linear analysis of the four landscape scenes that indicate multiple levels in Table 5.2 Goodland and Macon clearly show the pattern of alternating shallow slopes and slopes approaching -1.0. Greenville and West Palm Beach show the predicted pattern, but the steep line segments are far from -1.0. The piecewise linear analysis confirms the results obtained by the sliding window analysis: There are line segments that alternate between shallow (correlated or patterned) and steep (uncorrelated or random) slopes.

Discussion

The present study demonstrates the importance of analyzing heterogeneity in ecological systems. The analysis permits a test of the prediction of hierarchy theory that complex systems, such as landscapes, are often structured into discrete levels. There is evidence of a multilevel structure on

Table 5.3. Piecewise Linear Regression of Landscape Pattern

Landscape	Line Segments			Slope	
	Begin	End	No.	Shallow	Steep
Goodland	7.14	8.32	8	− 0.11	
	8.32	9.02	7		− 0.76
	9.02	9.26	3	− 0.27	
	9.26	9.52	5		− 0.95
	9.52	9.91	5	− 0.27	
	9.91	10.00	5		− 0.96
West Palm Beach	7.81	8.42	6	− 0.003	
	8.42	8.89	5		− 0.52
	8.89	9.52	10	− 0.018	
	9.52	9.91	7		− 0.65
	9.91	10.14	5	− 0.039	
Macon	7.14	8.20	8	− 0.22	
	8.20	8.97	8		− 1.13
	8.97	9.81	13	− 0.005	
Greenville	7.33	8.31	8	− 0.10	
	8.31	8.62	3		− 0.60
	8.62	8.97	4	− 0.11	
	8.97	9.70	11		− 1.20
Natchez	7.95	8.89	10		− 0.92
	8.89	9.12	3	− 0.13	
	9.12	9.26	2		− 1.33
	9.26	9.40	2	− 0.14	

four of the six landscapes examined, and the evidence appears using two methods for analyzing the data. The conclusion that the levels are not artifacts is confirmed by the fact that the levels do not appear in such land uses, as barren ground (Fig. 5.1), which is unlikely be hierarchically structured.

The theory does not predict that every landscape must be hierarchically structured, and indeed two of the landscapes examined do not show convincing evidence of a multilevel structure. At the present time we are unable to identify any simple criterion that predicts a priori whether a landscape scene will or will not show a hierarchical pattern. This question seems to be an important one to be addressed by hierarchy theoreticians.

The current analysis raises additional questions for hierarchy theory. For example, it would be interesting to have a prediction of how many hierarchical levels to expect over a given range of scales, e.g., 1,257 to 25,447 hectares. If the theoretical expectation was that the levels should be separated by an order of magnitude, it is not surprising that some of our scenes showed only one level.

One of the most unsatisfying aspects of the current study is that the analysis gives no hint of what is causing the level structure on the landscapes. In nested hierarchies, dynamics and pattern at larger scales are the result of interactions among lower level systems. Thus the level structure is generated by dynamics intrinsic to the nested system. There is some evidence that larger-scaled patterns in vegetation can result from such lower level interactions (Anderson, 1971). However, this interaction is not the only way a level structure can be produced. In the case of landscape ecology, it is not at all clear that higher level structure is due to intrinsic dynamics. Each level may represent a new level of constraints imposed on the system from outside. In the present study one suspects, at least, that larger scale patterns are being imposed by topography, coastlines, or patterns in geological variables such as soil. The question of what is causing the level structure detected in our analyses must await further research for resolution.

Summary

Ecologists often focus on controlled experiments in which the central tendency of a data set is of primary interest. However, patterns in the heterogeneity of the data can reveal much about the structure of the system. For example, hierarchy theory predicts that variance will decrease slowly at scales corresponding to hierarchical levels and more rapidly at intermediate scales. This prediction is tested on data for cover types at the landscape scale. On several landscapes, the analysis of heterogeneity revealed the underlying spatial structure of the system.

Acknowledgments. This research was supported in part by the Office of Health and Environmental Research, Department of Energy under contract DE-AC05-840R21400 with Martin Marietta Energy Systems, Inc.; in part by the Ecosystem Studies Program, National Science Foundation under grant BSR-8614981 to the University of New Mexico and Interagency Agreement NSF-BSR-8315185; and in part by an Alexander Hollaender Distinguished Postdoctoral Fellowship, administered by Oak Ridge Associated Universities, to M.G. Turner. (Publication No. 3612, Environmental Sciences Division, Oak Ridge National Laboratory.)

References

Allen TFH, Starr TB (1982) *Hierarchy: Perspectives for Complexity.* Chicago University Press, Chicago
Allen TFH, O'Neill RV, Hoekstra T (1987) Interlevel relations in ecological research and management: some working principles from hierarchy theory. J Appl Syst Anal 14:63–79
Anderson DJ (1971) Spatial patterns in some Australian dryland plant communities. In Patil GP, Pielou EC, Waters WE (eds) *Statistical Ecology, Vol 1. Spatial Patterns and Statistical Distributions.* Pennsylvania State University Press, University Park, pp 271–285

Eldredge N (1985) *Unfinished Synthesis: Biological Hierarchies and Modern Evolutionary Thought*. Oxford University Press, New York

Fegeas RG, Claire RW, Guptill SC, Anderson KE, Hallam CA (1983) *Land Use and Land Cover Digital Data*. Geological Survey Circular 895-E. US Geological Survey, Cartographic Information Center, Reston, VA

Greig-Smith P (1957) *Quantitative Plant Ecology*. Butterworth, London

Levin SA, Buttel L (1986) *Measures of Patchiness in Ecological Systems*. Publication ERC-130. Ecosystem Research Center, Cornell University, Ithaca

MacMahon JA, Phillips DL, Robinson JV, Schimpf DJ (1978) Levels of biological organization: an organism-centered approach. Bioscience 28:700–704

McIntosh RP (1985) *The Background of Ecology*. Cambridge University Press, Cambridge

O'Neill RV (1989) Perspectives in hierarchy and scale. In May RM, Roughgarten J (eds) *Ecological Theory*. Princeton University Press, Princeton, pp 140–156

O'Neill RV, DeAngelis DL, Allen TFH, Waide JB (1986) *A Hierarchical Concept of Ecosystems*. Princeton University Press, Princeton

Overton WS (1972) Toward a general model structure for forest ecosystems. In Franklin JF (ed) *Proceedings of the Symposium on Research on Coniferous Forest Ecosystems*. Northwest Forest Range Station, Portland

Salthe S (1985) *Evolving Hierarchical Systems: Their Structure and Representation*. Columbia University Press, New York

Smith HF (1938) An empirical law describing heterogeneity in the yields of agricultural crops. J Agric Sci 28:1–3

Webster JR (1979) Hierarchical organization of ecosystems. In Halfon E (ed) *Theoretical Systems Ecology*. Academic Press, Orlando, pp 119–131

Wiegert RG (1962) The selection of an optimum quadrat size for sampling the standing crop of grasses and forbs. Ecology 43:125–129

6. Communities in Patchy Environments: A Model of Disturbance, Competition, and Heterogeneity

Hal Caswell and Joel E. Cohen

All landscapes are to some extent patchy. The biological heterogeneity of communities on patchy landscapes reflects the time scales of local biotic interactions and abiotic disturbance, the time and space scales of dispersal, and (especially) the interaction of these scales. To investigate these factors, we examine here a simple model that provides a framework for building models of patchy communities directly from hypotheses about time scales. The model has numerous applications (Caswell and Cohen, 1991, in preparation); here we focus on the interplay of competition and disturbance as well as the kinds of biological heterogeneity that can be maintained by that interplay.

Our model describes a landscape composed of an effectively infinite set of effectively identical patches. Species colonize these patches, interact, are affected by abiotic disturbance, and eventually become locally extinct. Each of these processes has a characteristic temporal scale, in terms of which we describe the stochastic dynamics of individual patches and the resulting statistical properties of the landscape. Because we assume that all patches are identical, we are providing only the bare minimum of *environmental heterogeneity*—that produced by the independence of the patches. Our focus is on heterogeneity generated by the biological processes.

Consideration of the interaction of competition and disturbance has led to two important ecological concepts. The first is the idea of *fugitive species*

(Hutchinson, 1951), which persist regionally even though they are excluded locally by nonfugitive or *equilibrium species*. Fugitive species rely on disturbance to perturb the process of local competitive exclusion and cannot persist in its absence. The existence of fugitive species makes it impossible to infer competitive dominance from relative abundances; depending on the rates of disturbance, dispersal, and within-patch interactions, fugitive species may be much more common than competitive dominants.

The second concept is the maximization of species diversity at an intermediate disturbance frequency (Connell, 1978a,b; Huston, 1979). If disturbance is too rare, local competition proceeds to equilibrium and fugitive species are eliminated. If disturbance is too frequent, it eliminates all species and produces a desert. At intermediate frequencies, the combination of fugitive species and equilibrium species produces a maximum in species diversity.

The relation between disturbance and diversity is likely to be affected by resource supply rates or productivity. Huston (1985), in a discussion of coral reef communities, proposed that at high light intensities rapid coral growth should lead to rapid competitive exclusion and require a higher frequency of disturbance to maintain diversity than at lower light intensities. Discussions of the possible role of disturbance in maintaining high species diversity in deep-sea benthic communities (e.g., Dayton and Hessler, 1972; Grassle and Sanders, 1973; Rex, 1981, 1983) have emphasized the apparent low frequency of disturbance in the deep sea. However, it is known that growth rates and productivities in these communities (hydrothermal vents excluded) are low. Thus it is entirely possible that the time scale on which disturbance frequency must be measured is different in the deep sea.

Diversity enhancement is at odds with the concept of *biotic impoverishment* (Woodwell, 1983): that disturbance leads to shortened food chains, decreased diversity, and dominance by the handful of species hardy enough to resist the disturbance. Biotic impoverishment is a well documented response to radiation, pollution, and overgrazing.

At least two resolutions exist to this apparent contradiction. First, the disturbances that lead to biotic impoverishment tend to be chronic, whereas those leading to enhancement tend to be transient at any one spot. Whether disturbance leads to diversity enhancement or biotic impoverishment might depend on disturbance frequencies. Second, some disturbances, especially anthropogenic ones, delay recolonization for a long time. Levin and Smith (1984), for example, found that whereas defaunated sediment in the Santa Catalina basin was recolonized slowly, as is usually the case in the deep sea, defaunated sediment enriched by kelp was essentially uncolonized by macrofauna for the duration of their study. Such delays in recolonization may keep disturbance from enhancing diversity. Evolutionarily novel disturbances such as toxic waste are more likely to have long-lasting residual effects, which may make them more likely to produce biotic impoverishment than to enhance diversity.

Model Structure

Consider a patchy landscape inhabited by N species, $S_1, S_2, \ldots, S_N$. The landscape consists of an infinite set of physically identical patches. The state of a patch is determined by the presence or absence of each of the N species; there are thus 2^N possible states for each patch. These states can be conveniently numbered from 1 to 2^N by adding 1 to the binary representation of the presence and absence of each species. Consider two competing species, of which S_1 is locally superior to S_2, and let 0 denote absence and 1 denote presence. We then have

Species 2 (loser)	Species 1 (winner)	State
0	0	X_1
0	1	X_2
1	0	X_3
1	1	X_4

We denote the possible states of a patch by X_s, $s = 1, 2, \ldots, 2^N$. The proportion of all patches that are in X_s, is denoted by x_s; the state of the landscape is given by a vector $\mathbf{x}$, the elements of which give the proportion of patches in each of the states. This definition of the state of the landscape assumes that the spatial arrangement of the patches is unimportant, so that any landscape with a specified proportion of its patches in each state is effectively identical to any other such landscape.

The dynamics of the community are described by a nonlinear, discrete-time Markov chain

$$\mathbf{x}(t + 1) = \mathbf{A_x}\mathbf{x}(t) \tag{1}$$

where $\mathbf{A_x}$ is a column-stochastic matrix whose elements may depend on the vector $\mathbf{x}$. An element $a_{rs}(\mathbf{x})$ of $\mathbf{A_x}$ gives the transition probability from X_s, to X_r.

The transition probabilities a_{rs} are calculated from hypotheses about the time scales of interspecific interactions, dispersal, and disturbance. In general, the rate of a process is given by the inverse of the time scale on which that process occurs.

Colonization plays an important role in these models because any satisfactory description of the colonization process renders the model (1) nonlinear. The probability that a patch is colonized by S_i in $(t, t + 1]$ depends on the proportion of patches in the landscape in which S_i is present. Thus the entries a_{rs} that depend on colonization are functions of the current state vector **x**.

We make the following hypotheses about competition, dispersal, and disturbance.

1. Disturbance follows a Poisson process, with a time scale (mean time between disturbances) given by τ_d. The expected number of disturbances per unit time is τ_d^{-1}, and the probability of at least one disturbance in the interval $(t, t+1]$ is

$$p_d = 1 - e^{-\tau_d^{-1}} \tag{2}$$

2. The rate of competitive exclusion is specified by τ_c, the mean time required for S_1 to exclude S_2. If the probability of exclusion during the interval $(t, t+1]$ is p_c, the time required for exclusion (t_c) follows a zero-truncated geometric distribution, with

$$P(t_c = k) = p_c(1 - p_c)^{k-1} \qquad k = 1, 2, \ldots \tag{3}$$

The mean time required for exclusion is then $\tau_c = E(t_c) = p_c^{-1}$. Thus

$$p_c = \tau_c^{-1} \tag{4}$$

3. Colonization follows a Poisson process. The mean number of colonists of species i, $i = 1, \ldots, N$, arriving in a vacant patch in $(t, t+1]$ is proportional to the frequency of occurrence of species i. The constant of proportionality (the *dispersal coefficient*) d_i combines the effect of the production of offspring by populations in the occupied patches and the success at dispersal of those offspring. Let f_i denote the frequency of S_i; e.g., in our two-species example $f_1 = x_2 + x_4$ and $f_2 = x_3 + x_4$. The conditional probability of at least one colonist arriving in a patch, given that the patch is vacant, is then

$$C_i = 1 - e^{-d_i f_i} \tag{5}$$

4. Either or both species may colonize a vacant (X_1) patch; the winning species S_1 may also colonize a patch (X_3) containing S_2, but the losing species S_2 may not colonize a patch (X_2) that contains the winning species.

5. Disturbance affects only patches containing species. Thus empty patches (X_1) are not subject to disturbance while colonization is in progress.

Each of these hypotheses may be modified, but this simple model allows us to demonstrate the approach in the context of disturbance and competition.

A directed graph showing the possible transitions among the community

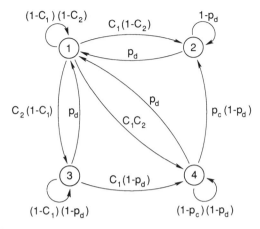

Figure 6.1. State transition graph for the two-species competition model. See text for definitions of states and notation.

states is shown in Figure 6.1. The corresponding transition matrix $\mathbf{A_x}$, which determines the dynamics in equation (1) is

$$
\begin{pmatrix}
(1 - C_1)(1 - C_2) & p_d & p_d & p_d \\
C_1(1 - C_2) & 1 - p_d & 0 & (1 - p_d)p_c \\
(1 - C_1)C_2 & 0 & (1 - C_1)(1 - p_d) & 0 \\
C_1 C_2 & 0 & C_1(1 - p_d) & (1 - p_d)(1 - p_c)
\end{pmatrix}
\tag{6}
$$

where the C_i values are given by equation (5).

Analysis

The analysis of this class of models is challenging only because of the non-linearity introduced by the dependence of colonization probabilities (equation 5) on the current state $\mathbf{x}(t)$, through f_i. If not for that nonlinearity, the model would be a linear, finite-state Markov chain, and standard methods would tell us everything there is to know about it.

The nonlinear model has several potential classes of dynamic behavior. The simplest is convergence of $\mathbf{x}(t)$ to a stable fixed point $\hat{\mathbf{x}}$. Once the community reaches such a fixed point, the matrix $\mathbf{A}$ is constant, and standard Markov chain methods can be applied. The nonlinear difference equation (1) may in principle also possess periodic, quasiperiodic, and chaotic attractors. We have no a priori reason to rule out these possibilities, but after extensive numerical investigation we have never found any of them. We conjecture that this model always has at most a single fixed point in the interior of the unit simplex, and that this fixed point attracts all initial vectors in the interior of the simplex.

Fixed Points and Stability

The existence of at least one fixed point in the closed unit simplex (including its boundaries) is guaranteed by the Brouwer fixed point theorem. Depending on the parameter values, some of these fixed points are on the boundaries, corresponding to landscapes in which one or more species are absent. Conditions guaranteeing the existence of an interior fixed point are difficult to obtain.

The stability of fixed points is surprisingly difficult to analyze. The simplest possible model, with one species and two states, can be reduced to a one-dimensional map of the interval (Caswell and Cohen, in preparation). In this model, if the dispersal rate is less than the disturbance rate, 0 is the only fixed point. If the dispersal rate is greater than the disturbance rate, there is an interior fixed point that is always globally stable. However, we have been unable to extend this proof to higher-dimensional models.

As an alternative, we carried out a numerical search for *un*stable equilibria, randomly sampling parameters from the sets $d_1 \in (0,10)$, $d_2 \in (0,10)$, $p_d \in (0,1)$, and $p_c \in (0,1)$. For each parameter set, we used a nonlinear equation-solving routine (part of the GAUSS software package, Aptech Systems, Inc., Kent, WA 98064) to find as many fixed points as possible. The local stability of these points was evaluated by calculating the linear approximation to the system at that fixed point and evaluating the eigenvalues of the resulting Jacobian matrix. If all the eigenvalues are less than 1 in magnitude, the fixed point is stable.

Examination of several thousand samples from the parameter space revealed no unstable interior fixed points. It is not a proof but it is the best we can do now.

Model Output

The output of the four-state model (or its generalizations) is surprisingly copious, considering its simplicity. It includes the following.

1. *State frequencies*. The immediate output of the model is the vector of equilibrium state frequencies $\hat{x}$, which provides the basic description of the landscape. The other output variables are calculated from $\hat{x}$.

2. *Species frequencies*. The frequency (or prevalence) of species i is given by the sum of the state frequencies for the states in which that species is present.

3. *Alpha diversity*. Local (or alpha) species diversity is measured by the mean number of species per patch. For our two-species example,

$$\alpha = x_2 + x_3 + 2x_4 \tag{7}$$

The variance in alpha diversity provides one measure of spatial heterogeneity in community structure. For the two-species example it is

$$V(\alpha) = x_2 + x_3 + 4x_4 - \alpha^2 \tag{8}$$

4. *Beta diversity*. Beta diversity measures spatial heterogeneity in species composition. It can be measured in a number of ways, one of the simplest being the entropy of the state frequency vector.

$$\beta = -\sum_{s=1}^{2^N} x_s \log x_s \tag{9}$$

A value of $\beta = 0$ corresponds to a homogeneous landscape. In practice, beta diversity is usually considered only in terms of the biological component of heterogeneity, with empty patches being ignored. We refer to it as biological beta diversity, and calculate it as

$$\beta_b = -\sum_{s=2}^{2^N} \left(\frac{x_s}{\sum_{j=2}^{2^N} x_j} \log \left(\frac{x_s}{\sum_{j=2}^{2^N} x_j} \right) \right) \tag{10}$$

A value of $\beta_b = 0$ corresponds to a landscape in which all occupied patches are homogeneous, although there may be some heterogeneity due to the presence of empty patches.

5. *Species–area slope*. Increasing the number of patches in a sample increases the number of species collected. The shape of the curve relating the number of species to the number of patches, or area, is sometimes used as a measure of diversity. It is possible to calculate the expected number of species found in any specified number of patches by a simple extension of the calculation of alpha diversity above. There are $(2^N)^k$ possible combinations of states in a sample of k patches (e.g., for $k = 2$, the two patches are respectively in states X_1 and X_1 with probability x_1^2, X_1 and X_2 with probability $x_1 x_2$, . . . , X_4 and X_4 with probability x_4^2), each of which yields a specific number of species. The average taken over this distribution gives the expected number of species in a sample of k patches.

6. *Interspecific association patterns*. The equilibrium state vector $\hat{x}$ contains all the information necessary to calculate the association between any pair of species. A number of association indices are available. One of the simplest in the two-species case is the log odds ratio

$$\log \left(\frac{x_1 x_4}{x_2 x_3} \right) \tag{11}$$

which is positive when the species tend to occur together, zero when they are independently distributed, and negative when they tend to occur alone. Association indices are discrete analogues of spatial niche overlap measures, indicating the extent to which the species co-occur in space.

7. *Rates of community change.* Along with the equilibrium state vector $\hat{\mathbf{x}}$, the model provides the equilibrium Markov matrix $\mathbf{A}_{\hat{\mathbf{x}}}$, which describes the dynamics of the landscape. Because these dynamics are described by a homogeneous Markov chain, it is easy to calculate a variety of measures of rates of community change. An ecologist following the fate of patches within this landscape might find some types of patch to be ephemeral, whereas others persist for long times. Certain developmental pathways (e.g., an empty, disturbed patch changing to a patch occupied by only the competitive dominant) might be traversed rapidly, others slowly. Insight into these dynamic patterns can be obtained from the following measures.

  a. *Turnover rates.* Although the landscape reaches an equilibrium, characterized by $\hat{\mathbf{x}}$, the individual patches change state continually. The probability of remaining in state s from t to $t + 1$ is given by a_{ss}. Thus the residence time in state s is geometrically distributed with mean $(1 - a_{ss})^{-1}$; the turnover rate is the inverse of this mean residence time. The mean turnover rate for a randomly selected patch is then

$$\sum_s \hat{x}_s(1 - a_{ss}) \tag{12}$$

  b. *First passage times.* Consider a patch in state j. How long, on the average, does it take before this patch first reaches state i? This first passage time provides some insight into the apparent rates of "succession" in this landscape. For example, a landscape in which the mean time required to go from state X_1 (recently disturbed) to state X_2 (occupied by the dominant competitor) is short appears to undergo a much more rapid succession than a landscape in which this time is long. Let m_{ij} denote the mean first passage time from state j to state i. This value is given by the (i,j) entry of a matrix $\mathbf{M}$ given by

$$\mathbf{M} = (\Pi_{dg})^{-1}(\mathbf{I} - \mathbf{Z} + \mathbf{Z}_{dg}\mathbf{E}) \tag{13}$$

where $\Pi = $ a matrix each of whose columns is $\hat{\mathbf{x}}$, $\mathbf{Z} = [\mathbf{I} - (\mathbf{A}_{\hat{\mathbf{x}}} - \Pi)]^{-1}$, and $\mathbf{E} = $ a matrix of ones (Iosifescu, 1980, theorem 4.7). Π_{dg} and $\mathbf{Z}_{dg}$ denote matrices containing the diagonal elements of Π and $\mathbf{Z}$.

The mean recurrence time for a state is the mean first passage time from that state to itself and is given by the diagonal elements m_{ss} of $\mathbf{M}$. The formula simplifies for these elements to $m_{ss} = 1/\hat{x}_s$. It is also possible to calculate the mean passage time to state j from a randomly selected patch by taking the mean of m_{ji} over the stationary distribution $\hat{\mathbf{x}}$ (Iosifescu, 1980, p. 135).

$$\sum \hat{x}_i m_{ji} = \frac{z_{jj}}{\hat{x}_j} \qquad (14)$$

c. *Smoluchowski recurrence time.* The recurrence times m_{ss} may be heavily influenced by the fact that the patch may *stay* in state s from t to $t + 1$, in which event the recurrence time is 1. The Smoluchowski recurrence time θ_s of state X_s is the time elapsing between *leaving* state X_s, and the next return to state X_s. Its mean is given by the following (Iosifescu, 1980, p. 135).

$$E(\theta_s) = \frac{1 - \hat{x}_s}{\hat{x}_s(1 - a_{ss})} \qquad (15)$$

Like the recurrence time, this index gives some insight into the rates of community development. $E(\theta_1)$ gives the mean time elapsing between colonization of a patch and its return to the disturbed state.

The Smoluchowski recurrence time can also be calculated for *sets* of states, defined as the mean time between leaving that set and the next return to it. This measurement is particularly useful for sets of states defined by the presence of a species; the recurrence time for such a set of states gives the mean time elapsing between the local extinction of a species (for whatever reason) and its next reappearance. Let χ denote the set of states under consideration. Then, as noted by Iosifescu (1980)

$$E(\theta_\chi) = \frac{1 - \sum_{s \in \chi} \hat{x}_s}{\sum_{s \in \chi} \hat{x}_s \sum_{r \notin \chi} a_{rs}} \qquad (16)$$

d. *Rate of convergence.* The dominant eigenvalue λ_1 of $\mathbf{A}_{\hat{x}}$ is 1; the corresponding eigenvector is proportional to $\hat{x}$. An idea of the rate of convergence to the landscape equilibrium can be had from the largest subdominant eigenvalue λ_2 of $\mathbf{A}_{\hat{x}}$. The smaller $|\lambda_2|$, the more rapidly the landscape converges to the equilibrium $\hat{x}$. This analysis applies only to arbitrarily small perturbations from $\hat{x}$; the rate of convergence from larger perturbations reflects the nonlinearity of the full model.

Relation to Other Approaches

A variety of complementary approaches have been used to study community dynamics in space and time. Each has its advantages and disadvantages. Our models lie at one end of a spectrum: Space, time, and the state of individual cells are all discrete. At the other extreme are partial differential equation models (e.g., Dubois, 1975; Wroblewski, 1977; Okubo, 1978), in which space, time, and state are continuous. Such models, especially when coupled with fluid mechanics for description of aquatic systems, can provide detailed insight into specific situations (Wroblewski, 1977), but they are complex, analytically intractable, and computationally difficult. It is not easy to obtain general theoretical insight from them.

Reaction-diffusion models (Levin, 1974, 1976; Yodzis, 1978), in which space is discrete but each patch contains a continuous model for species abundances, are intermediate in complexity. However, their detail can still make their results difficult to interpret (Levin, 1976).

Our models are closely related to deterministic differential equation models for patch state frequencies (e.g., Levins and Culver, 1971; Horn and MacArthur, 1972; Vandermeer, 1973; Slatkin, 1974; Hastings, 1977, 1978, 1980; Crowley, 1979; Acevedo, 1981; Greene and Schoener, 1982; Hanski, 1983, 1985; Caraco and Whitham, 1984). The state variables in these models are the proportions of patches in each of several states. Analysis typically focuses on the conditions for stability of the equilibria of the resulting equations, with an emphasis on conditions favoring coexistence.

Some of the early models in this category (Levins and Culver, 1971; Horn and MacArthur, 1972; Vandermeer, 1973) are written as differential equations in species frequencies. Unless each patch can be occupied by only a single species, or species occupy patches independently (which usually contradicts the hypothesis that the species interact) such patch state frequencies do not define a probability distribution and the models cannot correspond to a Markov chain (Slatkin, 1974).

More recently—certainly since Slatkin (1974)—these models have been written in terms of patch state frequencies, which sum to 1. The resulting models are nonlinear continuous time Markov processes. With the exception of Caraco and Whitham (1984), however, these studies have by and large failed to take advantage of the stochastic formulation underlying the deterministic models to get additional insight into community structure and dynamics.

The analysis of patchy community dynamics in terms of Markov chains was introduced by Cohen (1970). He considered a linear Markov chain describing transitions between four states defined by the presence and absence of two species at the equilibrium of Lotka-Volterra competition equations. Our nonlinear formulation of community models in terms of Markov chains defined by the time scales of within-patch processes is a combination of that proposed by Cohen (1970) and Caswell (1978).

Results

To examine the behavior of this model, we conducted a numerical experiment. The model parameters were varied over all 120 combinations of the elements of the following sets.

$d_1 \varepsilon \{1,10\}$
$d_2 \varepsilon \{1,10\}$
$p_c \varepsilon \{0.01, 0.1, 1\}$
$p_d \varepsilon \{10 \text{ values, log-uniformly spaced between } 0.001 \text{ and } 1\}$

Relying on the presence of a single stable fixed point for the model, we iterated the model (1) until $\mathbf{x}(t)$ converged to a fixed point $\hat{\mathbf{x}}$. The elements of $\hat{\mathbf{x}}$ give the proportions of the various states to be expected in a large landscape; the corresponding matrix $\mathbf{A}_{\hat{\mathbf{x}}}$ describes the transition dynamics of patches in such a landscape. The results of this experiment are shown as a series of figures. Most plot the response variable as a function of disturbance probability for several values of some other variable.

State Frequencies

The state frequencies $\hat{x}_s$, $s = 1, \ldots, 4$, are plotted as a function of p_d, in Figure 6.2. The frequency $\hat{x}_1$ of empty patches is nearly proportional to p_d, with only a small amount of variation generated by the different values of d_1, d_2, and p_c used in the experiment. This finding suggests that the fre-

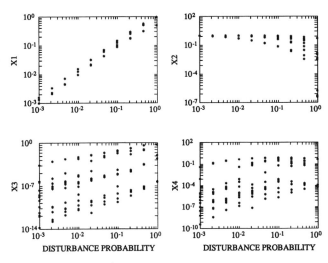

Figure 6.2. Equilibrium state frequencies x_s, $s = 1, \ldots, 4$, plotted as functions of the disturbance probability for the results of the numerical experiment described in the text.

quency of empty patches could be used as an index of disturbance probability in natural systems.

At low disturbance frequencies, $\hat{x}_2 \approx 1$ and almost all patches are occupied by the dominant competitor. As p_d increases, $\hat{x}_2$ decreases and becomes more variable as the dispersal rates and competitive exclusion rates come to play a part in determining community structure.

At any given disturbance frequency, the frequencies $\hat{x}_3$ and $\hat{x}_4$ of patches containing the inferior competitor vary over six to eight orders of magnitude, depending on the rates of competitive exclusion and dispersal. This variability is reflected in the variety of community patterns generated by the model.

Species Frequencies

The frequency f_1 of the winning competitor is independent of the losing competitor. Thus the dynamics of S_1 can be described by the one-species, two-state model described above: $f_1 > 0$ if and only if $d_1 > p_d$, and f_1 declines with increasing disturbance frequency independently of p_c (Fig. 6.3).

The losing competitor S_2 is a genuine fugitive species. In the absence of disturbance, it is eventually excluded by S_1 in every patch. In the presence of disturbance, however, it can persist and reach high frequencies. Figure 6.4 shows some results. When $d_1 = d_2$, f_2 is usually maximized at an intermediate frequency of disturbance. Numerically, it appears that f_2 begins to increase above 0 when

$$p_d > \frac{p_c}{d_2} \tag{17}$$

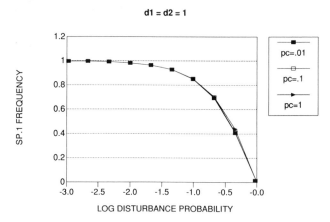

Figure 6.3. Equilibrium frequency f_1 of the winning competitor as a function of the disturbance frequency and the rate of competitive exclusion.

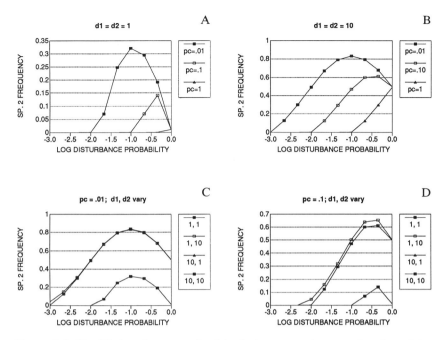

Figure 6.4. Equilibrium frequency f_2 of the losing competitor as a function of the disturbance frequency, dispersal rates, and rate of competitive exclusion. In (A,B) dispersal rates are fixed and competitive exclusion rates vary. In (C,D) competitive exclusion rates are fixed and dispersal rates vary.

Thus the minimum disturbance frequency required to maintain the fugitive species is directly proportional to the rate of competitive exclusion and inversely proportional to the dispersal rate of the fugitive. The disturbance frequency at which f_2 is maximized increases with p_c.

It is apparent from Figure 6.4A,B that the dispersal rate has an important positive influence on f_2. Fugitive species are often characterized as producing large numbers of propagules, but it is often also assumed that a fugitive species must have a dispersal advantage over the superior competitor. This assumption is not true (Fig. 6.4C,D). If d_2 is high enough, the relative values of d_1 and d_2 have no impact on f_2. If d_2 is lower ($d_2 = 1$ in our calculations), a dispersal advantage on the part of the superior competitor prevents the fugitive from persisting.

Diversity Enhancement

The interaction of competition, disturbance, and dispersal in this model is capable of enhancing local species diversity at intermediate disturbance frequencies. Figures 6.5 and 6.6 show the results.

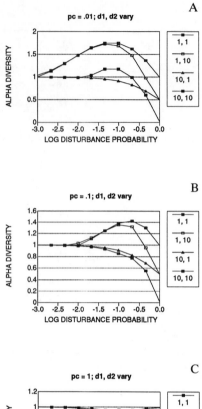

Figure 6.5. Mean alpha diversity as a function of the disturbance probability and dispersal rates (d_1, d_2) for three rates of competitive exclusion.

When dispersal rates are low and equal $(d_1 = d_2 = 1)$, alpha diversity is enhanced only when competition is slow $(p_c = 0.01)$ (Fig. 6.6A). When the fugitive species has a dispersal advantage $(d_1 = 1, d_2 = 10)$, diversity is enhanced (by as much as 80%) for $p_c = 0.01$ and $p_c = 0.1$ (Fig. 6.6B) but not for the fastest rate of exclusion. The same pattern holds when $d_1 = d_2 = 10$, so it apparently depends more on an adequate dispersal rate for the fugitive species than on the dispersal advantage per se. When the winning com-

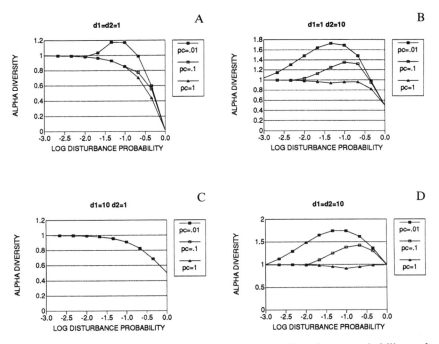

Figure 6.6. Mean alpha diversity as a function of the disturbance probability and rate of competitive exclusion for four combinations of dispersal rates.

petitor also has a dispersal advantage ($d_1 = 10$, $d_2 = 1$) there is no trace of diversity enhancement, regardless of p_c (Fig. 6.6D).

Spatial Heterogeneity

In the absence of disturbance, this model converges to a uniform landscape, with every patch occupied by S_1. Disturbance and the processes of colonization and interaction that follow disturbance create spatial heterogeneity in local community structure. The three measures of spatial heterogeneity—beta diversity (equation 9), biotic beta diversity (equation 10), and the variance in alpha diversity are highly correlated with each other and show similar patterns. Here we show the results for β_b (Fig. 6.7).

When dispersal is low ($d_1 = d_2 = 1$) (Fig. 6.7A), β_b generally increases with disturbance probability and decreases with the rate of competitive exclusion. At the other extreme, when $d_1 = d_2 = 10$ (Fig. 6.7D), β_b is maximized at a disturbance frequency $p_d \approx p_c$. The patterns when d_1 and d_2 differ are more complex. When the fugitive species has a dispersal advantage ($d_1 = 1$, $d_2 = 10$) (Fig. 6.7B), β_b shows a bimodal pattern for slow exclusion rates and a peak at high disturbance rates for faster exclusion rates.

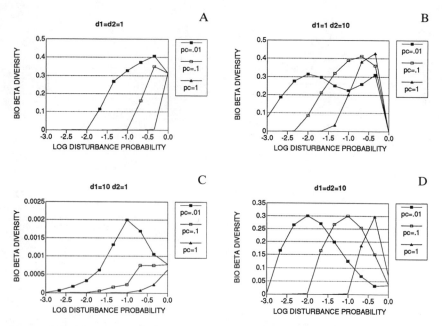

Figure 6.7. Biotic beta diversity (β_b) as a function of the disturbance frequency and the rate of competitive exclusion for different combinations of dispersal rates.

When the fugitive has a dispersal disadvantage, β_b is low (Fig. 6.7C), reflecting the nearly complete absence of S_2 in this situation.

Interspecific Association

The landscape-level pattern of association of two species tells remarkably little about their local interaction. In this case, the two species are pure competitors, and their association, were it to reflect that fact, would be negative. Figure 6.8 shows the results, and it is clear that association may be either positive or negative, depending on the parameters.

In general, association is more positive when the winning competitor has a high dispersal coefficient. Association tends to increase with p_d (Fig. 6.8A,B), although when the winning competitor has a dispersal advantage it decreases again at high values of p_d (Fig. 6.8C). When d_1 is high, association is positive over the entire range of disturbance probabilities. When d_1 is low, the association appears negative or positive, depending on the rate of disturbance.

Rates of Community Change

An observer following the fate of a randomly selected set of patches within the landscape described by this model would see continual change as

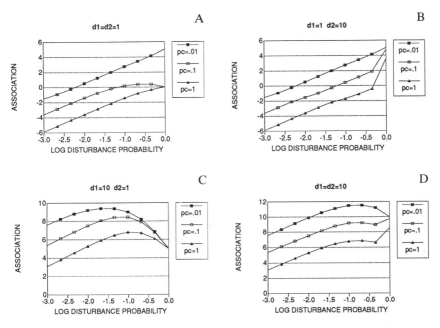

Figure 6.8. Interspecific association, measured by the log odds ratio, as a function of disturbance frequency and competitive exclusion rates for various combinations of dispersal rates.

patches are disturbed and colonized and as species interact with and exclude each other. The rates of this process of change can be measured in several ways.

Mean Turnover Rates

The mean turnover rate is an overall measure of the rate of community change, regardless of direction. In this model, it is determined almost completely by disturbance rate and in fact is nearly equal to the disturbance rate except when disturbance is so frequent relative to dispersal ability that the species begin to be driven to regional extinction (Fig. 6.9).

First Passage Times

The entries m_{ij} of the first passage time matrix **M** give the expected time required for a patch to reach state X_i for the first time, given that it starts in state X_j. Here we examine m_{i1}, $i = 1, \ldots, 4$: the mean first passage times from the disturbed state.

The mean first passage time from X_i to itself (*recurrence time*) is $\hat{x}_i^{-1}$. Thus $m_{11} = \hat{x}_1^{-1} \approx p_d^{-1}$. The mean first passage times from X_1 to X_2, X_3, and X_4 are shown in Figures 6.10 and 6.11.

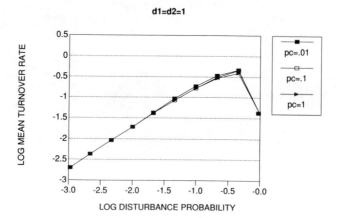

Figure 6.9. Log of the mean turnover rate as a function of disturbance frequency and competitive exclusion rate for low dispersal rates. The patterns for other dispersal rates are almost identical, except that they lack the drop in turnover rate at the highest disturbance probability. This drop reflects the fact that both species are being driven to extinction because their dispersal rates cannot match the disturbance rate.

State X_2 is the local equilibrium for this system, the "climax" of the two-species successional process described by the model. The mean first passage time m_{21} is thus a measure of how long an empty patch should take to reach the climax. At low dispersal rates ($d_1 = d_2 = 1$), m_{21} is low and insensitive to p_d, except when competitive exclusion is slow (Fig. 6.10A). In that case, there is a sudden increase in m_{21} when p_d exceeds p_c (thus permitting the persistence of S_2). Finally, m_{21} increases again, regardless of p_c, as p_d increases enough to begin to eliminate both species.

At high dispersal rates ($d_1 = d_2 = 10$) the pattern is similar (Fig. 6.11A). There is a sudden increase in m_{21} when p_d exceeds p_c. Then there is a plateau where m_{21} is inversely proportional to p_c, reflecting the fact that most patches arriving in X_2 must go through X_4, and that the transition from X_4 to X_2 is determined by p_c. Finally, when p_d is large, m_{21} increases again as the reduction in f_1 caused by the disturbance rate inhibits colonization by species 1.

The mean first passage time m_{31} to X_3 decreases with increasing p_d (Figs. 6.10B and 6.11B), as this state can be reached only by colonization of empty patches by the fugitive species. It also tends to increase with increasing p_c because when competition is more rapid the frequency, and hence the colonization probability, of the losing competitor is less.

Finally, the mean first passage time to X_4 also decreases with increasing p_d and with decreasing p_c. Like X_3, X_4 is reachable only by colonization of empty patches (by both species in this case). Thus it is reached more rapid-

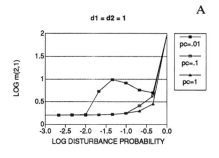

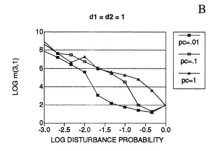

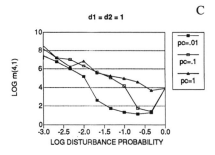

Figure. 6.10. Mean first passage times from state X_1 to states X_2, X_3, and X_4, for low dispersal rates ($d_1 = d_2 = 1$).

ly when empty patches are more common and when colonization is rapid enough to fill them with both species. The colonization rate increases with decreasing p_c and with increasing dispersal rates for both species (Figs. 6.10C and 6.11C).

This survey of mean first passage times reveals a wealth of patterns. The most reasonable candidate for a measure of the "rate of succession," m_{21}, suggests that there is a relatively sharp threshold value of disturbance frequency that, once exceeded, switches the community from one in which

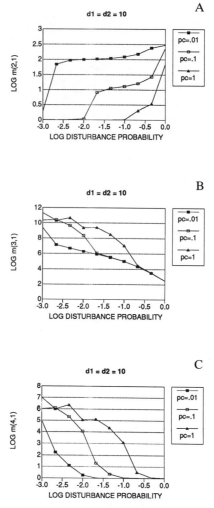

Figure 6.11. Mean first passage times from state X_1 to states X_2, X_3, and X_4 for high dispersal rates ($d_1 = d_2 = 10$).

succession proceeds almost immediately to its climax to one in which succession is much slower. In the latter case, a patch may cycle through disturbed and preclimax states many times before it reaches the climax state.

Smoluchowski Recurrence Time

The Smoluchowski recurrence time θ_i for state X_i measures the expected time between leaving X_i and returning to it for the first time. It is thus a measure of the rate at which a patch state recurs but without counting patches that remain in their present state.

In this experiment, $\theta_i \approx m_{i1}$, $i = 1, \ldots, 4$, which implies that the process of returning to a given state includes a passage through X_1, and that most of the variation in θ_i reflects variation in the time required to get from X_1 back to X_i. That time, of course, is measured by m_{i1}.

Convergence Rate

The rate of convergence of community structure, as measured by the second eigenvalue of $\mathbf{A_{\hat{x}}}$, shows a remarkable lack of variation compared to the variation in the other parameters. This fact suggests that the likelihood of a community being near its equilibrium distribution is independent of the parameter values, and hence observed differences between communities reflect differences in equilibria (at the landscape level) rather than differences in the rate of approach to equilibria.

Discussion

The results of this chapter demonstrate that the time scales of disturbance, dispersal, and competitive exclusion can interact to produce a variety of patterns in patchy communities. Studies of diversity, spatial heterogeneity, interspecific association, and rates of community change must take these time scales into account in order to make any sense of observed patterns of community structure.

Our results allow us to examine the relation between disturbance, rate of competitive exclusion, and alpha diversity, as discussed by Huston (1985). Figure 6.12 shows alpha diversity as a function of disturbance frequency and rate of competitive exclusion. In our model, diversity is maximized at intermediate disturbance frequencies and at low rates of competitive exclusion. At high enough disturbance rates, the competitive exclusion rate effect disappears. The value of p_d, which maximizes diversity, increases in log-log fashion with p_c. Huston (1985) proposed a somewhat different relation, in which diversity is maximized at intermediate values of p_d and p_c.

These results have important implications for discussions of diversity in low disturbance environments (e.g., of deep-sea benthic diversity). Alpha diversity is determined not by the disturbance rate per se but by that rate in relation to the rate of competitive exclusion. In environments characterized by low rates of competitive exclusion (e.g., low productivity, low growth rates), even very low rates of disturbance can significantly enhance species diversity and community heterogeneity. Comparisons between environments differing in disturbance frequency must also take into account differences in the rate of competitive exclusion.

We can also examine the apparent dichotomy between biotic impoverishment and diversity enhancement. The rate at which a disturbed patch

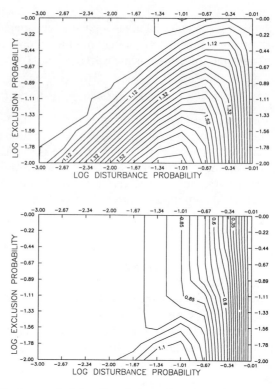

Figure 6.12. Contours of alpha diversity as a function of disturbance probability and the rate of competitive exclusion for high dispersal ($d_1 = d_2 = 10$) (top) and low dispersal ($d_1 = d_2 = 1$) (bottom) communities. In both graphs, diversity is at a minimum in the upper right corner.

recovers and becomes available for colonization influences whether the main effect of disturbance is diversity enhancement or biotic impoverishment. To illustrate this point we augment the basic model by adding a fifth state, corresponding to disturbed patches that are unavailable for colonization.

Unavailable	S_2	S_1	State
0	0	0	1
0	0	1	2
0	1	0	3
0	1	1	4
1	0	0	5

The graph for this model is shown in Figure 6.13, with the corresponding matrix $\mathbf{A_x}$.

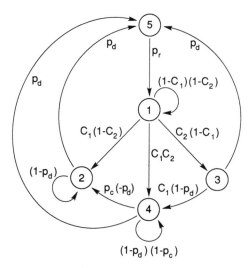

Figure 6.13. Transition graph for the residual effects model. Disturbance produces patches in X_5 that are unavailable for colonization until a recovery period (characterized by the recovery rate p_r) has passed.

$$\begin{pmatrix} (1 - C_1)(1 - C_2) & 0 & 0 & 0 & p_r \\ C_1(1 - C_2) & 1 - p_d & 0 & p_c(1 - p_d) & 0 \\ C_2(1 - C_1) & 0 & (1 - p_d)(1 - C_1) & 0 & 0 \\ C_1 C_2 & 0 & C_1(1 - p_d) & (1 - p_d)(1 - p_c) & 0 \\ 0 & p_d & p_d & p_d & 1 - p_r \end{pmatrix} (18)$$

The rate of recovery of disturbed patches is given by p_r; all other parameters are as in equation (6).

The rate of recovery determines the response of diversity to disturbance (Fig. 6.14A). Even when dispersal is high and competitive exclusion slow, diversity is enhanced by disturbance only if the rate of recovery is high enough. If recovery is slow enough ($p_r = 0.01$ in this example), disturbance leads only to biotic impoverishment.

This biotic impoverishment reflects the increasing proportion of patches in X_5 as disturbance becomes more frequent (Fig. 6.14B). At least in this case, in which dispersal rates are high ($d_1 = d_2 = 10$), it appears that $x_5 \approx p_d/p_r$ for low disturbance frequencies. As p_d increases, x_5 eventually approaches 1, the sooner the lower the value of p_r.

Summary

We have presented a simple modeling framework for studying the interaction of the rates of disturbance, dispersal, and competitive exclusion

A

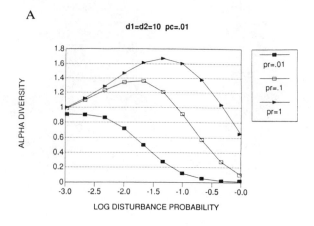

B

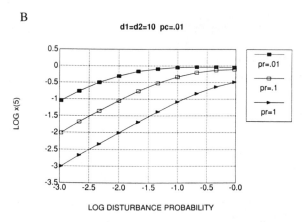

Figure 6.14. Alpha diversity (A) and log x_5 (B) for the residual effects model as a function of the disturbance rate and the rate of recovery (p_r) of disturbed patches.

in a patchy environment. The model is a nonlinear Markov chain, the transition probabilities of which are derived from hypotheses about the aforementioned rates. The state of a patch is defined by the presence and absence of the two competing species and the state of the landscape by the probability distribution of patch states. Numerical studies strongly suggest that the model has, at most, one stable fixed point in the interior of the unit simplex; this fixed point represents an equilibrium community at the landscape level, although equilibrium is never attained at the level of the individual patch. The output of the model includes species frequencies, alpha and beta diversity, interspecific association patterns, and measures of community change.

The model shows that disturbance can generate heterogeneity, both spatially and temporally, in patchy communities. The resulting patterns depend strongly on the relation between the time scales of the interacting processes.

The results are used to examine the relation between productivity and the phenomenon of diversity enhancement by disturbance. A single model can account for both diversity enhancement and biotic impoverishment, depending on the relations among its parameters.

Acknowledgments. This research was supported by National Science Foundation grants OCE85-16177, BSR86-9395, BSR87-4936, and BSR87-05047. JEC is grateful for the hospitality of Mr. and Mrs. William T. Golden. HC acknowledges support of a John Simon Guggenheim Memorial Fellowship. This work is Woods Hole Oceanographic Institution Contribution 7214.

References

Acevedo MF (1981) On Horn's Markovian model of forest dynamics with particular reference to tropical forests. Theor Pop Biol 19:230–250

Caraco T, Whitham TH (1984) Immigration-extinction competition on islands: associations among three species. J Theor Biol 110:241–252

Caswell H (1978) Predator-mediated coexistence: a nonequilibrium model. Am Nat 112:127–154

Caswell H, Cohen JE (1991) Disturbance and diversity in metapopulations. Biological Journal of the Linnean Society (in press)

Caswell H, Cohen JE (in preparation) Models for patchy communities: disturbance, dispersal, and interspecific interaction.

Cohen JE (1970) A Markov contingency-table model for replicated Lotka-Volterra systems near equilibrium. Am Nat 104:547–560

Connell JH (1978a) Diversity in tropical rain forests and coral reefs. Science 199:1302–1310

Connell JH (1978b) Tropical rain forests and coral reefs as open nonequilibrium systems. In Anderson RM, Taylor LR, Turner B (eds) *Population Dynamics* Blackwell, Oxford

Crowley PH (1979) Predator-mediated coexistence: an equilibrium interpretation. J Theor Biol 80:129–144

Dayton PK, Hessler RR (1972) Role of biological disturbance in maintaining diversity in the deep sea. Deep Sea Res 19:199–208

Dubois DM (1975) A model of patchiness for prey-predator plankton populations. Ecol Model 1:67–80

Grassle JF, Sanders HL (1973) Life histories and the role of disturbance. Deep Sea Res 20:643–659

Greene CH, Schoener A (1982) Succession on marine hard substrata: a fixed lottery. Oecologia 55:289–297

Hanski I (1983) Coexistence of competitors in patchy environment. Ecology 64:493–500

Hanski I (1985) Single species spatial dynamics may contribute to long-term rarity and commonness. Ecology 66:335–343

Hastings A (1977) Spatial heterogeneity and the stability of predator prey systems. Theor Pop Biol 12:37–48

Hastings A (1978) Spatial heterogeneity and the stability of predator-prey systems: predator-mediated coexistence. Theor Pop Biol 14:380–395

Hastings A (1980) Disturbance, coexistence, history and competition for space. Theor Pop Biol 18:363–373

Horn HS, MacArthur RH (1972) Competition among fugitive species in a harlequin environment. Ecology 53:749–752

Huston M (1979) A general hypothesis of species diversity. Am Nat 113:81–101

Huston M (1985) Patterns of species diversity on coral reefs. Annu Rev Ecol Syst 16:149–178

Hutchinson GE (1951) Copepodology for the ornithologist. Ecology 32:571–577

Iosifescu M (1980) *Finite Markov Processes and Their Applications*. Wiley, New York

Levin LA, Smith CR (1984) Response of background fauna to disturbance and enrichment in the deep sea: a sediment tray experiment. Deep Sea Res 31:1277–1286

Levin SA (1974) Dispersion and population interactions. Am Nat 108:207–228

Levin SA (1976) Population dynamic models in heterogeneous environments. Annu Rev Ecol Syst 7:287–310

Levins R, Culver D (1971) Regional coexistence of species and competition between rare species. Proc Natl Acad Sci USA 68:1246–1248

Okubo A (1978) *Ecology and Diffusion*. Springer-Verlag, New York

Rex MA (1981) Community structure in the deep-sea benthos. Annu Rev Ecol Syst 12:331–353

Rex MA (1983) Geographic patterns of species diversity in the deep-sea benthos. Sea 8:453–472

Slatkin M (1974) Competition and regional coexistence. Ecology 55:128–134

Vandermeer JM (1973) On the regional stabilization of locally unstable predator-prey relationships. J Theor Biol 41:161–170

Woodwell GM, Cole J, Hartman J (1983) Biotic impoverishment: A review of experience in terrestrial and aquatic communities. The Ecosystems Center, Marine Biological Laboratory, Woods Hole MA.

Wroblewski JS (1977) A model of phytoplankton plume formation during variable Orgeon upwelling. J Marine Res 35:357–394

Yodzis P (1978) *Competition for Space and the Structure of Ecological Communities*. Springer-Verlag, New York

7. Stochastic Population Models

Peter Chesson

Deterministic models dominated theory in ecology for much of its history despite recognition of the role of unpredictable environmental factors in population dynamics (Hutchinson, 1951, 1961; Andrewartha and Birch, 1954, 1984; Grubb, 1977, 1986; Sale, 1977; Wiens, 1977, 1986; Connell, 1978; Hubbell, 1979, 1980; Murdoch, 1979; Connell and Sousa, 1983; Sale and Douglas, 1984; Strong, 1984, 1986). Mathematical techniques for stochastic modeling were poorly developed and poorly understood. As a consequence, most ecological thinking about the role of stochastic factors was purely intuitive. Progress in stochastic population and community models has now allowed rigorous deduction to replace intuition. This progress has shown more complicated and intricate roles for stochastic factors than previously invisaged; but as is shown in this chapter, once elucidated, these roles can be understood intuitively.

The key to understanding stochastic factors is recognizing that even though they involve unpredictability they nevertheless have definable properties. For example, a stochastic factor such as rainfall has a frequency distribution characteristic of a locality. Moreover, there are a variety of ways in which any given stochastic factor may influence a population. For example, a stochastic factor might primarily affect the birth rate with little effect on the death rate; or it might not affect a given species directly but does affect the species' competitors or predators, thereby leading to an indirect effect on the species. Just as birth, death, competition, and predation are biologically distinct, each with their associated roles and theories, we have to expect that the role of a stochastic factor depends on which of these or other population processes it affects.

Another aspect of stochastic factors is scale. A stochastic factor may first appear or originate on one scale but have effects transmitted to other scales. We need to distinguish local phenomena, such as local thunderstorms, from regional phenomena, such as weather patterns over a large area. We must ask: Does this stochastic factor originate with individuals, local populations, or regional populations? Is its effect through time or over space? These questions of scale are basic when assessing stochastic factors.

However, the major scientific interest in stochastic factors is their effects on scales larger than their scales of origin. For example, one asks: Do stochastic effects seen at the level of an individual have an effect at the level of the total population? This question has two parts. First, do stochastic effects at one level (e.g., the individual level) lead to stochastic fluctuations at a higher level (e.g., the population)? Second, and more significantly, do stochastic effects at one level have systematic effects at a higher level? For example, do stochastic fluctuations on a short time scale lead to reductions in average population density on a long time scale, as has been suggested by some density-dependent population models (May, 1973)?

As used in ecology, the word "stochastic" has unfortunately taken on a rather mystical connotation. In mathematics, from which its ecological use is derived, a stochastic process is a function of time that is chosen with some probability from a set of possible functions of time. A stochastic process then is the appropriate description of population growth if we wish to consider the uncertainty of population trajectories. However, describing uncertainty is the least important attribute of a stochastic model. Rather, variation, which can be described by a frequency distribution but can be entirely deterministic in origin, is often key. From the population dynamic perspective of this review, stochastic factors are important because they generate variability. In contrast, in an evolutionary setting the unpredictability of stochastic factors can be the focus because it provides a problem for adaptation (Colwell, 1974).

Types of Variability

In 1978 I devised a classification of ecologically relevant variability based on its scale of origin. Any such classification is artificial to the extent that it imposes disjoint scales on a continuum. Nevertheless, such classifications have been useful. Three types of scale may be distinguished: spatial scale, temporal scale, and "population scale" (Chesson, 1982). Population scale relates to how large a population is or how many individuals are being considered. There may be few or many individuals in a given area, but the number of individuals increases as the spatial scale increases. Thus although population and spatial scale are logically distinct, they are not independent in nature.

Within-Individual Variability

Even if two individuals have identical phenotypes, their future longevity and reproduction can be different. Chance processes may affect different individuals independently. For example, chance effects may lead to death by predation for some individuals but not others, mates for some but not others, and foraging success for only some individuals. There is no doubt that an individual's phenotype influences its success or failure, but variation remains after the effects of an individual's phenotype have been factored out. This residual variation is called within-individual variability (Chesson, 1978), or demographic stochasticity in earlier classifications (May, 1973). Within-individual variability involves purely chance phenomena that have nothing to do with an individual's phenotype.

Within-individual variability was the first kind of stochastic variation to be considered in population models and is the only variation included in the classic stochastic population models of mathematics (Feller, 1971). Within-individual variability works on the scale of an individual, because each individual experiences variation independently and on a temporal scale shorter than or equal to the lifetime of an individual.

A special case of within-individual variability is the sampling variation considered in population genetics, which leads to some copies of an allele, but not others, being transferred to the next generation. In the context of population genetics, within-individual variability leads to genetic drift (Roughgarden, 1979). In a similar way, it has been incorporated in community models to describe the possible drifting of species composition of a tropical forest in the case where it is hypothesized that species have identical niches and are average demographic equals (have equal average birth rates, longevities, and so on). In this setting, it has been possible to generate species abundance relations that resemble those of a real forest (Hubbell, 1979).

Within-individual variability also has an important role in discussions of population persistence: When populations are small, within-individual variability can lead to chance extinction, as discussed in more detail below in the section on population persistence. More generally, within-individual variability contributes to local population fluctuations and therefore contributes to the interesting effects of within-patch variability, also discussed below.

Between-Individual Variation

Individuals in a population usually vary phenotypically. Such variation has received only a little attention in ecological models, primarily within the predator–prey and host–parasitoid framework. Variation in prey selection among individual predators gives frequency-dependent predation at the population level (Chesson, 1984b). On the other hand, between-individual variation in susceptibility of host to parasitism is able to stabilize the community-level interaction between parasitoids and hosts (Chesson and Murdoch, 1986). For systems of competitors, between-individual variation is a component of a species' niche breadth (Roughgarden, 1974). Milligan (1986) has taken this point further to investigate how heritable between-individual variation affects invasion and coexistence of competitors.

Between-individual variation has been found to have an important three-way interaction with competition and temporal environmental fluctuations (see below) and can play an important role in determining the outcome of interspecific competition in a variable environment. Between-individual variation occurs on the population scale of a single individual and on a temporal scale equal to or greater than the lifetime of an individual. The extent to which between-individual variation has a genetic basis determines the extent to which a phenotype is transmitted to an individual's offspring and therefore the temporal scale to which between-individual variation applies.

Within-Patch Variability

Within-patch variation is the analogue at the level of a local population of within-individual variability. Thus it is at a population scale greater than the individual but less than a closed population. It concerns stochastic fluctuations in population densities and environmental variables over time, locally in space (Chesson, 1981). If a population had only this sort of variation, and if there were no migration, each local population and the relevant environmental variables would fluctuate over time independently in different patches. Thus, the fluctuations in different spatial locations would be asynchronous. In nature, however, several sources of variation are present at any one time, and there is migration between patches. Rather than total asynchrony, there is some common element to the fluctuations. This common element is discussed below (see Pure Temporal Variation).

In the context of analysis of variance, within-patch variability is the space–time interaction, which may be partitioned out from other sorts of variation. Within-patch variability leads to asynchronous fluctuations in local population densities and local environmental parameters. For example, disturbance may affect some patches, but not others, in a way that is not predictable on the basis of the physical properties of a patch.

Asynchronous population fluctuations are at the heart of the verbal ideas of Andrewartha and Birch (1954) in their general theory of population regulation. General reasons for expecting within-patch variability to change the nature of population dynamics on the scale of a regional population consisting of many local populations are discussed by Chesson

(1981). Models and ideas about disturbance (Connell, 1979; Hastings, 1980) provide good illustrations. In these disturbance theories, catastrophes occur randomly in time and space, eliminating all organisms locally in space. Destruction of the organisms on a patch permits a successional process in which organisms with good colonizing abilities establish and grow for a while in the absence of later establishing competitive dominants.

Within-patch variability is especially important in these ideas on disturbance, as other patches are an important source of propagules for recolonization of a disturbed patch. The fact that only a portion of the patches is disturbed at any given time is critical.

Temporal fluctuations in mortality at a locality, which are a feature of disturbance, are critical too. One can imagine the same amount of mortality being spread uniformly through time. Can that additional mortality by itself promote coexistence without fluctuations in time or space? The first thing to note is that mortality does not eliminate linear hierarchies of competitive ability in resource exploitation (Armstrong and McGehee, 1980). At best, it changes the relative rankings of species favoring species with higher intrinsic rates of increase and so can merely alter the identity of the winner in competition. Thus mortality alone does not promote diversity in such a linear hierarchy of competitors.

Environmental fluctuations may be the cause of the local catastrophes involved in disturbance, or disturbance may be the result of predation. Discovery of patches by predators may vary in time and space, and so predators may disturb different patches independently. Although predation is more complicated than the simple effects of environmental variation, it nevertheless can work in principle in the same ways (Caswell, 1978; Hastings, 1978).

Spatiotemporal variation of discovery of local patches of habitat by different species of competitors, without invoking disturbance or predation, has also been regarded as having an important role in the maintenance of diversity on regional scales (Yodzis, 1978). More generally, one could simply consider within-patch variability of migration rates, which leads to spatially asynchronous fluctuations in the number of individuals colonizing different patches.

If fluctuating migration rates are less than perfectly correlated among species, chance fluctuations in the relative densities of competitors result. As a consequence, on average, individuals experience higher densities of conspecifics than heterospecifics. This statement is true for all species however, only when all patches are considered or when all time in a given patch is considered: It cannot be true for more than one species in a given patch at a given time. Thus the effect depends on spatially asynchronous temporal fluctuations, i.e., on within-patch variability.

This sort of fluctuating partial spatial segregation of species has the effect of increasing average intraspecific competition at the expense of interspecific competition on the spatial scale of many patches. The classic consequence of this situation still applies: Species diversity is promoted (Atkinson and Shorrocks, 1981; Chesson, 1985; Comins and Noble, 1985; Ives and May, 1985; Ives 1988).

Within-patch variability in migration rates or parasitoid search rates can have a stabilizing role in host–parasitoid systems (May, 1978; Chesson and Murdoch, 1986). In this situation, variation in the numbers of parasitoids discovering patches has the effect of creating partial refuges for the prey. On a spatial scale consisting of many patches, this condition reduces overshoot of the equilibrium and stabilizes the host–parasitoid interaction.

More generally, in predator–prey systems it has long been known that within-patch variability has a role stabilizing dynamics on a regional spatial scale. The time lag inherent in the dynamics of the predator–prey interaction in theory often leads to unstable oscillations. Predation that strikes different patches independently can stabilize such interactions and eliminate oscillations for the system as a whole (e.g., Maynard Smith, 1974; Hastings, 1977). More recently, Crowley (1981) and Reeve (1988) have explored spatially asynchronous population fluctuations that result from spatiotemporal environmental fluctuations interacting with unstable local population dynamics. Within-patch variability arising in this way has also been found to lead to regional stability.

Between-Patch Variation

Different places may have essentially permanent environmental differences. Such between-patch variation has long been appreciated in community ecology, where in essence it allows niche differentiation and therefore promotes coexistence. An example is Tilman's (1982) discussion of how between-patch variation in relative abundances of resources may permit high diversity in plant communities. Other discussions of between-patch variation explicitly taking into account migration between patches are those of Pacala and Roughgarden (1982), Shigesada and Roughgarden (1982) Shigesada (1984), Iwasa and Roughgarden (1986), and Pacala (1987).

Between-patch variation may also promote diversity without any involvement of habitat segregation when it is combined with pure temporal variation, as discussed in the next section. In predator–prey and host–parasitoid systems, between-patch variation can provide complete or partial refuges for prey or hosts, adding stability to the interaction between the species (Bailey et al., 1962; St Amant, 1970; Hassell, 1978).

Pure Temporal Variation

The weather, which is responsible for much temporal environmental variation, is correlated over large areas of space. Such spatially correlated patterns are not covered by any form of variation discussed so far. Pure temporal variation is variation over time on a spatial scale sufficiently large to contain essentially closed communities, i.e., communities in which migration has a negligible effect. Pure temporal variation is the variation remaining once all effects of space have been accounted for. Discussed here is the theory of temporal environmental variation in communities of competitors. This theory has been developed to the stage where broad generalizations are available. Some space is devoted to discussing them, as they may indicate the sorts of development that can be expected in other areas.

Stochastic Competition Models

When do stochastic fluctuations promote coexistence, and when do they hasten competitive exclusion? Two sorts of circumstance have been investigated. In the first case, population growth rates of different species are different nonlinear functions of the amount of competition the species experience: Thus the species differ in the range of competition to which they are most sensitive. For example, a species that exploits resources well when they are abundant but not so well when they are in short supply feels the effects of competition strongly while competition is still relatively weak. A species that is adapted to the situation of shortage of resources and that cannot increase its uptake rate even when resources are abundant would not respond to competition until it is severe. These nonlinear responses to competition are referred as negative and positive nonlinearities, respectively.

Fluctuations can promote coexistence of such species provided the species with the greater positive nonlinearity experiences smaller fluctuations in competition when it is at low density than the other species does when it is at low density. These ideas come mostly from models of deterministically varying factors (Levins, 1979; Armstrong and McGehee, 1980), but they hold up also in the stochastic case (Ellner, 1987a, Chesson, in preparation).

Models of disturbance (Chesson and Huntly, manuscript) can be formulated in this context. Disturbance was considered in the section on within-patch variability where asynchrony of disturbances on different patches was an important factor. Disturbance that is spatially synchronous is properly considered to be pure temporal variation. Naturally, such a disturbance must lead to less than 100% mortality. It can promote coexistence, provided organisms have complementary life histories as defined by Ellner (1987a).

Another broad class of models focuses not on differences between species in their response to competition but differences in their responses to environmental fluctuations. Most theoretical models do not deal with the environment directly, for example, temperature and rainfall usually are not variables. Population parameters that are presumed to depend on the environment serve instead and are referred to as environmentally dependent parameters. Examples are density-independent birth rates (Chesson

and Warner, 1981), survival rates (Chesson and Warner, 1981), seed germination rates (Ellner, 1984), and resource uptake rates (Abrams, 1984).
Indeed, any parameter in a model that is not a function of density can be
made an environmentally dependent parameter and a function of time.

Although environmentally dependent parameters themselves are not
functions of density, it is nevertheless to be expected that their values affect
the amount of competition that occurs in a system through their effects on
population densities. For example, in models of annual plants the seed
germination rate may be an environmentally dependent parameter that,
together with the size of the seed bank, determines the density of plants
that grow during a given year. Thus competition for resources needed by
growing plants depends on the environmentally dependent germination
fraction (Chesson, 1988).

In some models (e.g, Abrams, 1984) the environmentally dependent
parameter is a resource uptake rate, and the involvement with competition
is even clearer. The environmentally dependent parameter invariably has
an indirect effect on population growth by altering the amount of competition, and it can be expected to have a direct effect as well. The indirect
effect, however, depends on population density, whereas the direct effect
is, by assumption, density-independent.

Variation of competition with the environmentally dependent parameter can be measured by a covariance, and this covariance is an important
factor in the long-term dynamics of competing species. The direct and indirect effects of environment work in opposition, and the covariance between environment and competition indicates the extent to which environmental fluctuations are canceled out by opposing competitive effects. For
example, in single-species models of organisms competing for space,
fluctuations in the birth rate are exactly opposed by corresponding fluctuations in competition whenever space becomes saturated. Reflecting this
situation, the measure of covariance between environment and competition is equal to the variance of the environmentally dependent parameter
when the two are measured in the same standard units (Chesson, 1988). In
several species models of space limitation (Chesson, 1984a), however, the
covariance is usually less than the variance of the birth rate; and, as a
consequence, a species' share of space fluctuates.

More generally, when a species is competing with others, the covariance
between its environmentally dependent parameter and competition depends on the correlations between the environmentally dependent parameters of the different species and the absolute and relative densities of
these species. For example, consider the case where the environmentally
dependent parameters of different species are independent. When a given
species approaches zero density, it experiences only interspecific competition, which is uncorrelated with its environmentally dependent parameter.
The covariance between environment and competition is then zero.

When environmentally dependent parameters of different species are positively correlated, the covariance between environment and competition does not drop to zero as a species approaches zero density but remains at some positive value. In general, however, this value is less than when the species is at high density unless there is a perfect correlation between the environmentally dependent parameters of different species or unless interspecific competition is stronger than intraspecific competition for all species pairs.

The final case, that of negative correlations between environmentally dependent parameters of different species, means that as population density decreases the covariance between environment and competition decreases from a positive to a negative value. *Density dependence* of the covariance between environment and competition means that the growth rate of a species at low density fluctuates more than that of a species at high density. Intuitively, we might expect that this fact is bad news and should hasten extinction. However, many organisms have traits that reduce the magnitude of negative fluctuations in the growth rate while allowing advantage to be taken of positive fluctuations (Chesson and Huntly, 1988). To see how it occurs we must consider the response of population growth rates to different environmental and competitive situations.

Because population growth is multiplicative over time, we take logs and define the growth rate as the change in log population size per unit time. This method is equivalent to considering the growth rate parameter, "r" of demography. On this log scale, changes in population size are additive over time. In certain simple circumstances the growth rate itself, applying for a given period of time (e.g., 1 year), is an additive function of the effects of environment and competition. It occurs if the effect of competition on survival and reproduction of an individual is independent of how that individual has been affected by the environment. For example, if environmental conditions lead to 60% survival, and of those individuals remaining competition permits only 40% survival, total survival is the product of these figures, or 24%. Taking logs, this product becomes a sum: log $0.24 = \log 0.6 + \log 0.4$. We say that the growth rate is additive over the effects of environment and competition.

The additive case seems to occur in only the simplest situations, although approximate additivity may be common. To see how deviations from additivity arise, consider a population subdivided into two types of individual, e.g., robust versus fragile, and for simplicity let them be equally abundant. Let environmental conditions yield 90% survival of the robust type but only 30% survival of the fragile type; competitive conditions lead to, respectively, 60% and 20% survival of the remaining individuals. Then total survival of the population as a whole is $\frac{1}{2}(90\% \times 60\%) + \frac{1}{2}(30\% \times 20\%) = 30\%$.

To see that environment and competition are not additive in this second case, note that when acting alone the average mortality due to the environment would be 60% as in the first example; and considered alone (i.e., without prior action of the environment), competition would lead to an average mortality that is also the same as the first example. Hence if we combine these average survival rates by taking a simple product (i.e., assuming additivity on a log scale) we get 24%, which is too low. Thus we see that the effects of environment and competition in this subdivided population are not additive.

In biological terms, the presence of "fragile" and "robust" individuals in the second example provides a buffer against the joint negative effects of environment and competition. Thus the growth rate of the population is not as severely affected by unfavorable environmental and competitive effects as predicted on the basis of the sum of their separate effects. Consequently, this sort of situation is referred to as *subadditive*.

Subadditivity due to population subdivision can arise in a number of ways. The example using robust and fragile individuals shows that it can result from between-individual variation. It can also arise from between-patch variation, as environmental and competitive factors may be more important in some patches than in others (Chesson and Huntly, 1988). For subadditivity to occur, some individuals must be more susceptible to environmental factors than others, and these same individuals must be more susceptible to competition than others. In the earliest models incorporating such population subdivision, the various classes of individuals were defined by stages of their life cycle, e.g., juveniles and adults (Chesson and Warner 1981) or dormant seeds and growing plants (Ellner, 1984). Differences in sensitivity to environment and competition in such cases can be expected to be large.

Superadditivity, the opposite of subadditivity, arises in situations where sensitivity to competition usually means insensitivity to the environment. For example, a species may occupy a mosaic of habitats, some of which provide benign environmental conditions but competition among many individuals, whereas other habitats may be exposed to harsh and fluctuating environmental conditions but have lower densities of individuals and little competition.

Subadditivity has the benefit of providing protection against unfavorable combinations of environmental and competitive events arising temporally. However, the preceding discussion of the covariance between environment and competition implies that the density of a species determines the extent to which it can take advantage of subadditivity. At low density, a species experiences more extreme fluctuations in environmental and competitive conditions. Subadditivity dampens the unfavorable extremes while permitting advantage to be taken of favorable extremes. As a consequence, a species at low density has an advantage when the results of all these fluctuations are combined over time. Competitive exclusion is thus opposed and species diversity promoted by environmental fluctuations.

These effects depend critically on subadditivity and the argument that a species at high density has positive covariance between environment and competition. If positive covariance is combined with superadditivity, environmental fluctuations promote competitive exclusion (Chesson, 1989). Additive growth rates are neutral to coexistence in a fluctuating environment regardless of the covariance between environment and competition. Finally, if we consider negative covariance between environment and competition, all of the above conclusions about subadditivity and superadditivity are interchanged.

Models incorporating these effects have been reviewed elsewhere (Chesson and Huntly, 1988, 1989) together with various applications. There is much need for the development of models that apply in specific applications. Tools useful for developing and analyzing such models are also discussed elsewhere (Chesson, 1988).

Scale Transition

In all the models reviewed above, the interest is how variability on one scale leads to population and community phenomena on another. For example, questions of coexistence in the community models above involve density-dependent effects that appear when the results of fluctuations over many years are combined. Similarly, in the discussion of disturbance, we considered how large fluctuations locally in space contribute to a stable coexistence regionally.

Variability on one scale may or may not lead to significant variability on some larger scale. Variability on one scale, however, leads to different mean effects on a scale above it in most nonlinear systems (Chesson, 1981). Most of the results discussed above depend on this fact. To gain a better understanding of the subject, consider a single-species model exposed to pure temporal variation, such as that discussed by Turelli and Petri (1980).

$$X(t+1) = X(t)e^{r[1-X(t)/K(t)]} \tag{1}$$

This equation is the Moran-Ricker model, which is a discrete time version of the logistic model. $X(t)$ = the population density at time t, and $K(t)$ = the carrying capacity, also a function of time representing stochastic factors. Assume for the purpose of this illustration that the values of the carrying capacity are independent from one time to the next and fluctuate within some finite range above 0. It is not difficult to see that this assumption implies that the fluctuations in $X(t)$ also are bounded within some finite range above 0.

Of the many interesting questions one may ask about this system, let us focus on the average of population density over time. The average over a time interval of length T is

$$\bar{X}_T = \frac{1}{T} \sum_{t=s+1}^{T+s} X(t) \qquad (2)$$

It is shown in the Appendix that as we increase T (i.e., increase the temporal scale on which we are taking the average) the variance vanishes and $\bar{X}_T$ approaches a constant value $\bar{X}_\infty$ equal to the harmonic mean, $H(K)$, of $K(t)$, i.e.,

$$\bar{X}_T \rightarrow \bar{X}_\infty = H(K) = 1/E[1/K] \qquad (3)$$

where E = the theoretical average or expected value of the random variable in the brackets. The scale or the value of T on which $\bar{X}_T$ is reasonably approximated by $\bar{X}_\infty$, which we call "long term," depends on the rate of decline of the variance of $\bar{X}_T$ as T increases. In general, the variance, $(V(\bar{X}_T))$ of $\bar{X}_T$ is given by the approximation

$$V(\bar{X}_T) = \sigma^2 c/T \qquad (4)$$

where σ^2 = the variance of $X(t)$, measuring the magnitude of fluctuations on a yearly time scale, and the constant c is expressed in terms of the correlation $\rho(s)$ between the population size, $X(t)$, at time t, and the population size, $X(t+s)$, at time $t+s$, according to the formula

$$c = 1 + 2 \sum_{s=1}^{\infty} \rho(s) \qquad (5)$$

Thus for T sufficiently large, the variance given by formula (4) is small. If we look at averages of population density over successive intervals of time of length T, we find that all of those averages are in fact close to $\bar{X}_\infty$. Thus on a time scale defined by this value of T, variation from year to year no longer causes fluctuations in population density on that scale.

In this model with $r > 2$, unstable deterministic dynamics can also cause fluctuations (May and Oster, 1976), but formula (4) still applies in that case, but with c as the limit of the Cesàro mean in formula (5), and indicates the time scale on which short-term fluctuations no longer propogate to long-term fluctuations.

Although we do not expect short-term fluctuations to cause fluctuations on a long time scale, they can nevertheless have an effect in the long run: They can affect the value of $\bar{X}_\infty$. Indeed, we found above that $\bar{X}_\infty$ is equal not to the arithmetic mean of $K(t)$ but to the harmonic mean. It is a general theorem of mathematics that the harmonic mean is less than the arithmetic mean (see below). Thus

$$\bar{X}_\infty < EK \qquad (6)$$

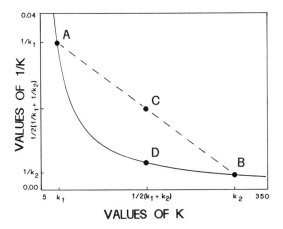

Figure 7.1. Plot of $1/K$ against values of K.

In other words, in a temporally stochastic environment, the long-term average population density is less than the ordinary mean or arithmetic average of the carrying capacity. We conclude that random fluctuations in the environment over time about a given mean carrying capacity lower the average population density.

This lowered carrying capacity comes from a generic property of non-linear stochastic models that is fundamental to many of the interesting conclusions derived from them. That the harmonic mean is less than the arithmetic mean can be rephrased as $E[1/K] > 1/EK$; i.e., the average of a reciprocal is greater than the reciprocal of the average. More generally, we can conclude that under most circumstances the average of a nonlinear function, $Ef(K)$, is different from the nonlinear function of the average, $f(EK)$ (Levins, 1979).

These ideas are illustrated in Figure 7.1, where the graph of $1/K$ is plotted. If K takes on just the two values k_1 and k_2 with probability $\frac{1}{2}$ each, EK is just $\frac{1}{2}(k_1 + k_2)$. $E[1/K]$ can be located similarly on the vertical axis at $\frac{1}{2}(1/k_1 + 1/k_2)$. It can be seen that the value of $E[1/K]$ is at point C on the straight line joining A and B; the value of $1/EK$ is at the point D below it. More generally, when k_1 and k_2 have different probabilities, $E[1/K]$ always lies on line AB, and $1/EK$ lies on the curve below it. With more than two values for K, the geometrical construction is more complicated, but the same principles apply. Moreover, the same construction can be done with any nonlinear function $f(Y)$ of a random variable Y, not just the reciprocal. When the function always curves in the same direction, as it does in the figure, the resulting difference between $Ef(Y)$ and $f(EY)$ is known as Jensen's inequality (Feller, 1971).

The example we have been studying involves a nonlinear function of a parameter. More generally, nonlinear functions of population density arise. For example, consider the following modification of the Moran-Ricker model (Turelli and Petri, 1980).

$$X(t+1) = X(t)e^{r\{1 - [X(t)/K(t)]^\theta\}} \tag{7}$$

In this case, there is no simple equation for the long-term average of $X(t)$, but there is an equation for the long-term average of $[X(t)]^\theta$.

$$\overline{[X_\infty]^\theta} = 1/E[1/K^\theta] \tag{8}$$

which means that

$$\left(\overline{[X_\infty]^\theta}\right)^{1/\theta} = \{1/E[1/K^\theta]\}^{1/\theta} \tag{9}$$

For θ more than 1, Jensen's inequality shows that $\bar{X}_\infty$ is less than the left-hand side of equation (9). Jensen's inequality also implies that the right-hand side of equation (9) is a decreasing function of θ. It follows that the long-term average density must decrease as a function of θ, at least for θ more than 1, although it need not do so monotonically.

This result depends on the presence of variation in K, as K is still the equilibrium value, regardless of the value of θ, and is stable for small $r\theta$ (Turelli and Petri, 1980). In this stochastic version, however, the value of θ affects the long-term average density in a way that depends on the variance of K, as the right-hand side of equation (9) can be expected to decrease more strongly as a function of θ for distributions with larger variance.

Although this second result follows from the same principle, i.e., the existence of differences between averages of nonlinear functions and non-linear functions of averages, this result differs qualitatively from the first. It shows how fluctuations can link properties of an equation (in this case the parameter θ and the outcome $\bar{X}_\infty$) that previously were independent. Such qualitative changes represent some of the most important effects of variability in ecological systems and are at the heart of many of the results discussed in this review.

A particularly clear example that reveals this effect is Ellner's (1987b) discussion of the evolution of dormancy in annual plants. In some equilibrium models, he found no advantage to between-year dormancy, where only a fraction of the seeds of an individual may germinate during any year. In the presence of fluctuations in the yield of germinating seeds, there is a selective advantage to between-year dormancy. Thus fluctuating yield in Ellner's models changes qualitatively the relation between fitness and dormancy.

This general mechanism by which short-term fluctuations translate into systematic effects on a longer time scale, if not treated carefully, can also be a source of confusion in stochastic models (Hastings and Caswell, 1979).

We illustrate this point by reparameterizing the Moran-Ricker model so that the carrying capacity (K) is replaced by its reciprocal, the intraspecific competition coefficient (α). Thus we get the equation

$$X(t+1) = X(t)e^{r[1 - \alpha(t)X(t)]} \tag{10}$$

Both α and K are meaningful biologically, and it is not a priori obvious which one is best used in the analysis. However, a naive approach to the analysis gives strikingly different results. Because $\alpha = 1/K$, it follows that $E\alpha = E[1/K]$; and equation (6) now implies that

$$\bar{X}_\infty = 1/E\alpha \tag{11}$$

Because in a constant environment the equilibrium value of this model is $1/\alpha$, we might be tempted to conclude from equation (11) that fluctuations in α have no effect on the long-term average of $X(t)$, a result that appears to contradict the previous conclusions that fluctuations in the equivalent quantity K lower $\bar{X}_\infty$. In fact, both conclusions are correct; the key is to appreciate that they are based on different assumptions.

Parameterizing by K only leads to the conclusion that the fluctuations lower the mean because of the implicit assumption that EK, the arithmetic mean of K, remains constant as we alter the variance. However, if the mean changes with the variance, the conclusion that fluctuations in K lower the value of $\bar{X}_\infty$ is not valid. Similarly, the conclusion that fluctuations in α do not alter $\bar{X}_\infty$ depends on the assumption that $E\alpha$ remains constant as we increase its variance. Jensen's inequality implies that keeping $E\alpha$ constant is different from keeping EK constant. Indeed, keeping $E\alpha$ constant and increasing its variance almost always increases EK.

Another way of looking at this problem is to note that $H(K) = 1/E\alpha$. If when parameterizing with K we decide to use $H(K)$ rather than EK as the measure of location and keep it constant as we introduce variability, we would conclude that variation in K does not affect long-term average population sizes. Using EK as the measure of location of the distribution of K leads to the opposite conclusion. Thus the effects of fluctuations of a parameter, holding its location constant, may depend on the actual measure of location used.

These problems are not special difficulties with stochastic models. They simply reflect that determining the effect of some parameter in a model involves deciding whether it is linked with other parameters or can be varied independently in a meaningful way. In stochastic models, however, linkages among parameters are not always appreciated. Thus, if $V(K)$ means the variance of K, it is tempting to treat $V(K)$ and EK as equivalent to $V(\alpha)$ and $E\alpha$, with $EK = 1/E\alpha$. As we have seen, however, this assumption is not correct. Indeed, to a rough approximation, $EK = 1/E\alpha + V(\alpha)/(E\alpha)^3$.

A difficulty with stochastic models arises because it is not always easy to provide a rationale for keeping $E\alpha$ constant, EK constant, or some other mean value constant while a variance is changed. Rational criteria can be developed in some cases, however. For example, suppose the carrying capacity, $K(t)$, is proportional to food production during year t. Then keeping EK constant while changing its variance keeps the total food productivity in the system constant but shifts the way it is apportioned over time. Thus we can examine the effect of fluctuating food productivity without changing the overall amount. If we do not have a justification such as this one, we cannot say that fluctuations in the carrying capacity per se decrease long-term average population densities.

In the second model involving θ, more robust conclusions were obtained that did not require decisions about how to measure the location of the fluctuating parameter. That fluctuations in K make long-term average population densities dependent on θ is true regardless of whether we keep $E\alpha$ constant or EK constant. Such interactions between variability and some other factor are common and often lead to similarly robust results. Most of the conclusions from stochastic models we discussed in previous sections are of this nature. Some of the remaining difficulties can be dealt with by choosing standard parameterizations that have a common biological meaning in different models (Chesson, 1988). In general, however, one must be careful to avoid drawing conclusions from stochastic models that depend on arbitrary parameterizations (Hastings and Caswell, 1979; Bulmer, 1985).

Species Persistence

Most of our discussion has dealt with the way in which variability on one scale changes mean trends on a larger scale. Variability on one scale can also lead to variability on a larger scale. For example, Lewontin and Cohen (1969), using a density-independent population model, showed how pure temporal environmental variability on a short time scale translates into large population fluctuations on a long time scale. This situation contrasts with the density-dependent population model of the previous section in which such fluctuations on a short time scale do not lead to population fluctuations on a long time scale. Similarly, the competition models discussed in the section on pure temporal variation do not show variability on a sufficiently long time scale: There is always some finite period of time T such that the average density over such a period shows little fluctuation through time.

Some of the earliest concerns with stochastic models involve the expectation that stochastic variation may lead to extinction (May, 1973; Ludwig, 1976). Some discussions of biological conservation focus on this same idea (Leigh, 1981; Gilpin, 1987; Lande and Barrowclough, 1987; Pimm et al., 1988). It is self-evident that large population fluctuations about a given mean must increase the likelihood of extinction because lower populations will be more frequent. Lower populations mean that it is easier to arrive exactly at zero—extinction—or that serious loss of genetic variability may occur, reducing long-term survival potential (Lande and Barrowclough, 1987).

Within-individual variability, which is usually referred to as demographic stochasticity in this literature, may be the cause of population fluctuations for populations consisting of relatively few individuals (May, 1973; Leigh, 1981; Pimm et al., 1988). For any reasonably large population, though, such variation generated on the individual scale does not lead to significant fluctuations on the scale of the whole population. Temporal environmental variability is different, however. Environmental variation generated on a large spatial scale affects all individuals in a population in a correlated way and therefore leads to population fluctuations on a log scale that are independent of population size or, more precisely, independent of population scale as defined in Chesson (1982). Thus it is generally believed that temporal environmental variability is more important than within-individual variability in causing population fluctuations.

It is tempting to conclude from this discussion that environmental variability is detrimental to species persistence. There are several reasons why this proposal is not likely to be true. First, Gillespie (1978) has pointed out that environmental fluctuations themselves, apart from causing population fluctuations, may play an important role in maintaining the genetic diversity of a population.

Second, in communities of interacting species, environmental fluctuations may be the mechanism of coexistence. Environmental fluctuations mean in this context that all species have times when they perform well. In this regard it is interesting to note that in some community models (Hatfield and Chesson, 1989), and perhaps fairly generally in the subadditive competition models discussed above, population fluctuations are not very sensitive to environmental fluctuations and approach an asymptote as environmental fluctuations become large. This seemingly paradoxical situation arises because environmental variability increases average low density growth rates and therefore increases the rate of recovery from low density. This situation opposes the intuitively disruptive effects of environmental variability at higher densities.

None of this discussion is to suggest that population fluctuations are not an important concern in biological conservation, especially when populations are small. It must be emphasized, however, that the results of this review show that population fluctuations, environmental fluctuations, and stochastic processes occurring on other scales have important functional roles in communities that often go beyond their seemingly disruptive aspects. Understanding their sometimes subtle, sometimes counterintuitive, effects is critical to understanding community function and ultimately to the conservation of species.

Acknowledgments. Discussions with Michael Turelli over the years have greatly improved my appreciation of the "scale transition." This work benefited from the stimulating academic environment and the hospitality of my hosts during my sabbatical at the University of Arizona. It was supported by National Science Foundation grants BSR 8615028 and BSR 8615031.

References

Abrams P (1984) Variability in resource consumption rates and the coexistence of competing species. Theor Pop Biol 25:106–124

Andrewartha HG, LC Birch (1954) *The Distribution and Abundance of Animals.* Chicago University Press, Chicago

Andrewartha HG, Birch LC (1984) *The Ecological Web: More on the Distribution and Abundance of Animals.* Chicago University Press, Chicago

Armstrong RA, McGehee R (1980) Competitive exclusion. Am Nat 115:151–170

Atkinson WD, Shorrocks B (1981) Competition on a divided and ephemeral resource: a simulation model. J Anim Ecol 50:461–471

Bailey VA, Nicholson AJ, Williams EJ (1962) Interactions between hosts and parasites when some hosts are more difficult to find than others. J Theor Biol 3:1–18

Bulmer MG (1985) Selection for iteroparity in a variable environment. Am Nat 126:63–71

Caswell H (1978) Predator-mediated coexistence: a nonequilibrium model. Am Nat 112:127–154

Chesson PL (1978) Predator-prey theory and variability. Annu Rev Ecol Syst 9:323–347

Chesson PL (1981) Models for spatially distributed populations: the effect of within-patch variability. Theor Pop Biol 19:288–325

Chesson PL (1982) The stabilizing effect of a random environment. J Math Biol 15:1–36

Chesson PL (1984a) The storage effect in stochastic population models. Lect Notes in Biomath 54:76–89

Chesson PL (1984b) Variable predators and switching behavior. Theor Pop Biol 26:1–26

Chesson PL (1985) Coexistence of competitors in spatially and temporally varying environments: a look at the combined effects of different sorts of variability. Theor Pop Biol 28:263–287

Chesson PL (1988) Interactions between environment and competition: how fluctuations mediate coexistence and competitive exclusion. Lect Notes Biomath 77:51–71

Chesson, PL (1989) A general model of the role of environmental variability in communities of competing species. Lect Math in the Life Sci 20, 97–123

Chesson PL, Huntly N (1988) Community consequences of life-history traits in a variable environment. Ann Zool Fenn 25:5–16

Chesson PL, Huntly N (1989) Short-term instabilities and long-term community dynamics. Trends Ecol Evol 4:293–298

Chesson PL, Murdoch WW (1986) Aggregation of risk: relationships among host-parasitoid models. Am Nat 127:696–715

Chesson PL, Warner RR (1981) Environmental variability promotes coexistence in lottery competitive systems. Am Nat 117:923–943

Colwell RK (1974) Predictability, constancy and contingency of periodic phenomena. Ecology 55:1148–1153

Comins HN, Noble IR (1985) Dispersal, variability and transient niches: species coexistence in a uniformly variable environment. Am Nat 126:706–723

Connell JH (1978) Diversity in tropical rainforests and coral reefs. Science 199:1302–1310

Connell JH (1979) Tropical rainforests and coral reefs as open non-equilibrium systems. In Anderson RM, Turner BD, Taylor LR (eds) *Population Dynamics*. Blackwell, Oxford, pp 141–163

Connell JH, Sousa WP (1983) On the evidence needed to judge ecological stability or persistence. Am Nat 121:789–824

Crowley PH (1981) Dispersal and the stability of predator-prey interactions. Am Nat 118:673–701

Ellner SP (1984) Stationary distributions for some difference equation population models. J Math Biol 19:169–200

Ellner S (1987a) Alternate plant life history strategies. Vegetatio 69:199–208

Ellner S (1987b) Competition and dormancy: a reanalysis and review. Am Nat 130:798–803

Feller W (1971) *An Introduction to Probability Theory and Its Applications*. Vol 2, 2nd ed. Wiley, New York

Gillespie JH (1978) A general model to account for enzyme variation in natural populations. V. The SAS-CCF model. Theor Pop Biol 14:1–45

Gilpin ME (1987) Spatial structure and population vulnerability. In Soule ME (ed) *Viable Populations for Conservation*. Cambridge University Press, Cambridge, pp 125–139

Grubb PJ (1977) The maintenance of species richness in plant communities: the regeneration niche. Biol Rev 52:107–145

Grubb PJ (1986) Problems posed by sparse and patchily distributed species in species-rich plant communities. In Diamond J, Case T (eds) *Community Ecology*. Harper & Row, New York, pp 207–225

Hassell MP (1978) *The Dynamics of Arthropod Predator-Prey Systems*. Princeton University Press, Princeton

Hastings A (1977) Spatial heterogeneity and the stability of predator-prey systems. Theor Pop Biol 12:37–48

Hastings A (1978) Spatial heterogeneity and the stability of predator-prey systems: predator mediated coexistence. Theor Pop Biol 14:380–395

Hastings A (1980) Disturbance, coexistence, history and competition for space. Theor Pop Biol 18:361–373

Hastings A, Caswell H (1979) Role of environmental variability in the evolution of life history strategies. Proc Natl Acad Sci USA 76:4700–4703

Hatfield J, Chesson PL (1989) Diffusion approximation and stationary distribution for the lottery competition model. Theor Pop Biol 36:251–266

Hubbell SP (1979) Tree dispersion, abundance, and diversity in a tropical dry forest. Science 203:1299–1309

Hubbell SP (1980) Seed predation and the coexistence of tree species in tropical forests. Oikos 35:214–299

Hutchinson GE (1951) Copepodology for the ornithologist. Ecology 32:571–577

Hutchinson GE (1961) The paradox of the plankton. Am Nat 95:137–145

Ives A (1988) Covariance, coexistence and the population dynamics of two competitors using patchy resource. J Theor Biol 133:345–361

Ives AR, May RM (1985) Competition within and between species in a patchy environment: relations between macroscopic and microscopic models. J Theor Biol 115:65–92

Iwasa Y Roughgarden J (1986) Interspecific competition and among metapopulations with space-limited subpopulations. Theor Pop Biol 30:194–214

Lande R, Barrowclough G (1987) Effective population size, genetic variation and their use in population management. In Soule ME (ed) *Viable Populations for Conservation*. Cambridge University Press, Cambridge, pp 87–123

Leigh EG Jr (1981) The average life time of a population in a varying environment. J Theor Biol 90:213–239

Levins R (1979) Coexistence in a variable environment. Am Nat 114:765–783

Lewontin RC, Cohen D (1969) On population growth in a randomly varying environment. Proc Natl Acad Sci USA 62:1056–1060

Ludwig D (1976) Persistence of dynamical systems under random perturbations. Soc Ind Appl Math Am Math Soc Proc 10:87–104

May RM (1973) *Stability and Complexity in Model Ecosystems*. Princeton University Press, Princeton

May RM (1978) Host-parasitoid systems in patchy environments: a phenomenological model. J Anim Ecol 47:833–844

May RM, Oster GF (1976) Bifurcations and dynamic complexity in simple ecological models. Am Nat 110:573–599

Maynard Smith J (1974) *Models in Ecology*. Cambridge University Press, Cambridge

Milligan BG (1986) Invasion and coexistence of two phenotypically variable species. Theor Pop Biol 30:245–270

Murdoch WW (1979) Predation and the dynamics of prey populations. Fortschr Zool 25:245–310

Pacala SW (1987) Neighborhood models of plant population dynamics. 3. Models with spatial heterogeneity in the physical environment. Theor Pop Biol 31:359–392

Pacala SW, Roughgarden J (1982) Spatial heterogeneity and interspecific competition. Theor Pop Biol 121:92–113

Pimm SL, Jones HL, Diamond J (1988) On the risk of extinction. Am Nat 132:757–785

Reeve JD (1988) Environmental variability, migration, and persistence in host-parasitoid systems. Am Nat 132:810–836

Roughgarden J (1974) Niche width: biogeographic patterns among *Anolis* lizard populations. Am Nat 108:429–442

Roughgarden J (1979) *Theory of Population Genetics and Evolutionary Ecology: An Introduction*. Macmillan, New York

Sale PF (1977) Maintenance of high diversity in coral reef fish communities. Am Nat 111:337–359

Sale PF, Douglas WA (1984) Temporal variability in the community structure of fish on coral patch reefs and the relation of community structure to reef structure. Ecology 65:409–422

Shigesada N (1984) Spatial distribution of rapidly dispersing animals in heterogeneous environments. Lect Notes Biomath 54:478–501

Shigesada N, Roughgarden J (1982) The role of rapid dispersal in the population dynamics of competition. Theor Pop Biol 21:253–373

St Amant J (1970) The mathematics of predator-prey interactions. MA thesis, University of California, Santa Barbara

Strong DR (1984) Density vague ecology and liberal population regulation in insects. In Price PW, Slobodchikoff CN (eds) *A New Ecology: Novel Approaches to Interactive Systems*. Wiley, New York, pp 313–327

Strong DR (1986) Density vagueness: abiding the variance in the demography of real populations. In Diamond J, Case TJ (eds) *Community Ecology*. Harper & Row, New York, pp 257–268

Tilman D (1982) *Resource Competition and Community Structure*. Princeton University Press, Princeton

Turelli M, Petri D (1980) Density dependent selection in a random environment. Proc Natl Acad Sci USA 77:7501–7505

Wiens JA (1977) On competition and variable environments. Am Sci 65:590–597

Wiens JA (1986) Spatial and temporal variation in studies of shrubsteppe birds. In Diamond J, Case T (eds) *Community Ecology*. Harper & Row, New York, pp 154–172

Yodzis P (1978) *Competition for Space and the Structure of Ecological Communities*. Lect Notes Biomath 25:1–288

Appendix

To determine the behavior of the Moran-Ricker equation discussed in the text, we first take logs in equation (1) to get

$$\ln X(t+1) - \ln X(t) = r[1 - X(t)/K(t)] \tag{12}$$

Summing this equation from $t = 0$ to $t = T - 1$, and dividing by T, we obtain

$$\frac{\ln X(T) - \ln X(0)}{T} = r[1 - \frac{1}{T} \sum_{t=0}^{T-1} X(t) \cdot \left[\frac{1}{K(t)}\right]] \tag{13}$$

Because the fluctuations in $X(t)$ are bounded away from 0 and ∞ as $T \to \infty$, the left-hand side of equation (13) must converge to 0. In other words

$$\frac{1}{T} \sum_{t=0}^{T-1} X(t) \cdot \left[\frac{1}{K(t)}\right] \to 1 \tag{14}$$

Boundedness of $K(t)$ and $X(t)$ from 0 and ∞ imply that the convergence in relation (14) is mean square. The independent fluctuations of $K(t)$ over time mean that $X(t)$ and $1/K(t)$ are statistically independent, even though it is clearly not true of $X(t+1)$ and $1/K(t)$. Using this fact, routine calculations show that

$$\frac{1}{T} \sum_{t=0}^{T-1} X(t) \cdot K(t)^{-1} - EK^{-1} \tag{15}$$

coverges in mean square to 0; combining this result with equation (14) we see that $\bar{X}_T$ converges in mean square to $H(K)$, proving equation (3) in the text.

Equation (9) in the text has essentially the same derivation as equation (3).

8. Managing and Monitoring Ecosystems in the Face of Heterogeneity

Craig Loehle

Problems of air and water pollution have become sufficiently acute and extensive that piecemeal solution of the problems is no longer adequate. It has become clear that even if each city meets clean air and water standards for human health the regional ecosystems may be degraded. It is thus necessary to monitor and manage the environment on a regional or even global basis. At this scale, however, heterogeneity of all types becomes almost overwhelming. Spatial heterogeneity, multilevel phenomena, multiple management activities, and multiple time scales interfere with the abilities of managers and regulatory agencies to detect and prevent environmental deterioration. This chapter addresses strategies for dealing with complexity and heterogeneity, largely in the context of environmental protection.

A characteristic of large-scale environmental problems is that they occur slowly and in a piecemeal fashion. To a person on the ground and on a year-to-year time scale the problems may be imperceptible. An overall perspective is needed to integrate many types and sources of data before these problems can be demonstrated to policymakers and the public. Furthermore, without good regional monitoring data and ways to analyze such data, it is impossible to assess the economic impact of adverse changes, to recommend corrective actions, or to recognize improvement after intervention.

Managing ecosystems on a regional basis is not a straightforward extension of, for example, municipal water pollution control. New issues arise when regional problems are addressed because of spatial complexity, multiple media, multiple variables of interest, and synergy of impacts. We can recognize both unresolved scientific issues and difficulties of implementation.

Scientific Issues

Ecosystem Health

One of the basic questions, when moving from human health concerns (city air pollution) to ecosystem health concerns, is what exactly constitutes ecosystem health. A daunting problem is the lack of referents. In many places the entire natural community has been disturbed by pollution, logging, hunting, fire, fishing, and agriculture, sometimes to the extent that it is difficult to ascertain the original composition of the system before man. In the case of historical disturbances by preindustrial man (e.g., controlled burning) we may never know what the ecosystem was like in man's absence. Trying to ascertain the characteristics of ecosystem health by studying perturbed ecosystems is analogous to the efforts of early psychologists to define human behavior largely from studying the mentally ill. We are no longer satisfied with a definition of mental health that is stated in terms of the absence of hallucinations.

Even when reasonably unaltered referents are available, it is still not easy to define ecosystem health. If a whole forest dies, it is clear to all that the ecosystem is in bad shape. For monitoring purposes, however, we would prefer to be able to detect ill effects before a catastrophe occurs, so that action can be taken. Thus detection of degrees of ill health in the ecosystem is necessary. Detection of nonlethal impacts is not easy even when a single variable such as tree growth increment is used (Loehle, 1988). If overall system health is to be defined, the difficulties are even greater. Early monitoring studies took the simplistic view that diversity (species richness) is good and natural systems are diverse. Biotic diversity (species richness) thus became a key variable in most environmental impact statements. Others have taken an approach analogous to human health (Regier and Rapport, 1978; Rapport and Regier, 1980; Rapport, 1984; Rapport et al., 1985). I propose a successional framework for specifying ecosystem health because so many ecosystems today are perturbed or regenerating and are undergoing succession. This basic concept has been used before, but it is expanded here to show that certain indicators of stress are available that are not currently recognized. A successional context is particularly important for detecting regional trends and gradual habitat destruction because the systems change naturally over the period of measurement. It also enables us to develop specific null models and appropriate statistical tests.

It has long been recognized that ecosystems go through a sequence of stages during their development. This successional process that we may call progression has been characterized for a multitude of systems, and generalizations have been drawn about the characteristics distinguishing early versus late successional stages. Although certain stress agents affect only portions of the ecosystem (e.g., DDT and bird eggs), many affect the overall system (system health). In a classic paper Woodwell (1965) showed that chronic stress due to ionizing radiation resulted in a process that is in many ways the reverse of progression. In his experiment, forest reverted to shrubs and then grasses and forbs as radiation dose increased. The general process of retrogression in response to chronic stress has since been documented for many ecosystems (Rapport et al., 1985).

This situation suggests criteria for determining ecosystem health. If a system is going through progression along the expected trajectory, it is healthy. If its progression is slowed or halted, it is stressed. If retrogression is occurring, stress is severe. Note that in rare cases (e.g., early successional species that are highly susceptible) pollution could actually speed up progression. The particular variables useful for measuring progression and retrogression vary by system type, and sometimes a multivariate approach is necessary to characterize changes in community composition. A key to using this approach, however, is characterization of normal successional trajectories. Stressed systems can then be compared to unstressed ones of similar type and productivity (Loehle and Smith, 1990; Loehle et al., 1990). For example, for detecting acid rain stress on forests, recent tree growth rates have been compared to prior growth rates. A proper null model, however, takes into account endogenous natural changes in tree or stand growth rates with age (Loehle, 1988). Failure to use the proper null model has led to hasty conclusions of both effect and no effect.

Once trends in system variables with time have been established, it is possible to develop valid statistical tests. First, the path and time variance of the null model must be defined. Consider a stochastic autoregressive process, such as logistic growth perturbed by noise. Taking a number of replicate systems perturbed by independent noise vectors, the time paths of the systems when plotted together define a stochastic envelope for the nominal system. At each point in time this gives an expected value and a variance. Given this information, a time series for the test (stressed) system can be compared to the expected trend, and trend or deviation statistics can be developed (Loehle and Smith, 1990; Loehle et al., 1990). Note that standard references (Green, 1979; Gilbert, 1987) do not even address such successional contexts, in part because many monitoring studies are looking for evidence of pollution (e.g., heavy metal tissue concentrations), which is not necessarily related to overall successional status. By this time-series method, lack of progression due to stress would quickly show up as a deviation from the expected trend, whereas the system might not be visibly stressed and measures of living matter or production rates might not indicate a problem. On the other hand, in many ecosystems growth processes peak and then decline. For example, after fire, herbaceous biomass increases to a peak and then declines somewhat after several years (e.g., Kovacic et al., 1985). Measuring only the test system over time would indicate a "problem" decline in productivity, whereas comparison to an expected trajectory would not, in direct analogy with the forest decline problem. This same statistical approach can be applied to systems in equilibrium, where stress could be indicated by increased variance around the equilibrium or a retrogressive trend. Certain types of stress, e.g., DDT-induced thin egg shells, may show up at the species level much sooner than at the ecosystem level, of course, and such information should be utilized.

Not all systems behave in such a manner that the above approach can work. Some systems or species (e.g., grasshoppers on grassland) are so variable from year to year that no "nominal" successional trend can be discerned. In other cases a disturbance (e.g., introducing a herbivore) may change the system in such a way that it is incommensurate with that which previously existed. It might even still be classified as "healthy" using measures of system structure and function. It is thus necessary to use this approach appropriately.

Knowledge of expected trajectories is also useful for experimental design. Rates of change influence how frequently samples are required. In general, samples must be taken more often when rates are rapid or when it is necessary to detect a peak or inflection point. The current state of the system can be matched with typical trajectories, and location on the curve can be used to assess optimal sampling frequency for upcoming sample periods. For example, based on this information one would not attempt to detect yearly changes in forest growth because the rates of change are too slow.

The concept of stochastic trajectories is closely associated with the scale of disturbances and the nature of patchiness. At the small scale of habitat patches, succession is a stochastic process. Plotting individual patches versus time yields the stochastic envelope discussed above. In this context, much of the heterogeneity in a habitat is spurious, reflecting stochastic processes and patch age. Some amount of habitat heterogeneity may thus be reduced by putting it in a stochastic successional context. Detection of impact on such patches would need to consider rates of change in this context of an endogenous successional process. At the larger scale, a forest with patches of all ages is in a sense at equilibrium when considered as a whole and can be compared to another such forest. Most forests near human activities are far from this equilibrium condition, so in general the whole process of defining "plots," "patches," and so on should be in a successional context. For further details on statistical methodology, see Loehle and Smith (1990).

In addition to being able to detect incipient human health risks and ecosystem damage, it is also useful to be able to ascertain the point of no return (PNR). For many systems, such activities as fishing, farming, and forestry are accepted even though they perturb the system. It is assumed that consequences of such management are reversible, but when activities lead to irreversible damage, a PNR has been reached.

For a population, the PNR is that population size at which irreversible decline or random extinction is almost certain, even when protection from human disturbance is complete. The California condor and carrier pigeon are examples. The PNR can result from effects of stochastic fluctuations on small populations, a population density so low that mating success declines, habitat fragmentation, inbreeding suppression, or other causes.

There is often a PNR for ecosystems as well. Physical alteration of habitat can result in permanent degradation of productive capacity. Areas around copper smelters in several parts of the world have been irreparably degraded. The area around the smelter typically was logged for fuel, after which fumes killed the remaining vegetation and erosion removed the topsoil. These devastated landscapes remain today, only slowly revegetating even though copper smelting ceased long ago. Another example is laterite formation in a tropical jungle. Small clearings are much less subject to laterite formation than are large clearings. Yet another example is from the American West. When sagebrush grassland is overgrazed just once, cheatgrass (*Bromus tectorum*), a fast growing annual, invades and becomes a permanent dominant. Cheatgrass has little forage value (hence its name). Below a certain patch size, remnant forest stands may degenerate (Franklin and Forman, 1987; Loehle, 1989), and species of wildlife may not be able to persist. In the case of marsh disappearance, a PNR can be reached at a certain level of fragmentation due to canal dredging. Finally, old, severely stressed (by pollution, for example) or suppressed trees may fail to recover even when the pollution is removed or competition is reduced (Loehle, 1988).

It is imperative to identify PNRs before they are reached; afterward it may not be possible to undo the damage even with intensive management efforts. Enough human-created disasters have been generated to compile an extensive set of PNRs as a guide to policymakers. Table 8.1 lists a few PNRs and their indicator variables. Note that patch size and fragmentation are often indicator variables. Predictive models of succession or physiological response to stress can be used to anticipate PNRs. Ecological theory in conservation biology, community stability, e.g., catastrophe theory (Loehle, 1989), and the ecology of invasions can also contribute to this base of knowledge. For more examples of PNRs, see Munn (1988).

Spatial Considerations

When regional impacts are being studied, spatial aspects play an important role. For species preservation, the size and spatial contiguity of habitats is important. For air pollution studies, it is important to determine origins of pollutants and how they are transported and transformed regionally. Habi-

Table 8.1. Indications of Approach to a Point of No Return

Point of No Return	Indicators
Species extinction	Small population size, high hunting pressure, small habitat size, reproductive failure, low genetic diversity
Habitat destruction	Fragmentation, small remnant patch size, exotic invasion, rare species losses, large clearing sizes
Marsh disappearance	Deposition—settling—sea level rise < 0, fragmentation, salt water intrusion
Desertification	Browse lines, plant reproductive failure, lack of litter for soil cover, increased stream flashiness and erosion
Soil salinification	Salt crystals evident, changes in soil structure, reduced drainage

tat fragmentation can itself cause further fragmentation and destruction. In old-growth stands of trees, windthrow may increase when the forest is fragmented (Franklin and Forman, 1987). The spatial placement of canals and dikes in a marsh can dramatically influence its rate of disappearance (Sklar et al., 1985). An emerging realization is that extrapolation of laboratory toxicity data to the field cannot be based on average field conditions. Laboratory studies, even studies utilizing microcosms, maintain uniform conditions for the test system. In the field, however, spatial and temporal heterogeneity prevail. Breck et al. (1988) have shown that when organisms can detect and avoid disadvantageous locations (e.g., owing to salinity or pH) within their home range or can temporarily tolerate them, survival increases substantially. Thus spatial variance, by providing spatial or temporal refuges, can enable populations to survive under chronic stress conditions when an average or mean of sampled values compared to laboratory data would predict mortality. For acute stressors, the converse may be true. If a pollutant is rapidly toxic at high doses, a spatial average value that laboratory studies would predict to be safe would actually result in high mortality if the spatial variance were high because part of the population would pass through "hot spots" and perish. Hot water from power plants is such a stressor. Thus how spatial variables are averaged is a nontrivial problem.

   The systematic treatment of such spatial factors is an emerging field of ecology (Forman and Godron, 1986). An unresolved scientific problem is the inadequacy of our descriptive and quantitative tools for studying problems of fragmentation, transport, and spatial structure. There are fairly good descriptors for spacing of points on a plane (e.g., tree stems), particularly regarding degree of randomness, but descriptors for spacing of patches or linear features have been lacking. Promising progress has been made using fractal measures of patchiness and spatial complexity as well as concepts such as percolation (Burrough, 1981; Loehle, 1983; Gardner et al., 1987; Krummel et al., 1987; Turner, 1987; plus others this volume), but much more needs to be done.

A difficulty that arises in spatial studies is rapidly increasing sample sizes. Even when the goal is to compare a single ecological treatment with a control, adequate sample sizes are difficult to obtain because of inherent variability. Expense goes up even faster if multiple variables must be measured. For this reason, most studies of ecological scaling have utilized remote-sensing data of landscapes because it has already been collected and computerized. Similar studies done on the ground would be prohibitively expensive. When the goal is to detect a spatial trend in pollution or stress (e.g., with distance from a smelter), to create contour maps, or to locate "hot spots," standard methods for determination of sample size are inadequate. This is because the variance is not merely a matter of sample size (power) but of fineness of the sample grid in relation to the scaling properties of the phenomenon. Inadequate sampling can result in the inability to detect spatial patterns.

One method that can help is the survey/census method (Flatman and Yfantis, 1984), which involves a preliminary survey on a grid, with composite samples at each point. A semivariogram is used to test for adequacy of the grid. If certain areas are inadequately covered, more intermediate grid points are sampled. This method ensures the minimum number of points for adequate coverage. An isopleth map can then be drawn. The method works best when single or composite samples at each grid point are representative (i.e., within plot variability is not high).

For highly variable or complex systems, more effort should perhaps go into a search for the best variables to sample rather than the optimal grid. For example, total biomass on a grassland is much more closely related to precipitation than are individual species values because the species dominants on a plot are determined by succession, growing history, and chance in addition to precipitation. As another example, certain remote sensing signatures may be good indicators of physiological stress, regardless of the species. In conclusion, the question of adequate sampling for spatial problems involves determining the minimum number of grid points, samples at each grid point, and variables in each sample. Further refinements are needed on these issues.

Synergistic and Multidimensional Processes

Human impacts are rarely single factor. Extinction of a species may be due to the combined effects of lack of nest sites, hunting pressure, and chemical pollutant stress. Many polluted bodies of water resemble a chemical soup. Synergistic effects become important on a regional spatial scale because of multiple types and sources of pollutants, multiple land management activities, and the presence of migratory species. Cumulative impacts are supposed to be considered under the National Environmental Protection Act, but the means for doing so are not available. Animals that move around may encounter different stressors on any particular day, as may even a particular spot of ground as pollution levels fluctuate. Typical toxicology tests (e.g., for a new industrial chemical) consider only one chemical at a time at a fixed dose using only one or a few species (often species not native to the system of concern). Testing synergistic relations of chemicals is currently impractical, yet nevertheless important. A breakthrough is needed in this area, such as a reliable microbial test organism that can facilitate testing of multiple combinations of pollutants. A first step in this direction is community toxicity testing (Cairns, 1986). The ability to consider time and space variability effects is still almost nonexistent. Synergy of toxic materials with environmental alterations also needs more exploration. Graphic overlays of localities with several particular toxic or ecomanagement problems could highlight those regions with a spectrum of problems, so that concentrated attention can be focused there. Perturbation modeling is also a useful tool for studying synergistic or heterogeneous impacts, though model reliability becomes a significant issue.

Management Issues

Institutional Aspects

Given the need for regional environmental monitoring and management, it would seem that setting up a program should be straightforward: Define the data to be collected and assign the project to an agency. Unfortunately, it is not so simple. First, there are already a plethora of agencies, municipalities, and institutions doing monitoring for their own purposes (e.g., enforcement, research). A regional program should make use of these existing operations and conversely. Arranging cooperative relations between multiple institutions is a difficult task, particularly between state and federal levels or between countries (Bandurski et al., 1986). A related problem is maintaining continuity of funding in such an environment. Finally, the relation between monitoring programs and environmental research programs needs better definition. Answering certain questions and even obtaining certain data requires a research project. Measuring community diversity or breeding pairs of birds, for example, is relatively straightforward, but other measures of ecosystem integrity require in-depth study and experimentation, tasks better suited to university research centers. For example, "measuring" pollution-induced growth suppression of trees is far removed from routine air pollution data recording, and proper methods are still under debate (Loehle, 1988).

Because ecosystems are variable seasonally and over longer time periods, there is broad agreement on the need for long-term ecological data sets. Without such data it is difficult to assess if an observed change is man-induced. Unfortunately, long-term data are difficult to collect without institutional support.

There are three basic types of long-term data. An obvious one is continuously collected ecological data, an example of which is the Audubon Society Christmas bird counts, repeated now for many years. Another significant source of such data is the recently established network of Long-Term Ecological Research sites of the National Science Foundation. Biological field stations and government agencies (e.g., the US Forest Service) also compile useful data sets. Long-term data collection programs face problems of continuity of variables measured, funding continuity, comparability of data as analytical methods change, and lack of professional prestige for the scientists involved.

A second source of long-term data is archived samples. Archived water, filter, and biological samples can be useful for asking retrospective questions, such as when a particular toxicant first appeared in the environment. Archiving programs face problems of preservation and space. Volatile man-made organic substances may be difficult to preserve. Other samples are bulky, and space becomes limiting quickly (try finding freezer space for 100 ducks each year). Long-term maintenance of archived samples or even data requires better record keeping than most researchers are willing to perform or institutions to reward. For example, large historical museum collections in England face an uncertain future because of lack of curators and space. Remote sensing data storage and management at NASA is underfunded.

A third type of long-term data is contained in nature preserves, which can act as artificial "before impact" sites (i.e., time-control substitutes). They provide an opportunity for a better understanding of how nature works in the absence of human perturbation. Unfortunately, as air pollution has become regionally dispersed, fewer nature preserves are completely pristine, especially near industrial regions where baseline data are most desperately needed.

Data Management

There are numerous technical problems associated with implementing a regional monitoring program. Any such program generates mountains of data, which introduces the problem of data management; fortunately, modern computers and database software alleviate this problem, so no conceptual barrier to implementation exists. Interpretation of complex, spatially oriented data can be difficult, however, and more effort should be spent in this area than usually is. Expert system technologies (e.g., Loehle and Osteen, 1990) and geographic information systems (Johnston et al., 1988) can be utilized to help integrate data with regulatory requirements. Continuity of database updating may be a continuing problem, however. Flexibility should be built into the design of any database system to allow addition of sample locations and variables.

Serious problems are quality assurance and data comparability. When data from one site or time measured to parts per million (ppm) are combined with data measured to parts per billion, (ppb), a zero in the ppm data is not necessarily a zero on a ppb scale. Such censored data cause statistical problems (Taylor and Stanley, 1985). Similar problems arise when studies vary in the level of taxonomic detail. If data are collected by various agencies using a variety of laboratory and field techniques, data comparability is even more difficult to achieve. In fact, some data collection methods are actually invalid, but who is equipped to test all the methods used or to enforce standards? Long-term studies are frequently plagued by changes in methods or experimental design (Gore et al., 1979; Taylor and Stanley, 1985). Quality assurance also involves such issues as sample custody, sample archiving, double checking entered data, and checking for outliers (Taylor and Stanley, 1985). When a variable such as diversity is examined, it must be remembered that there are several incommensurate indices, so that one index must be chosen and the index being used indicated in the database for each site. Collection of data at different frequencies or different scales also creates problems for database management and data interpretation.

Given that data can be managed and data quality, continuity, and comparability achieved, the important step of interpretation can become a major bottleneck. Trying to discern patterns over several time and spatial scales for several media, species, and measures of ecosystem structure is a daunting task but is exactly what is required in regional studies. The various issues of concern at the regional scale may operate at different time and space scales and may scale differently with the patch size sampled. The acid rain problem is symptomatic of the difficulties that arise. There has been a tendency to overgeneralize results of single studies because scientists in the beginning did not have experience over the various sites, species, and causes of dieback. Those who found pollution but no dieback tended to denigrate pollution as a mortality cause. Conversely, those living in areas with high tree mortality tended to view all diebacks as due to that cause. A more balanced, multicausal view of diebacks is now beginning to emerge (e.g., Loehle, 1988) but only after much work. Similar problems have historically existed with environmental impact statements, where complex issues of causation have often not been resolved because of inadequate experimental design, time scales, money, or expertise (Schindler, 1976; Thomas et al., 1978, 1981; Gore et al., 1979; Bandurski et al., 1986). In general, we find that the difficulty of solving a problem goes up multiplicatively with the number of interacting factors, rather than linearly. If our question is about the prevalence of n independent pollutants in the environment, the difficulty is simply a linear n-multiple of effort required. When we study ecosystems, however, species interact. One species does better after forest fires, whereas another does worse. One species of concern may prey on another of concern. An even more fundamental limitation is the difficulty people have thinking about more than two or three causal factors at once. Tools are needed that reduce the dimensionality of the problem to something tractable, without too much loss of detail.

Data on toxicants can be converted to information about human health risks in several defensible ways. Maps can be generated showing regions that exceed legal standards for carcinogens and other health risk agents, including areas with multiple types of risk. Information can also be generated on expected induced cancers, workdays lost, or premature deaths.

Interpretation of measures of ecosystem health is likely not straightforward, however. Certain aspects (e.g., counts of rare species) relate to statutory provisions (e.g., preservation of endangered species), and these data can be converted to fairly objective information on meeting these societal goals. Data on rare objects are, however, the most subject to sampling error and sampling inadequacy, especially in the context of a patchy habitat. Other measures of health are difficult to assess (how "bad" is a 10% loss of diversity?) in a value-free context. Use of the successional standard for assessing impact makes quantification of the degree of impact more robust, but the problem of the seriousness of a given degree of change remains. A partial solution to evaluating ecosystem damage is conversion to dollar terms. An $X\%$ decline in tree vigor in the Northeast from acid rain may not sound too bad until it is converted to the information that maple sugar production will be reduced, resulting in a loss of $\$Y$ of income. Such a measure of impact is crude but is relatively objective so long as huge dollar values are not placed on rare species or habitat preservation, for which a nondollar value system should be used. Gosselink et al. (1974) provided an example of determining the value of a natural system by thorough analysis of the "services" provided by it. Such an approach is limited by distortions in our economic system.

It is imperative at the regional or global levels to be able to aggregate information in a meaningful way. An acre by acre assessment of a region is simply impossible, though distributed samples may be partial answers. In many cases mere sampling is not adequate, however. For example, in studies of the Mississippi River delta (Sklar et al., 1985) proper characterization of spatial units (pixel size and definition of types) was critical. In the study of the global carbon dioxide budget, methods for classifying vegetation types and estimating areal extents of these types is under debate. Work in spatial modeling (above) and remote sensing (below) are making a dent in these problems, but much remains to be done.

Cost Containment

Cost containment should be a prime consideration in monitoring study design because such studies can be black holes for money without yielding desired answers (e.g., Schindler, 1976; Thomas et al., 1981). There are two types of cost: the cost of conducting an inconclusive study (dollars wasted and undesirable consequences not detected in a timely manner), and the cost of conducting an inefficient or overly intensive sampling program. Of the thousands of environmental impact studies conducted during the 1970s and 1980s, most fall into the first category. Current practice has not always improved on this record, though the tools now exist to do so. Two major innovations exist: efficient experimental designs and more cost-effective technologies.

In general, there is an internal conflict in experimental design goals for monitoring studies. To make sure nothing adverse is happening (e.g., for an Environmental Impact Statement), a survey approach often seems indicated, and a few data are collected on everything. This tendency is particularly marked at a regional level, when many habitats are being considered. After the fact, hard results (pro or con) are often requested, but sample size or time replication may turn out to be inadequate. Proper statistical methods can help avoid such dilemmas. First, power analysis (which shows, given inherent variability, how many samples are needed to detect a given degree of difference between treatments) indicates that we cannot usually afford to "have our cake and eat it too." Thus although a survey might measure many attributes with only a few replicates, sufficient power for statistical testing may require 15 to 50 replicates at each time/location. The difference in sample size is such that hard choices are required. With survey-type data it generally is not possible to squeeze the data sufficiently to extract rigorous conclusions. By making hard choices and limiting variables sampled to those that are most informative, e.g., microbial dynamics (Seki, 1986), the cost of doing an inconclusive study can be avoided. Such an approach may not, of course, appease those who insist that we must know the impact on every species before we draw any conclusions; but given funding limitations, some choices are necessary.

Statistical analysis can help when choosing the best variables to measure. Certain variables are so inherently variable that no reasonable sample size is capable of characterizing them. Such variables should be avoided, no matter how informative they might be in principle. Consideration of the expected time behavior for successional systems (discussed above) can also help when choosing variables. Properties of the system that relate closely to successional status and respond quickly to changing conditions make good indicator variables. The more rapidly a variable is changing, the easier it is to detect a change from one time period to the next (for more details see Loehle et al., 1990). Variables that may change owing to reasons other than human impact are not particularly useful (Thomas et al., 1981) unless they exhibit a predictable underlying trend due to, for example, succession (Loehle et al., 1990).

Given the above considerations, efficient experimental designs can be devised that provide adequate power at minimal cost (Green, 1979; Skalski and McKenzie, 1982; Alldredge, 1987; Gilbert, 1987). For regional monitoring of air quality, substantial cost savings can be achieved by carefully reducing the overlap of coverage between adjacent stations (Modak and Lohani, 1985). Even more dramatic savings can be achieved if extreme values are the item of interest. Consider a search for outcropping of contaminated groundwater. Samples are cheap to collect but expensive to analyze (e.g., for heavy metals and radionuclides). Uncontaminated sites have low or undetectable levels of contaminants, whereas hot spots contain levels several orders of magnitude higher. Using the pooling method of Casey et al. (1985), adjacent samples on a grid (5 to 20 at a time) are subsampled and the subsamples pooled. This pooled sample is analyzed, and if it comes up hot the original samples are analyzed to pinpoint the location of the problem. The sensitivity of the test decreases with the number pooled; but note that even if nine semples are clean and one is hot (e.g., 1 ppm) the pooled sample is likely to be detectably hot (in this case the pooled sample registers 100 ppb). Furthermore, when the original samples are analyzed, only those substances that showed up as hot in the pooled sample need be assayed. Such methods can help overcome the barriers posed by spatial patchiness and heterogeneity.

In addition to efficient experimental design, improvements in technology are needed to bring monitoring costs under control. In particular, labor costs are high, especially when PhD-level expertise is needed (e.g., for taxonomic identification of difficult species). We can look to several areas where costs are coming down. Standard ecological field instrument costs, for example, are decreasing (though slowly because of low volume). New devices offer in situ measurements that previously required handling and laboratory work. In situ devices combined with automated, computerized data-loggers can eliminate many man-hours of repeated site visits. Automatic cell counters comprise another labor-saver. Connecting laboratory equipment directly to a computer eliminates both a time-consuming manual step and a source of errors. There should be a trend for laboratory equipment (e.g., mass spectrometer, autoanalyzer) costs to come down, but the tendency is to replace old equipment with more sensitive, more expensive instruments. In some cases new equipment allows for automatic processing of large numbers of samples, thereby reducing labor costs. The tendency to measure to ever-smaller detection limits is not necessarily cost-effective, however. Although we need to know about the ubiquity of low-level contaminants, in routine monitoring studies we are interested mainly in levels near or above statutory limits. Using a less costly, less sensitive laboratory method could reduce both cost and the volumes of data that need to be interpreted and managed.

Remote sensing (e.g., Spitzer, 1986) provides truly a quantum leap in cost-efficiency. Initial costs of a satellite or set of aerial photographs can seem high, but spatial coverage is also high and costs are much lower than for ground studies. In some cases resolution is a problem, but in others it is excellent (e.g., infrared aerial photographs can detect individual stressed trees). In addition to biomass and stress measures, some new methods even allow the discrimination of nitrogen content of a forest canopy (Wessman et al., 1988).

By combining less costly technologies and those that reduce labor costs, efficient spatially oriented experimental designs, and large-scale measurement methods (e.g., remote sensing), it should theoretically be possible to reduce monitoring and impact study data collection costs such that regional studies can become more routine. A competing tendency, described by the paradox of the computer and the accountant, is at work, however. This paradox is evident when it is noted that large-scale computerization should have reduced the cost of routine accounting and inventory control. Instead, it has enabled more detailed control. Detailed inventory control may reach the point where every part in a warehouse has a unique number and can be traced. An individual's time and expenses may now be charged against several cost codes. Falling into this trap is easy with environmental work because as we learn more about ecosystems and pollution we can think of more things to measure. This point is particularly true when we try to handle spatial heterogeneity. The only antidote is to keep firmly in mind the statistical trade-off between the number of things measured and the sample size of any one item. Monitoring programs should reflect the minimum effort necessary to obtain information sufficient to determine an appropriate response. Occasional fishing expeditions that measure everything can act to counterbalance this narrow focus.

Conclusions

In the past, environmental assessments have tended to be mind-numbing collections of raw facts, maps, and species lists. Extending such an approach to regional problems is unacceptable. By consideration of the concept of ecosystem health and the existence of points of no return, regional monitoring can become more focused. Statistical consideration of scale, spatial complexity, suitability of variables, and sampling efficiency can lead to better experimental and monitoring designs. Without such an integrated approach, regional monitoring programs will be doomed to either inconclusive results or exponentially rising costs.

Acknowledgments. Manuscript preparation was supported by contract DE-AC09-76SR00001 with the US Department of Energy. Helpful reviews were provided by J. Bowers, C. Comiskey, M. Paller, and M.L. Scott.

References

Alldredge JR (1987) Sample size for monitoring of toxic chemical sites. Environ Monitor Assess 9:143–154.

Allen TFH, Starr TB (1982) *Hierarchy Perspectives for Ecological Complexity*. University of Chicago Press, Chicago

Bandurski BL, Haug PT, Hamilton AL (1986) *Toward a Trans-boundary Monitoring Network: A Continuing Binational Exploration*. Vols 1 and 2). International Joint Commission, Washington, DC

Breck JE, DeAngelis DL, van Winkle W, Christensen SW (1988) Potential importance of spatial and temporal heterogeneity in pH, Al, and Ca in allowing survival of a fish population: model demonstration. Ecol Model 41:1–16

Burrough PA (1981) Fractal dimensions of landscapes and other environmental data. Nature 294:240–242

Cairns J Jr (1986) *Community Toxicity Testing*. ASTM, Philadelphia

Casey D, Nemetz PN, Uyeno D (1985) Efficient search procedures for extreme pollutant values. Environ Monitor Assess 5:165–176

Flatman GT, Yfantis AA (1984) Geostatistical strategy for soil sampling: the survey and the census. Environ Monitor Assess 4:335–349

Forman RTT, Godron M (1986) *Landscape Ecology*. J Wiley, New York

Franklin JF, Forman RTT (1987) Creating landscape patterns by forest cutting: ecological consequences and principles. Landscape Ecol 1:5–18

Gardner RH, Milne BT, Turner MG, O'Neill RV (1987) Neutral models for the analysis of broad-scale landscape pattern. Landscape Ecol 1:19–28

Gilbert RO (1987) *Statistical Methods for Environmental Pollution Monitoring*. Van Nostrand Reinhold, New York

Gore KL, Thomas JM, Watson DG (1979) Quantitative evaluation of environmental impact assessment, based on aquatic monitoring programs at three nuclear power plants. J Environ Manag 8:1–7

Gosselink JG, Odum EP, Pope RM (1974) –The value of the Tidal Marsh. Center for Wetland Resources, Louisiana State University, Baton Rouge

Green RH (1979) *Sampling Design and Statistical Methods for Environmental Biologists*. Wiley, New York

Johnston CA, Detenbeck NE, Bonde JP, Niemi GJ (1988) Geographic information systems for cumulative impact assessment. Photogram Eng Remote Sens 54:1609–1615

Kovacic DA, Dyer MI, Cringan AT (1985) Understory biomass in ponderosa pine following mountain pine beetle infestation. Forest Ecol Manag 13:53–67

Krummel JR, Gardner RH, Sugihara G, O'Neill RV, Coleman PR (1987) Landscape pattens in a disturbed environment. Oikos 48:321–324

Loehle C (1983) The fractal dimension and ecology. Speculat Sci Technol 6:131–142.

Loehle C (1988) Forest decline: endogenous dynamics, tree defenses, and the elimination of spurious correlation. Vegetatio 77:65–78

Loehle C (1989) Forest-level analysis of stability under exploitation: depensation responses and catastrophe theory. Vegetatio 79:109–115

Loehle C, Osteen R (1990) Impact: an expert system for environmental impact assessment. AI Applications 4:35–43

Loehle C, Smith EP, (1990) An assessment methodology for successional systems. II. Statistical tests and specific examples. Environmental Management 14:259–268

Loehle C, Smith EP, Gladden J (1990) An assessment methodology for successional systems. I. Null models and the regulatory framework. Environmental Management 14:249–258

Modak PM, Lohani BN (1985) Optimization of ambient air quality monitoring networks (part 1). Environ Monitor Assess 5:1–19

Munn RE (1988) The design of integrated monitoring systems to provide early indications of environmental/ecological changes. Environ Monitor Assess 11: 203–217

Rapport DJ (1984) State of ecosystem medicine. In Cairns JW, et al (eds) *Contaminant Effects on Fisheries*. Wiley, New York pp 315–324

Rapport DJ, Regier HA (1980) An ecological approach to environmental information. Ambio 9:22–27

Rapport DJ, Regier HA, Hutchinson TC (1985) Ecosystem behavior under stress. Am Nat 125:617–640

Regier HA, Rapport DJ (1978) Ecological paradigms, once again. Bull Ecol Soc Am 59:2–6

Schindler DW (1976) The impact statement boondoggle. Science 192:509

Seki H (1986) Thresholds of eutrophication in natural waters. Environ Monitor Assess 7:39–46

Skalski JR, McKenzie DH (1982) A design for aquatic monitoring programs. J Environ Manag 14:237–1251

Sklar FE, Costanza R, Day JW Jr (1985) Dynamic spatial simulation modeling of coastal wetland habitat succession. Ecol Model 29:261–281

Spitzer D (1986) On applications of remote sensing for environmental monitoring. Environ Monitor Assess 7:263–271

Taylor JK, Stanley TW (1985) Quality assurance for environmental measurements. ASTM-867

Thomas JM, Mahaffey JA, Gore KL, Watson DG (1978) Statistical methods used to assess biological impact at nuclear power plants. J Environ Manag 7:269–290

Thomas JM, McKenzie DH, Eberhardt LL (1981) Some limitations of biological monitoring. Environ Inter 5:3–10

Tsai EC (1985) Statistical determination of the optimal sample size of secondary effluent BOD_5 and SS. Environ Monitor Assess 5:177–183

Turner MG (1987) Spatial simulation of landscape changes in Georgia: a comparison of 3 transition models. Landscape Ecol 1:29–36

Wessman CA, Aber JD, Peterson DL, Melillo JM (1988) Remote sensing of canopy chemistry and nitrogen cycling in temperate forest ecosystems. Nature 335:154–156

Woodwell GM (1965) Effects of ionizing radiation on ecological systems. In Woodwell GM (ed) *Ecological Effects of Nuclear War*. Publication 917. Brookhaven National Laboratory, Brookhaven, NY, pp 20–38

9. Biological Heterogeneity in Aquatic Ecosystems

John A. Downing

The study of heterogeneity in the biological components of aquatic eco-systems has been an important part of ecology for more than a century. Although many early ecologists (see review by Lussenhop, 1974) perceived especially the pelagic milieu to be uniform [hence the term plankton or "wanderers" for its inhabitants (Ruttner, 1953)], early quantitative limnol-ogists such as Birge (1897) found the aquatic habitat to be highly hetero-geneous in factors such as light, temperature, oxygen, and limiting nu-trients. The physical heterogeneity of the aquatic habitat has long been known to be reflected in the spatial patterns of aquatic populations.

It became clear early in the study of temperate aquatic habitats that they varied in composition temporally and spatially. The most important tem-poral variations were perceived to be seasonal developments of pelagic communities mediated by annual cycles of temperature, and the most im-portant spatial variation seemed to be related to depth. At the close of winter, lakes reheat and the renewed light availability and higher tempera-tures give rise to a rapid increase in plant and animal life. Reviews of clas-sic limnology suggest that the temporal development of the pelagic com-munity follows a predictable trajectory throughout summer and autumn until activity is decreased again during winter (Wetzel, 1983).

Seasonal variation in temperature is also a major source of spatial heter-ogeneity in lake ecosystems. Owing to the physical characteristics of water (Ruttner, 1953), temporal variation in climate yields important spatial vertical and horizontal gradients in physical, chemical, and biotic components of lakes. The seasonal succession of various communities (e.g., planktonic, benthic, fish) in lakes and the study of spatial and temporal gradients and their influence on lake ecosystems has made up a large fraction of the limnological studies performed over the last century. These studies are the backbone of limnological research and are reviewed in depth by several limnological texts (e.g., Ruttner, 1953; Cole, 1983; Goldman and Horne, 1983; Wetzel, 1983).

Superimposed on these spatial and temporal gradients is another kind of heterogeneity, called "stochastic variation" in temporal studies and "spatial aggregation," "contagion," or "patchiness" in spatial studies. Such heterogeneity, not arranged in perceptible gradients, has been studied much less completely than variation along aquatic clines. This review concentrates on the spatial heterogeneity (i.e., aggregation) of biotic components of lake ecosystems not related to gradients. First to be examined are some serious methodological problems associated with its quantification that have retarded the advancement of knowledge in this field. A comparative analysis of spatial heterogeneity in several biological components of lake ecosystems is then presented.

Measuring Spatial Heterogeneity in Aquatic Organisms

Many of the classic studies of nonclinal spatial heterogeneity were performed using large, slow-moving or sessile organisms such as trees, bushes, or grasses (e.g., Grieg-Smith, 1952; Pielou, 1977). It was thus easy to make interpretable and somewhat time-stable measurements of the spatial position of organisms within an ecosystem. In lakes, both the medium (e.g., water, sediments) and the organisms are highly dynamic; and wave action, turbulence, and currents render spatial patterns highly ephemeral. In aquatic ecosystems, the concept of "place" is illusory and dynamic. It is probably of greater importance to the organisms, however, how the members of the population are spaced relative to other members of the population, rather than to geographical space. Such relative spacing is important to tests of theoretical questions because aggregation or heterogeneity is believed to be important to interorganismal interactions such as reproduction (Waters, 1959; Dana, 1976; Jackson, 1977; Cowie and Krebs, 1979), competition (Ryland, 1972; Keen and Neill, 1980; Veresoglou and Fitter, 1984), and predation (Anscombe 1950; Gilinsky, 1984; Sih, 1984).

Another factor that renders the measurement of spatial heterogeneity difficult in aquatic ecosystems is the fact that organisms are usually small in lakes, with most plants and animals being less than a few centimeters long (fish and aquatic macrophytes are exceptions, but they are rarely numerically dominant taxa in lakes) . Because normal swimming speed scales as around $W^{0.14}$ (W = body mass) and length and mass are related as about

$W \propto L^3$ (Peters 1983), the number of body lengths moved at normal swimming speed scales as about $L^{-0.6}$. That is, smaller animals can move many more body lengths per unit time than can larger animals. This fact, combined with the difficulty of observing small, elusive animals in dark places such as deep waters or muddy sediments, has made direct measurement of the spacing of lake dwelling organisms impossible.

Limnologists, like many other ecologists, have therefore relied strongly on indices of spatial heterogeneity derived from measures of variations in numbers of organisms among replicate (i.e., same sized samples taken in the same habitat) samples of population density. When several replicate samples have been taken at random or regular points within the space occupied by the population, a frequency distribution of different numbers of organisms collected in replicate samples can be produced. Two types of information are derived from this approach. The central tendency of the frequency histogram indicates the density of the population. The shape of the frequency histogram indicates the degree of heterogeneity of the population over the sampling space, greater heterogeneity being implied by more skewed frequency distributions. Because of the difficulty of observing most aquatic organisms, tests of hypotheses about spatial heterogeneity have usually been based on synthetic indices of spatial heterogeneity derived through such replicate sampling, and interpretation of these indices has been founded on the shape of density frequency histograms that should be encountered in aquatic organisms exhibiting different spatial patterns.

Advances in our knowledge about factors influencing spatial aggregation and the ecological consequences of different levels of spatial heterogeneity have been hampered by two technical problems. First, the development of indices of spatial aggregation has paid more attention to their capacity to indicate lack of randomness than to measure differences in degree of aggregation among significantly aggregated populations. Organisms that are spatially heterogeneous or aggregated are those that show more variability in space than would be expected from either a random (Poisson: $s^2 = m$), i.e., variance equals the mean, or uniform ($s^2 < m$) distribution (Taylor et al., 1978; Mauchline, 1980; Lincoln et al., 1982; Taylor, 1984). Aggregation is inferred by demonstrating differences from the spatial variance expected based on the Poisson distribution that is known to result from randomly occurring events. Thus the relative size of their s^2 and m can be used to sort populations into three categories: uniform, random, or aggregated. A population with $s^2 > m$ is aggregated; and for more than one population of equal density, the population with the highest s^2 the most aggregated. The problem that has daunted ecologists is that few populations exist at exactly the same density. Several indices have been proposed to permit comparison of the spatial aggregation of populations of unequal densities (see Elliott, 1979, for review). There is reason to suspect that some of these indices contain mathematical artifacts that lead to systematic errors and misleading interpretations (Green, 1966; Elliott, 1979; Taylor 1984; Downing, 1986), but no systematic analysis of them has been presented.

The second technical difficulty has been that few indices have functioned generally enough to permit comparisons. The analysis of frequency histograms of spatially replicated population estimates (e.g., Polya, 1930; Neyman, 1939; Waters, 1959; Waters and Henson, 1959) is especially notorious because entirely different distribution functions are often found to fit population data collected on consecutive dates or with different treatments (Taylor, 1965, 1984). Similarly, methods based on measures of distance among individual organisms (e.g., Pielou, 1977) can be applied only under particular circumstances. Methods based on variation among spatially separated population counts can be applied much more frequently than those based on distance measures (Patil and Stiteler, 1974) because mobile organisms and those that cannot be observed for long periods can be examined only by measuring the variability among repeated population estimates made in slightly different locations.

Spatial aggregation is easy to define but has proved difficult to measure. A good index of spatial aggregation should be widely applicable, should provide measures that correspond to the definition of spatial aggregation, and should provide reliable values that can be compared among populations. This chapter concentrates on the universally applicable indices based on the variance (s^2) and mean (m) of spatially repeated population estimates and tests the simple hypothesis that mathematical constraints on measures and comparative analyses of spatial heterogeneity yield indices that are biased and confounded by variations in sampling design.

Indices of Spatial Aggregation Based on Replicate Counts

Although many indices of spatial aggregation have been proposed, only those that have been used most frequently in aquatic studies are discussed here (Table 9.1). Historically, the first indices used were simple ratios of either standard deviation or variance of replicate counts to the average of the replicates (s/m or s^2/m). Figures 9.1 and 9.2 show that s^2/m of replicate samples of marine and freshwater zooplankton (example data from Downing et al., 1987) increases with m, whereas s/m decreases with m. Several other indices attempt to compensate for systematic variation with m. The distribution coefficient of the negative binomial distribution, k (e.g., Elliott, 1979), describes the skewness of frequency histograms of population counts but also varies systematically with m (Fig. 9.3). In addition, k is meaningless for $s^2 < m$ and is difficult to calculate if $0 < k < 4$ (Anscombe, 1950). Lloyd's index of mean crowding ($\overset{*}{m}$) (Lloyd, 1967) is a variant of s^2/m (Table 9.1) and tends to increase with m (Fig. 9.4). Although not illustrated here, Lloyd's index of patchiness ($\overset{*}{m}/m$) is also correlated with

Table 9.1. Upper and Lower Limits to Indices of Heterogeneity and Their Dependence on Various Factors[a]

Index	Calculation	Limits to Indices	
		Low Heterogeneity	High Heterogeneity
s^2/m	s^2/m	$\dfrac{R(1-R/n)}{m(n-1)}$	nm
CV	s/m	$\dfrac{(R-R^2/n)^{0.5}}{m(n-1)^{0.5}}$	$n^{0.5}$
k^b	$m^2/(s^2-m)$	$\dfrac{m^2}{[R(1-R/n)(n-1)^{-1}]-m}$	$\dfrac{m}{(n-m^{-1})}$
$\overset{*}{m}$	$m+(s^2/m)-1$	$m+\dfrac{R(1-R/n)}{m(n-1)}-1$	$m+nm-1$
$\overset{*}{m}/m$	$1+s^2-m^{-1}$	$1+\dfrac{R(1-R/n)}{m^2(n-1)}-\dfrac{1}{m}$	$1+n-m^{-1}$
I_d	$\dfrac{n(\sum X^2-\sum X)^c}{(\sum X)^2-\sum X}$	$\dfrac{R(1-R/n)}{nm^2-m}-\dfrac{m-1}{m-n^{-1}}$	n

[a] Number of samples taken (n), mean number of organisms collected per sample (m), total number of organisms collected in the n samples ($\sum X$), and remainder of the division of $\sum X$ by n (R). CV = the coefficient of variation (Elliott 1979); k = the distribution coefficient of the negative binomial distribution (Pieters et al., 1977); $\overset{*}{m}$ = Lloyd's (1967) index of mean crowding and index of patchiness; and I_d = Morisita's (1954, 1962) index. Minimum and maximum variances are calculated after equations 2, 4, and 5.
[b] Larger k value denotes less heterogeneity.
[c] Undefined for $\sum X = 1$.

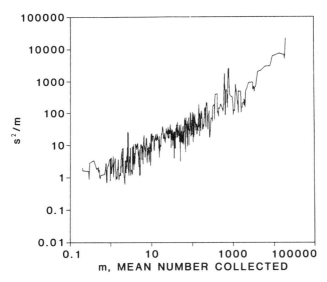

Figure 9.1 Variation in the variance mean ratio (s^2/m) with mean density (no. sample^{-1}) in 1200 sets of replicate zooplankton samples. Data are from marine and freshwater systems (Downing et al., 1987). Means and variances were calculated on a per-sampler basis and are not normalized to a common volume. The trend line is a simple moving average with a window size of 10 data points. Moving averages are shown instead of data points to facilitate examination of the trend in s^2/m with m.

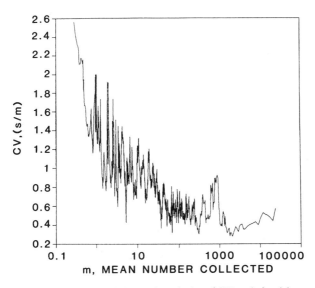

Figure 9.2 Variation in the coefficient of variation ($CV = s/m$) with mean density (no. sample^{-1}). Data and analyses are as in Figure 9.1.

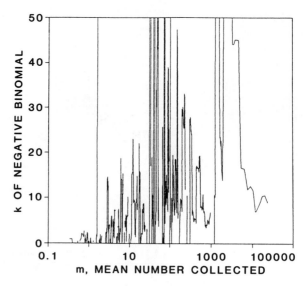

Figure 9.3 Variation in k of the negative binomial distribution with mean density (no. sample^{-1}). k was calculated by the product-moment method because it is the best method for small sample size (Pieters et al., 1977). Small values of k imply highest heterogeneity. Negative values of k were included in moving averages but were not plotted. Data and analyses are as in Figure 9.1.

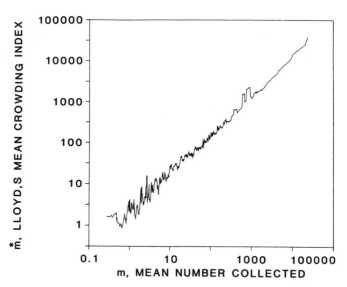

Figure 9.4 Variation in Lloyd's (1967) mean crowding index ($\overset{*}{m}$) with mean density (no. sample^{-1}). Data and analyses are as in Figure 9.1. Systematic variation with m is also found (but not shown) using Lloyd's index of patchiness ($\overset{*}{m}/m$).

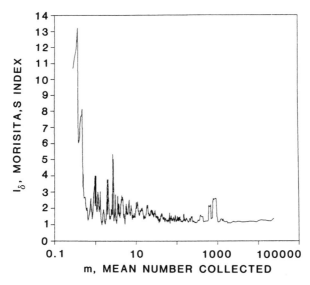

Figure 9.5 Variation in Morisita's (1954, 1962) index (I_d) with mean density (no. sample^{-1}). This index is least correlated with m because the average relation between s^2 and m in these data was $s^2 = am^{1.85}$ (Downing et al., 1987) and I_d tends to a constant where n and m are large and $s^2 = am^2$ (see Table 9.1). Data and analyses are as in Figure 9.1. The high values of I_d at low m are probably due to the high n necessary to detect low m (see Table 9.1).

m. Morisita's index (I_d) (1954, 1962) reduces to $I_d = s^2/m^2$ at large values of n and m (Table 9.1). Morisita's index tends to decrease with m in actual data (Fig. 9.5).

Some of these indices lead to directly conflicting interpretations. Such indices of spatial aggregation have often been used to make comparisons of spatial distributions of populations that occur at different densities. Using s/m, k of the negative binomial distribution, or I_d would lead to the conclusion that sparser aquatic populations are the most aggregated, whereas s^2/m, $\overset{*}{m}$, and $\overset{*}{m}/m$ lead to the contrary conclusion.

Mathematical Bounds to Variances

All of these indices of spatial heterogeneity are related to s^2 and m of population estimates made at different locations within a population. The parameter s^2 is estimated as follows.

$$s^2 = \frac{\sum X^2 - (\sum X)^2/n}{(n-1)} \tag{1}$$

where $\sum X^2$ = the summation over all samples of the squares of observations in each replicate sample; $\sum X$ = the summation of the observations found in all replicate samples; and n = the number of replicate samples. Downing (1989) has shown that variances of population counts have distinct, logical lower and upper limits. The minimum s^2 (s^2_{min}) for a set of population counts, where the total number of organisms counted ($\sum X$) in all samples is less than the number of samples taken (n) and so the mean number per sample (m) is less than unity, can be calculated as

$$s^2_{min} = \frac{\sum X - (\sum X)^2/n}{n-1} \qquad (2)$$

Because $\sum X$ is an integer, number theory proves that it can be expressed as

$$\sum X = nq + R \qquad (3)$$

where q = the integer part of m; and R = the remainder of the division of $\sum X$ by n. Equation (2) can thus be generalized to all values of m as

$$s^2_{min} = \frac{R - R^2/n}{n-1} \qquad (4)$$

because $m = q + R/n$, the variance associated with q is 0, the s^2_{min} associated with the remainder (R) is obtained by substituting R for $\sum X$ in equation (2), and the variance of the sum of the two values q and R/n is the sum of the variances of the two values where there is no covariance (Colquhoun, 1971). Equation 4 thus shows that s^2_{min} is a finite positive value except where m is an integer, yielding $R = s^2_{min} = 0$.

Downing (1989) also showed that the maximum possible variance (s^2_{max}) of a set of population counts is

$$s^2_{max} = nm^2 \qquad (5)$$

for all values of m. These bounds are illustrated for $n = 10$ in Figure 9.6.

Effect of Variance Bounds on Aggregation Indices

Because s^2 is bounded, indices of heterogeneity based on s^2 are also subject to logical limits. Columns 3 and 4 of Table 9.1 show that the upper and lower limits of various indices of aggregation must vary with the average population density (m), the number of replicate samples taken (n), and the remainder of the division of $\sum X$ by n. The upper limit of s^2/m increases linearly with m, whereas the lower limit decreases from 1 to 0 between

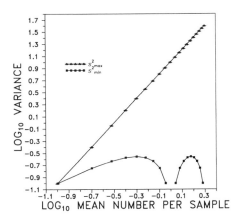

Figure 9.6 Trends in the minimum and maximum variances for data consisting of discrete counts. Predictions are from equations 2, 4, and 5 and are shown for $n = 10$ samples.

$\sum X = 1$ and $\sum X = n$. Empirical observations of s^2/m must increase with m (Fig. 9.1) because high value of s^2/m cannot occur at low m. The upper limit of the coefficient of variation (CV) is constant at $n^{0.5}$, whereas the lower limit decreases from $n^{0.5}$ to 0 as a function of m between $\sum X = 1$ and $\sum X = n$. Empirical measurements of CV must decrease with m (Fig. 9.2) because low values of CV cannot occur at low m. The limit to high values of k (lowest heterogeneity) is a rapidly increasing function of m between $\sum X = 1$ and $\sum X = n$, which allows empirically derived k values to appear positively correlated with m (Fig. 9.3). Above $\sum X = n$, high positive limits to k fall to negative, decreasing values. Values of k obtainable for maximum spatial aggregation generally increase with increased m. The upper limit to $\overset{*}{m}$ is a linear increasing function of m, whereas the lower limit tends to $m - 1$ as R approaches n. Empirical measurements of $\overset{*}{m}$ must therefore increase with increased m (see Fig. 9.4). I_d is bounded above by n, bounded below by 0 between $\sum X = 1$ and $\sum X = n$, and an increasing function of m where m is an integer of more than 1. These mathematical limits combined with the structure of the indices result in the trends seen in Figures 9.1 to 9.5. The maximum possible values of s^2/m, s/m, $\overset{*}{m}$, and I_d increase with increased n, whereas the range of k must decrease. The trend in I_d seen in Figure 9.5 is probably due to the fact that low average numbers per sample can be detected only using many samples, which can yield higher values of I_d (Table 9.1) and for the example data, $s^2 = am^{1.5}$ (Downing et al., 1987); therefore $I_d \propto m^{1.5}/m^2$ for large values of n and m. Not only do these indices result in conflicting interpretations of spatial aggregation of aquatic organisms, but because of their systematic variation with m they cannot be used for comparative purposes except when m is equal among treatments. In most cases, indices based on more replicate samples can indicate greater apparent aggregation (Table 9.1, column 4). These problems severely limit the use of any s^2-based index of spatial aggregation.

Suspicions about such deficiencies have led to a search for a better index of spatial variation. The literature has suggested that a perfect index should have five attributes (Taylor, 1984).

1. It should provide real and continuous values for the complete range of spatial distributions (Green, 1966).
2. It should be uninfluenced by the number of sample units, the size of the unit sample, and the number of individuals (Green, 1966).
3. Values appropriate to some theoretical expectation should be in central position (Lefkovitch, 1966).
4. Tests of the significance of differences between two or more such values should be available (Lefkovitch, 1966).
5. The descriptive function should be clearly separated from the theoretical justification (Taylor, 1984).

None of the indices in Table 9.1 pass these criteria because their upper and lower limits are logically correlated with m, n, or both.

Power Relation Between Variance and Mean

It has long been observed (e.g., Bliss, 1941) for many types of data (e.g., Taylor 1961; Finney, 1976; Downing and Peters, 1980; Real et al., 1982; Downing and Anderson, 1985) that s^2 tends to increase as a power function of m.

$$s^2 = am^b \tag{6}$$

where a and $b =$ fitted coefficients (Taylor, 1971). Taylor (1961, 1965) suggested on intuitive grounds that "b appears to be a true population statistic, an 'index of aggregation'" that is independent of m and n (Taylor, 1961), whereas a is largely or wholly a sampling or computing factor. The coefficient b is the relative rate at which s^2 increases with increased m. If m changes by a factor of 10, s^2 changes by a factor of 10^b. The coefficient b has been advanced as the index of spatial aggregation that most nearly satisfies all attributes of a perfect index (Taylor, 1984). It has further been suggested that b is a characteristic of species (Taylor et al., 1978, 1980) and is determined by evolution (Taylor and Taylor, 1977; Taylor et al., 1983) or demography (Anderson et al., 1982).

The fit of published relations between s^2 and m is precise. Coefficients of determination for such relations found by Taylor et al., (1983) were more than 0.8 in 80% of 444 species and more than 0.9 in more than 50% of the species. Values of b also show a high degree of central tendency. Exponents for specific taxa generally vary between 0 and 4, with average values

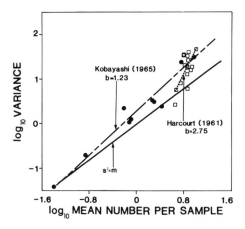

Figure 9.7 Relation between variance and mean for the spatial distribution of the eggs of *Pieris rapae* found by Harcourt (1961) and Kobayashi (1965). The broken line is the least squares fitted trend, and a *t* test shows that the *b* values are significantly different for the two data sets ($p < 0.01$) (Downing, 1986). The solid line represents the variance that would have been found if the eggs were distributed randomly in space.

somewhere around 2 for large assemblages of coefficients (Taylor et al., 1983). The strength of relations between s^2 and m and the small range of b values found for highly dissimilar species presents an imposing body of observation that has prompted some (e.g., Taylor, 1965; Southwood, 1966; Elliott, 1979) to call the empirical s^2 relation a natural law of spatial heterogeneity.

The use of b in equation (6) as an index of spatial aggregation is questionable. The relation between b and the accepted definition of spatial aggregation ($s^2 > m$) (Taylor et al., 1978; Mauchline, 1980; Taylor, 1980; Lincoln et al., 1982) is not clear. An illustration of the ambiguity of b is provided by the data of Harcourt (1961) and Kobayashi (1965) on the spatial distribution of eggs of the cabbage butterfly *Pieris rapae*. Downing (1986) showed that the b values found were 2.75 and 1.23, respectively, and a *t* test showed that these b values were significantly different ($p < 0.01$). Traditional interpretation of b (e.g., Taylor et al., 1983) suggested that Harcourt's eggs were more aggregated than those collected by Kobayashi because b was significantly higher. Inspection of Figure 9.7, however, shows that more than one-half of the variances observed by Harcourt fell between the $s^2 : m$ relation of Kobayashi and the $s^2 = m$ line. In contrast, according to the definition of spatial aggregation, s^2 closer to m indicates that spatial distributions are more nearly random. Throughout much of the range of m observed by Harcourt, the insects with the lower value of b actually had higher variances, indicating greater departures from randomness. Departure from randomness is our objective criterion for determining spatial aggregation, and Figure 9.7 shows that b alone is not a reliable measure of this.

Making Valid Comparisons of Spatial Aggregation of Aquatic Organisms

Valid comparative analyses of spatial aggregation can and must be made. Two important questions about spatial distributions must be resolved routinely by aquatic ecologists. First, like early ecologists (e.g., Hensen, 1884) we must determine whether organisms are uniformly, randomly, or heterogeneously distributed in nature. This need is fulfilled by comparisons of spatial distributions with expectations based on the Poisson distribution (Elliott, 1979), although the discovery of agreement with the Poisson distribution does not necessarily mean that chance is the only mechanism involved in the spatial arrangement of the organisms (Kac, 1983). Second, many important ecological questions demand comparisons of spatial heterogeneity among populations or treatment groups that may differ in average spatial density. Such comparisons cannot be made using current indices of spatial aggregation because they do not control properly for the effect of m and n on s^2.

Our problem is therefore to determine which of two or more populations is the most aggregated in space. In most aquatic populations, where actual organismal pattern and spacing cannot be measured, this question reduces to determining which of the populations yields the highest s^2 at given levels of m. The mean density cannot be ignored as a covariable because the s^2 indicating randomness varies directly with m. The simple solution to this problem is to compare entire $s^2:m$ relations established for different populations. This solution was suggested several years ago (e.g., Downing, 1980), but the allure of panaceistic indices has been great. The comparison of entire $s^2:m$ relations, like those in equation (6), can be performed easily using any of the standard methods for the comparison of two or more bivariate relations (e.g., Gujarati, 1978). Naturally, such analyses should be performed only on s^2 determined using the same number of replicate samples and employing ranges of means of similar magnitude among treatment groups or populations when possible. Examples of this type of analysis have been presented by Rasmussen and Downing (1988) and Pinel-Alloul et al. (1988).

Normative Data on Biological Heterogeneity

The relative degree of spatial aggregation cannot, therefore, be measured reliably by any existing index but can be measured by comparison of the degree of variation among samples observed at comparable values of m.

Such comparisons satisfy or avoid criteria 1, 2, 4, and 5 cited above. The salient question to answer then is whether aquatic populations exhibit a degree of spatial variation that is greater or less than other comparable populations. Implicit in such analyses is that we know what characteristics render populations comparable, i.e., what factors have a broad, general influence on the spatial heterogeneity of natural populations. Due to the biasing influence of indices of heterogeneity discussed above, we have little general knowledge about the subject. A good point of departure for attacking this question would be to know what $s^2:m$ relations have been observed for aquatic taxa; i.e., what are normal values of a and b?

Several authors have now assembled large numbers of observations on m and s^2 of aquatic populations that can serve as a basis of comparison. Table 9.2 summarizes the relation found between m and s^2 of replicate samples of aquatic organisms and characteristics. These data are derived from more than 18,000 sets of observations and include both marine and freshwater organisms to provide a broad basis for comparison. Fitted values of b (equation 6) for data based on counts of organisms ranged only from 1.11 to 2.52. The highest b values were obtained for a small set of data on oligochaetes from the lake benthos (Standen and Latter, 1977) and for freshwater phytoplankton (Gilyarov et al., 1979). Values for both a and b obtained for heterogeneity in noncount variables, e.g., biomass of macrophytes (Downing and Anderson, 1985) or concentrations of sediment components (Downing and Rath, 1988), diverged greatly from those found for heterogeneity of population counts (Table 9.2).

Figure 9.7 illustrates that b alone cannot be used to make comparisons of the spatial heterogeneity of populations, but information on both the elevation (a) and the rate of change in variance functions (b) must be employed. A group of organisms showing very low spatial heterogeneity would have both low a and low b, whereas a group showing very high spatial heterogeneity would have both high a and high b. With the exception of perhaps Morin's (1985) stream benthos data, no broad groups of organisms seem to be either strikingly or consistently more or less heterogeneous than any others (Fig. 9.8). Instead, estimated coefficients a and b tend to negatively covary ($r = -0.66$, $n = 27$, $p < 0.0001$), suggesting that variations in a and b are simply errors in the estimation of some single common relation between spatial s^2 and average number collected.

The similarity of $m:s^2$ relations for aquatic organisms is reiterated in Figure 9.9. When the fitted relations between log s^2 and log m are plotted throughout the range of m for which they were determined, there is little divergence in overall trend, even when such disparate organisms as bacteria, phytoplankton, and benthos are compared (Table 9.2). As suggested by several authors (reviewed by Downing, 1979) most estimates of s^2 and m suggest nonrandom spatial distributions (Fig. 9.9). Departures from $m = s^2$ are most acute when large numbers of organisms per sample are encountered in samples. Intersample s^2 therefore suggests that aquatic popula-

Table 9.2. Coefficients of the Relation Between m and s^2 in Published Studies (Equation 6)

Organism	Type[a]	log a	b	n'	Ref.[b]
Lake benthos					
All benthos	LB	0.38	1.36	—	1
All benthos	LB	0.03	1.40	1673	2
Phytofauna	LB	0.18	1.63	497	3
Amphipoda	LB	—	1.45	159	4
Chironomidae	LB	-0.17	1.39	—	1
Chironomidae	LB	0.96	0.84	22	5
Chironomidae	LB	0.75	1.34	1166	6
Bivalvia-Unionidae	LB	0.18	1.15	66	7
Bivalvia-Pisidiidae	LB	0.24	1.32	30	8
Bivalvia-Pisidiidae	LB	-0.35	1.71	22	9
Oligochaeta	LB	0.27	1.46	—	1
Oligochaeta	LB	0.23	1.11	25	10
Oligochaeta	LB	-0.55	1.37	25	10
Oligochaeta	LB	0.75	1.21	8	11
Oligochaeta	LB	-1.70	2.52	9	11
Stream benthos					
All benthos	SB	-0.28	1.60	1461	12
Simulidae (on tiles)	SB	0.58	1.84	50	13
Marine benthos					
All benthos	MB	0.22	1.22	3015	14
Colonial Ascidiacea	MB	0.09	1.50	52	15
Freshwater zooplankton					
All zooplankton	FZ	-0.07	1.64	714	16
All zooplankton	FZ	-0.50	1.69	10	17
All zooplankton	FZ	0.11	1.49	583	18
Cyclopoida	FZ	0.01	1.69	8	19
Marine zooplankton					
All zooplankton	MZ	0.47	1.67	475	16
All zooplankton	MZ	-0.22	1.24	16	20

Phytoplankton					
Freshwater	FP	0.02	1.58	20	21
Freshwater	FP	0.04	2.02	8	17
Bacteria					
Marine	B	−0.34	1.76	75	22
Non-count equations					
Sediment organic matter	FS	4.30	6.23	11	23
Sediment organic matter	FS	−4.75	−2.02	12	23
Sediment water content	FS	−5.52	−18.91	11	23
Sediment chlorophyll	FS	−0.16	0.90	20	23
Sediment phosphorus	FS	−2.07	1.94	30	23
Macrophytes (biomass)	FM	−0.19	1.61	1198	24
Macrophytes (biomass)	FM	−0.53	2.00	74	25

All m values are expressed on a per-sample basis, not normalized to common units of volume or surface area. n' = the number of m:s^2 pairs used in the regression estimation of a and b.

[a]Type of organisms: LB = freshwater benthos; SB = stream benthos; MB = marine benthos; FZ = freshwater zooplankton; MZ = marine zooplankton; FP = freshwater phytoplankton; B = bacteria; FS = freshwater sediment; FM = freshwater macrophytes.

[b]Literature sources; 1, Bakanov (1980); 2, Downing (1979); 3, Downing and Cyr (1985); 4, France (1987); 5, Rasmussen and Downing (1988); 6, Drake (1983); 7, Downing and Downing (in preparation); 8, Hornbach et al. (1982); 9, Way and Wissing (1982); 10, McElhone (1978); 11, Standen and Latter (1977); 12, Morin (1985; personal communication); 13, Morin (1987); 14, Vézina (1988); 15, Stocker and Bergquist (1987); 16, Downing et al. (1987); 17, Gilyarov et al. (1979); 18, Pinel-Alloul et al. (1988); 19, Pont (1986); 20, Frontier (1972); 21, Jones and Francis (1982); 22, Trousselier et al. (1986); 23, Downing and Rath (1988); 24, Downing and Anderson (1985); 25, France (1988).

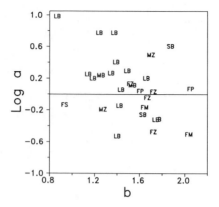

Figure 9.8 Relation between log a and b in the variance relation $s^2 = am^b$ (equation 6) for groups of aquatic organisms. Data and abbreviations are as in Table 9.2.

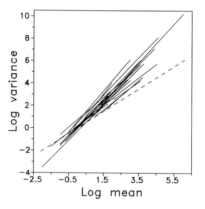

Figure 9.9 Regression relations between log s^2, and log m plotted throughout the ranges of observed m for each analysis. Data are from Table 9.2. The dashed line indicates $s^2 = m$.

tions are spatially aggregated, especially when they are dense. Overall, however, there is great similarity in the degree of spatial heterogeneity encountered in various categories of organisms.

The data in Table 9.2 suggest that there exists a universal normative relation between m and s^2 of samples of aquatic organisms. The average value of log a is not significantly different from 0 ($p > 0.05$), and the average value of b is 1.507 ($s = 0.32$, $n = 28$) . The normative relation between s^2 and m for aquatic organisms is therefore close to $s^2 = m^{1.5}$. Drawing inspiration from Fulton's condition index (Fulton, 1902; Doyon et al., 1988), a good empirically derived index of spatial aggregation for aquatic populations is

$$B = \log (s^2/m^{1.5}) \tag{7}$$

This index yields values where $B = 0$ indicates that the population is aggregated to an average degree for aquatic populations, $B = 1$ indicates that the population has s^2 10 times the average for aquatic populations, and $B = -1$ indicates that the population has s^2 one-tenth the average for aquatic populations. This index satisfies all five criteria for a perfect index presented by Green (1966), Lefkovitch (1966), and Taylor (1984), except that attention should be paid to the possible influence of n on s^2 (equations 2, 4, and 5) and unlikely values of $s^2 = 0$ render B undefined. Because $s^2 : m$ relations (i.e., equation 6) usually have residuals normal in their logarithms, values of B should be normally distributed and thus amenable to testing using parametric analyses of variance.

Progress in the understanding of biological heterogeneity in aquatic ecosystems has been hindered by the use of biased indices of spatial aggregation. The similarity of empirically observed relations between spatial s^2 and m found in diverse groups of aquatic organisms renders comparative analysis possible. The outlook for future progress in the understanding of spatial processes in aquatic ecosystems therefore seems bright.

Acknowledgments. I am grateful to A. Morin and A. Vézina for sharing raw data. Sophie Lalonde helped in the analysis of data and the preparation of the manuscript. Thanks go also to J. Kolasa for his patience.

References

Anderson RM, Gordon DM, Crawley MJ, Hassell MP (1982) Variability in the abundance of animal and plant species. Nature 296:245–248

Anscombe FJ (1950) Sampling theory of the negative binomial and logarithmic series distributions. Biometrika 37:358–382

Bakanov AI (1980) Dependence of statistical characteristics of benthos samples on area sampled. Akad Nauk SSSR Inst Biol Vnutr Vod 270:1–28

Birge EA (1897) Plankton studies on Lake Mendota. II. The Crustacea from the plankton from July, 1894, to December, 1896. Trans Wis Acad Sci Arts Lett 11:274–448

Bliss CI (1941) Statistical problems in estimating populations of Japanese beetles larvae. J Econ Entomol 34:221–231

Cole GA (1983) *Textbook of Limnology.* 3rd ed. Waveland Press, Prospect Heights, Illinois

Colquhoun D (1971) *Lectures on Biostatistics—An Introduction to Statistics with Applications in Biology and Medicine.* Clarendon Press, Oxford

Cowie RJ, Krebs JR (1979) Optimal foraging in patchy environments. In Anderson RM, Turner BD, Taylor LR (eds) *Population Dynamics.* Blackwell, Oxford, pp 183–205

Dana TF (1976) Reef-coral dispersion patterns and environmental variables on a Caribbean coral reef. Bull Mar Sci 26:1–13

Downing JA (1979) Aggregation, transformation and the design of benthos sampling programs. J Fish Res Board Can 36:1454–1463

Downing JA (1980) Precision vs. generality: a reply. Can J Fish Aguat Sci 37:1329–1330

Downing JA (1986) Spatial heterogeneity: evolved behaviour or mathematical artefact? Nature 323:255–257

Downing JA (1989) Precision of the mean and the design of benthos sampling programmes: caution revised. Mar Biol 103:231–234

Downing JA, Anderson MR (1985) Estimating the standing biomass of aquatic macrophytes. Can J Fish Aquat Sci 42:1860–1869

Downing JA, Cyr H (1985) The quantitative estimation of epiphytic invertebrate populations. Can J Fish Aquat Sci 42:1570–1579

Downing JA, Peters RH (1980) The effect of body size and food concentration on the in situ filtering rate of Sida crystallina. Limnol Oceanogr 25:883–895

Downing JA, Rath LC (1988) Spatial patchiness in the lacustrine sedimentary environment. Limnol Oceanogr 33:447–458

Downing JA, Pérusse M, Frenette Y (1987) Effect of interreplicate variance on zooplankton sampling design and data analysis. Limnol Oceanogr 32:673–680

Doyon JF, Downing JA, Magnin E (1988) Variation in the condition of northern pike, Esox lucius. Can J Fish Aquat Sci 45:479–483

Drake CM (1983) Spatial distribution of chironomid larvae (Diptera) on leaves of the bulrush in a chalk stream. J Anim Ecol 52:421–437

Elliott JM (1979) Some Methods for the Statistical Analysis of Samples of benthic invertebrates. 2nd ed. Scientific publication 25. Freshwater Biological Association, Ambleside

Finney DJ (1976) Radioligand assay. Biometrics 32:721–740

France RL (1987) Aggregation in littoral amphipod populations: transformation controversies revisited. Can J Fish Aquat Sci 44:1510–1515

France RL (1988) Biomass variance function for aquatic macrophytes in Ontario (Canada) shield lakes. Aquat Bot 32:217–224

Frontier S (1972) Calcul de l'erreur sur un comptage de zooplancton. J Exp Mar Biol Ecol 8:121–132

Fulton TW (1902) Rate of growth of sea fishes. Annu Rep Fish Board Scotland 20:326–446

Gilinsky E (1984) The role of fish predation and spatial heterogeneity in determining benthic community structure. Ecology 65:455–468

Gilyarov AM, Chekryzheva TA, Sadchikov AP (1979) The structure of horizontal plankton distribution in the epilimnion of a mesotrophic lake. Hydrobiol J 15:7–14

Goldman CR, Horne AJ (1983) Limnology. McGraw-Hill, New York

Green RH (1966) Measurement of non-randomness in spatial distributions. Res Popul Ecol 8:1–7

Grieg-Smith P (1952) The use of random and contiguous quadrats in the study of the structure of plant communities. Ann Bot 16:293–316

Gujarati D (1978) Basic Econometrics. McGraw-Hill, New York

Harcourt DG (1961) Spatial pattern of the imported cabbageworm Pieris rapae (L.) (Lepidoptera: Peridae) on cultivated Cruciferae. Can Ent 93:945–952

Hensen V (1884) Ueber die Bestimmung der Planktons oder des im Meer treibenden Materials an Pflanzen und Tieren. Bericht der Commission zur Wissenschaftlichen Untersuchungen der Deutschen Meere 5(2)

Hornbach DJ, Wissing TE, Burky AJ (1982) Life-history of a stream population of the freshwater clam Sphaerium striatinum Lamarck (Bivalvia: Pisidiidae). Can J Zool 60:249–260

Jackson JBC (1977) Competition on marine hard substrata: the adaptive significance of solitary and colonial strategies. Am Nat 111:743–767

Jones RI, Francis RC (1982) Dispersion patterns of phytoplankton in lakes. Hydrobiologia 86:21–28

Kac M (1983) What is random? Am Sci 71:405–406

Keen SL, Neill WE (1980) Spatial relationships and some structuring processes in benthic intertidal communities. J Exp Mar Biol Ecol 45:139–155

Kobayashi S (1965) Influence of parental density on the distribution pattern of eggs in the common cabbage butterfly, *Pieris rapae crucivora*. Res Popul Ecol 7:109–117

Lefkovitch LP (1966) An index of spatial distribution. Res Popul Ecol 8:89–92

Lincoln RJ, Boxshall GA, Clark PF (1982) *A Dictionary of Ecology, Evolution, and Systematics*. Cambridge University Press, Cambridge

Lloyd M (1967) Mean crowding. J Anim Ecol 36:1–30

Lussenhop J (1974) Victor Hensen and the development of sampling methods in ecology. J Hist Biol 7:319–337

Mauchline J (1980) The biology of Mysids and Euphausiids. Adv Mar Biol 18:1–637

McElhone MJ (1978) A population study of littoral dwelling Naididae (Oligochaeta) in a shallow mesotrophic lake in north Wales. J Anim Ecol 47:615–626

Morin A (1985) Variability of density estimates and the optimization of sampling programs for stream benthos. Can J Fish Aquat Sci 42:1530–1534

Morin A (1987) Unsuitability of introduced tiles for sampling blackfly larvae (Diptera: Simuliidae). Freshwat Biol 17:143–150

Morisita M (1954) Estimation of population density by spacing method. Mem Fac Sci Kyushu Univ Ser E 1:187–197

Morisita M (1962) I_d-index, a measure of dispersion of individuals. Res Popul Ecol 4:1–7

Neyman J (1939) On a new class of "contagious" distributions, applicable in entomology and bacteriology. Ann Math Stat 10:35–57

Patil GP, Stiteler WM (1974) Concepts of aggregation and their quantification: a critical review with some new results and applications. Res Popul Ecol 15:238–254

Perry JN, Taylor LR (1985) Adès: new ecological families of species-specific frequency distributions that describe repeated spatial samples with an intrinsic power-law variance-mean property. J Anim Ecol 54:931–953

Peters RH (1983) The ecological implications of body size. Cambridge University Press, Cambridge

Pielou EC (1977) *Mathematical Ecology*. Wiley, New York

Pieters EP, Gates CE, Matis JH, Sterling WL (1977) Small sample comparisons of different estimators of negative binomial parameters. Biometrics 33:718–723

Pinel-Alloul B, Downing JA, Pérusse M, Codin-Blumer G (1988) Spatial heterogeneity in freshwater zooplankton: variation with body size, depth, and scale. Ecology 69:1393–1400

Polya G (1930) Sur quelques points de la théorie des probabilités. Ann Inst Henri Poincaré 1:117–161

Pont D (1986) Structure spatiale d'une population du Cyclopide *Acanthocyclops robustus* dans une rizière de Camargue (France). Acta Oecol Gen 7:289–302

Rasmussen JB, Downing JA (1988) The spatial response of chironomid larvae to the predatory leech, *Nephelopsis obscura*. Am Nat 131:14–21

Real L, Ott J, Silverfine E (1982) On the tradeoff between the mean and the variance in foraging: effect of spatial distribution and color preference. Ecology 63:1617–1623

Ruttner F (1953) *Fundamentals of Limnology*. University of Toronto Press, Toronto

Ryland JS (1972) The analysis of pattern in communities of bryozoa. I. Discrete sampling methods. J Exp Mar Biol Ecol 8:277–297

Sih A (1984) The behavioral response race between predator and prey. Am Nat

123:143–150

Southwood TRE (1966) *Ecological Methods with Particular Reference to the Study of Insect Populations.* Chapman & Hall, London

Standen V, Latter PM (1977) Distribution of a population of *Cognettia sphagnetorum* (Enchytraeidae) in relation to microhabitats in blanket bog. J Anim Ecol 46:213–229

Stocker LJ, Bergquist PR (1987) Importance of algal turf, grazers, and spatial variability in the recruitment of a subtidal colonial invertebrate. Mar Ecol Prog Ser 39:285–291

Taylor LR (1961) Aggregation, variance and the mean. Nature 189:732–735

Taylor LR (1965) A natural law for the spatial disposition of insects. In *Proceedings of the 12th International Congress on Entomology*, London, 1964, pp 396–397

Taylor LR (1971) Aggregation as a species characteristic. In Patil GP, Pielou EC, Waters WE (eds) *Statistical Ecology*, Vol 1. Pennsylvania University Press, University Park, pp 357–372

Taylor LR (1980) New light on the variance/mean view of aggregation and transformation: comment. Can J Fish Aquat Sci 37:1330–1332

Taylor LR (1984) Assessing and interpreting the spatial distributions of insect populations. Annu Rev Entomol 29:321–357

Taylor LR, Taylor RAJ (1977) Aggregation, migration and population mechanics. Nature 265:415–420

Taylor LR, Woiwod IP, Perry JN (1978) The density dependence of spatial behaviour and the rarity of randomness. J Anim Ecol 47:383–406

Taylor LR, Woiwod IP, Perry JN (1980) Variance and the large scale spatial stability of aphids, moths and birds. J Anim Ecol 49:831–854

Taylor LR, Taylor RAJ, Woiwod IP, Perry JN (1983) Behavioural dynamics. Nature 303:801–804

Trousselier M, Baleux B, André P (1986) Echantillonnage de variables bactériologiques dans les milieux aquatiques. In GERBAM (ed) *Deuxième Colloque International de Bactériologie Marine*. Actes du colloque 3. CNRS/IFREMER, Brest, pp 23–33

Veresoglou DS, Fitter AH (1984) Spatial and temporal patterns of growth and nutrient uptake of five co-existing grasses. J Ecol 72:259–272

Vézina A (1988) Sampling variance and the design of quantitative surveys of the marine benthos. Mar Biol 97:151–155

Waters WF (1959) A quantitative measure of aggregation in insects. J Econ Entomol 52:1180–1184

Waters WE, Henson WR (1959) Some sampling attributes of the negative binomial distribution with special reference to forest insects. For Sci 5:397–412

Way CM, Wissing TE (1982) Environmental heterogeneity and life history variability in the freshwater clams, *Pisidium variabile* (Prime) and *Pisidium compressum* (Prime) (Bivalvia: Pisidiidae). Can J Zool 60:2841–2851

Wetzel RG (1983) *Limnology*. 2nd ed. Saunders College Publications, Philadelphia

10 Working with Heterogeneity: An Operator's Guide to Environmental Gradients

Paul A. Keddy

Environmental gradients are a powerful and largely overlooked research tool. They exist only because of heterogeneity. The objective of this chapter is to provide an overview of techniques for using environmental gradients to explore heterogeneous systems. The chapter begins by noting that any tool is helpful only if it is used with a well-defined goal. The goal of "assembly rules" and "response rules" for communities is suggested and briefly described. Because not all habitats have obvious gradients, techniques for finding gradients where they are not obvious are then introduced. Eight guidelines for using gradients as research tools are explored. These guidelines deal with the following topics: (1) choosing gradients to maximize generality; (2) selecting independent variables; (3) selecting dependent variables and the value of screening to create "trait matrices"; (4) importance of inferential statistics for detecting patterns; (5) incorporating experiments into natural gradients; (6) merits of locating two orthogonal gradients; (7) making gradients if they cannot be found; and (8) considerations of scale. The chapter concludes by discussing centrifugal organization, a model of community organization at the landscape scale that is built around species' responses to gradients.

Gradients and Heterogeneity

If you begin with a world view based on equilibrium models and homogeneous environments, heterogeneity can appear to be an obstacle to
scientific progress. However, situations that at first appear to be obstacles
can in fact be opportunities. Heterogeneity can be transformed from an apparent obstacle into a powerful research tool. Gradients organize environmental and biotic heterogeneity for generating and testing hypotheses.

The existence of predictable relations between organisms or communities and their environments is a fundamental concept of ecology, yet there
are few general quantitative rules that specify which kinds of organisms or
communities occur in which environments. Gradients organize nature in
such a way that such empirical relations among environmental conditions,
distributions, and abundances of species, and traits of these species can be
easily explored. Such patterns are often obvious, and we need only choose
the appropriate state variables for describing them (Lewontin, 1974; Rigler, 1982; Keddy, 1987). Gradients not only rapidly suggest hypotheses by
making patterns obvious, they also provide a tool for conducting experiments because they specify predictable and quantifiable changes in environmental states. The objective of this chapter is to provide some preliminary guiding principles or rules for making efficient use of gradients to
maximize the rate that we as scientists are able to pose and answer important questions.

Most of the examples come from vascular plants, particularly those in
wetlands. This choice is partly because it is these systems with which I am
most familiar, and partly because there is a good literature on gradients in
these systems. More importantly, however, wetlands were deliberately
chosen as a model system precisely because of the obvious heterogeneity
and strong environmental gradients in them. The following ideas therefore
grew out of attempts by myself, Scott Wilson, Bill Shipley, Irene Wisheu,
Dwayne Moore, and Connie Gaudet to make most efficient use of wetland
gradients to develop and test ideas.

Choosing Goals: Assembly and Response Rules

Two points must be addressed before going further. First, although while
gradients are one of the most powerful tools in ecology today, tools are not
an end in themselves. To effectively use a tool, the user must have clear
questions. The fundamental issues involved in selecting questions and designing research have been discussed elsewhere (e.g., Platt, 1964; Dayton,
1979; Peters, 1980a,b; Rigler, 1982; Saarinen, 1982; Quinn and Dunham,
1983; Simberloff, 1983a,b; Wiens, 1983; Leary, 1985; Loehle, 1987; Shipley and Keddy, 1987; Keddy, 1989a, 1990a), but it is worth briefly considering possible long-term goals for community ecology that could be
reached by using gradients as a tool. One current area of debate is whether

goals ought to be "understanding" or "prediction" (Peters, 1980a,b). The following suggested goal for community ecology includes elements of both.

The first goal would be community assembly rules. Although Diamond (1975) has been criticized (e.g., Connor and Simberloff, 1979) for his early attempts, it is important to distinguish between the goal he suggested and the approach he used to try to reach this goal. I accept this goal as a useful one but suggest that one way of attaining it is to use the three following steps. First, we would determine that the species pool is for a particular area. We then know that any specific community will have a subset of these species. Second, we would collect systematic data on ecological traits of species in the pool. Last, we would devise rules that tell us which traits (and therefore which subset of the pool) would be present under which set(s) of environmental conditions. We could then specify a community for any given environment. Some steps in this direction can be found in Haefner (1978, 1981), van der Valk (1981), Nobel and Slatyer (1980), Keddy (1990a) and Shipley et al. (1989).

A second goal would be response rules. These rules would predict how a specific perturbation would change the species composition of a given community. That is, they would transform one vector of species abundances into another vector of abundances given a specified environmental change. Response rules, like assembly rules, would begin with information on the species pool for the area and a matrix of life history traits of these species. The final step would use this information to predict how a specified perturbation would transform the community from one state to another. This prediction would have to permit both deletions from and additions to the subset of species forming a specific community. Examples have been reported by van der Valk (1981) and Nobel and Slatyer (1980); a less mechanistic approach has been tried by Nilsson and Keddy (1989).

Finding Gradients

It is essential to draw a distinction between two classes of situation: gradients that are spatially continuous and those that are not. The first class, spatially continuous gradients, consists of habitats with clear and obvious gradients and therefore conspicuous zonation patterns in the organisms: salt marshes (Pielou and Routledge, 1976), mountainsides (Whittaker, 1956; Terborgh, 1971), lakeshores (Spence, 1982; Keddy, 1983), rocky marine shorelines (Connell, 1972; Underwood and Denley, 1984), and alpine meadows (del Moral, 1983a). The second class, spatially discontinuous gradients, consists of habitats in which the gradients must be constructed by the researcher. This latter situation leads to a second dichotomy. Gradients can be constructed by either (1) ordering patches along one or more gradients defined by the investigator, or (2) using multivariate analyses. These two methods for constructing gradients have been called "direct" and "indirect" gradient analyses (Whittaker, 1967).

Let us first consider the distinction between continuous and discontinuous gradients, i.e., gradients in zoned communities and gradients that are constructed by the researcher. The important differences between them arise primarily from differences in (1) the ways in which propagules and adults disperse and (2) the distances over which groups of individuals can interact with each other. Although they are both real and important, from the point of view of this chapter they are merely refinements that are eventually necessary when comparing different types of gradient.

A comparison of the use of continuous and discontinuous gradients can be found in a series of studies that explored changes in vegetation properties such as species richness and heterogeneity through successional time (Morrison and Yarranton, 1973, 1974; Shafi and Yarranton, 1973a,b). Morrison and Yarranton studied vegetation changes along a strong natural gradient of primary sand dune succession; i.e., they used a naturally zoned community. In contrast, Shafi and Yarranton converted a mosaic of different ages of boreal forest to a gradient.

In environments lacking continuous gradients, we must work harder. The methods available to construct gradients in heterogeneous environments are direct and indirect gradient analysis (see also Wittaker, 1967; Gauch, 1982). In the first case, direct gradient analysis, the observer can choose to order the habitats along gradients that appear to be important based on intuition or theory. That is, one may choose a particular gradient because there is already evidence for its importance as a tool for testing hypotheses. An example of this process is Grime's (1973, 1974, 1979) introduction of above-ground biomass (standing crop) as a gradient along which species richness and life history strategies of plant communities can be arranged (Fig. 10.1). More recent studies have also examined biomass gradients and confirmed that attributes of plant communities can often be predicted from standing crop (e.g., Silvertown, 1980; Tilman, 1982; Wheeler and Giller, 1982; Wilson and Keddy, 1986; Moore et al., 1989; Wisheu and Keddy, 1989; Puerto et al., 1990). Another example at a larger scale is the arrangement of vegetation types along gradients of temperature and moisture (Fig. 10.2).

In situations where there is no obvious gradient or no initial hypothesis to be tested, multivariate techniques can be used to find gradients (e.g., Orloci, 1978; Gauch, 1982; Legendre and Legendre, 1983; Pielou, 1984; Digby and Kempton, 1987). In contrast with direct gradient analysis, where the independent variable for creating the gradient is chosen by the scientist, in indirect gradient analysis the gradients are selected based on mathematical criteria of multivariate statistical techniques. Unfortunately, such procedures tend to distort the real patterns in the community being studied. Even if this problem can be solved, most resulting indirect gradients are too complex to use in the field as they are functions of many species or environmental factors.

Direct gradient analysis is therefore a better tool for asking questions

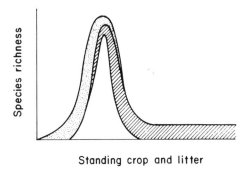

Figure 10.1. General model that organizes local heterogeneity in plant communities by plotting species richness (species per 0.25 m²) against standing crop and litter (above ground biomass) (grams per 0.25 m²). The stippled area represents stress tolerators or ruderals, and the hatched area represents competitive dominants. (After Grime, 1979).

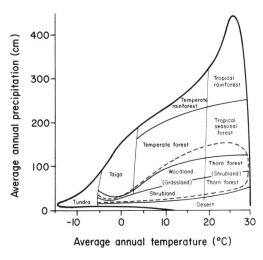

Figure 10.2. General model that organizes global heterogeneity by relating world vegetation types to gradients of temperature and moisture. It is direct gradient analysis (sensu Whittaker, 1967) at a global scale. (After Whittaker, 1975.)

and testing hypotheses. In fact, indirect gradient analysis often appears to be an end in itself. However, in some circumstances the detection of indirect gradients could be simply the first step in exploring a system for designing convincing field experiments. The reasons it is so rarely done probably have more to do with the habitual psychological patterns of scientists than anything else. For exceptions see Goldsmith (1973a,b, 1978) and del Moral (1983a,b, 1985).

Eight Guidelines for Using Gradients

Choosing the Gradient

Some gradients may be more useful than others. An important criterion for choosing a gradient for study is that it represents more than a single special case. Demonstrating that the pattern itself is of general occurrence is a key principle when designing field experiments with maximum generality (Keddy, 1989a). Before using field experiments to explore a pattern in one chalk grassland, Mitchley and Grubb (1986) reviewed nine published studies from chalk grasslands across Europe to show that their pattern was of general occurrence. Statistical tests for concordance (e.g., Siegel, 1956) would have made their work more rigorous. An example can be found in Wilson and Keddy (1985), who tested for concordance of plant distributions along lakeshore gradients before commencing experimental studies.

Temporal gradients are obviously of broad general interest for studies of succession and for developing predictive response rules for ecosystems. Careful selection of sites in heterogeneous landscapes can provide such gradients. For example, Morrison and Yarranton (1973, 1974) used a dated primary sand dune succession gradient to determine how properties such as diversity, evenness, and vegetational heterogeneity varied through time. Similarly, Shafi and Yarranton (1973a,b) used boreal forest areas dating from different fires to explore how the same properties varied through time in a different vegetation type. Jackson et al. (1988) have compared spatial patterns in wetland vegetation to successional trends reconstructed from paleoecological information and found them to be different. These authors emphasized that such reconstructions of temporal gradients assume that the areas being compared differ only in terms of age and degree of successional development.

Choosing Independent Variables to Measure

Tansley (1914) observed, "The mere taking of an instrument into the field and recording of observations, or the collection and analysis of soil samples, is no guarantee of scientific results." Given that there are nearly an infinite number of variables that could be measured, how do we simplify and choose the right ones? Theory could guide our choice, but as Lewontin (1974) observed, part of "the agony of community ecology" arises from lack of agreement as to the important state variables requiring measurement. However, once dependent and independent variables are agreed upon and are used in different studies, real empirical relationships in plant communities can be found (Keddy 1987).

The decision about what to measure therefore lies in the research question. A good question is framed in terms of measurable state variables such as biomass, species richness, size, competitive ability, etc. Once the question is posed, it is obvious what requires measurement. Environmental gradients then allow us to test for correlations between specified dependent and independent variables.

The choice of independent variables presents a special problem, since it is often difficult to translate measurements upon specific environmental factors that vary along gradients to physiological responses of organisms. This problem was carefully considered by Clements, and he concluded (Weaver and Clements, 1929; Clements, 1935) that there was little point in making complex physical measurements of the environment when it is difficult to translate such measurements into plant performance; instead he suggested using transplanted individuals as "phytometers" and "zoometers" to integrate the effects of environmental gradients upon organisms. The advent of miniature sensors and automatic data loggers has made his advice providential—but perhaps because phytometers and zoometers do not seem sufficiently technological, they are used rarely, and there is still no reference work on their selection and use (for some examples see Antonvics and Primack, 1982; del Moral, 1983a,b; Al-Farraj et al., 1984; Wilson and Keddy, 1986; Gaudet and Keddy, 1988; Keddy, 1989a). Obviously, the concept of phytometers and zoometers can be extended to other organisms, suggesting the more general term biometer or bioassay. Another possible extension is to use physiological measurements on individuals already present along the gradient as measures of position along a stress gradient (e.g., Hickman, 1975; Levitt, 1980; Fonteyn and Mahall, 1981, Weldon and Slauson, 1986).

Choosing Dependent Variables and Trait Matrices

The choice of dependent variable also rests with the question being posed. Presently, many questions are posed using species nomenclature rather than species traits; traits can be generalized from one situation to another more easily than species nomenclature. An important series of variables for developing ecological theories therefore includes the traits of organisms such as life-spans, sizes, reproductive outputs, competitive abilities, and so on (e.g., Grime, 1977; Harper, 1977; Tilman, 1988; Keddy, 1989a; Shipley et al., 1989). By testing for predictable relations between such traits and environmental gradients, it may be possible to find general empirical relations between environment and life history to test current theories (e.g., Grime, 1979; Southwood, 1988).

The dependent variables for testing whether such traits are related to environmental gradients would consist of a trait matrix of s species by t traits (Fig. 10.3). Traditionally, ecologists have studied rows of this matrix and accumulated details of the autecology and population biology of selected species (e.g., Harper, 1977, 1980). With more than a million spe-

Traits

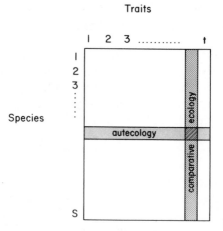

Figure 10.3. Trait matrix of *s* species by *t* traits. Few such matrices currently exist, in part because ecologists have emphasized studying species autecology (rows) rather than comparing traits (columns). These materices are an essential first step to exploring the interrelations of traits and the distributions of traits along environmental gradients.

cies of organisms on the planet (Colinvaux, 1986), it may be a long time before such studies provide general theory that is of use to community ecologists (Keddy, 1990a).

At this point attention is directed to exploring the columns of this matrix (as opposed to the rows) and the way the traits in the columns vary along natural environmental gradients. We now have floras that organize the species of the world and periodic tables for looking up the basic properties of elements, but there are no standard reference works one can consult to look up traits of organisms. The compilation of such traits would be an important step forward for the development of general theory. An essential step in this direction is the technique of "screening"—measuring the same trait on large numbers of species grown under standardized environmental conditions—introduced by Grime and his coworkers (e.g., Grime and Hunt, 1975; Grime et al., 1981). The flora near Sheffield is perhaps the most extensively studied anywhere in the world. Grime et al. (1986) compiled a volume on the essential traits of the common species of the area. Such compendia would be valuable tools for other ecosystems. Compiling trait matrices requires changes in attitude and research strategy; instead of making a great many precise measurements on one species, we would ask physiological ecologists to make a few general measurements on many species. Such a project would require some healthy debate about which traits are the important ones for constructing population and community models and which environments are appropriate for measuring these traits (Keddy, 1990b).

In the short run, researchers will probably have to do their own screening. A matrix including 20 traits has been constructed for wetland plants (Shipley, 1987; Shipley et al., 1989) and is now being expanded by incorporating traits such as competitive ability (Gaudet and Keddy, 1988) stress tolerance (Shipley and Keddy, 1988) and palatability (McCanny et al., in press). The ease of organizing and exchanging data bases on computers suggests that we are now at a stage where we could be doing it for other major systems on the planet. Until such data are available, there is the risk that studies using gradients will continue to be based on species nomenclature and therefore will be of limited general use.

Using Inferential Statistics

One of the obvious features of gradients is the changes in species abundances along them. It has led to questions about the arrangement of species distributional limits along gradients in zoned communities (e.g., Whittaker, 1956, 1967, 1975). Beyond merely describing such distribution patterns, there are now tests for determining if the distribution patterns of species along gradients deviate significantly from null models (e.g., Pielou, 1977a; Underwood, 1978; Shipley and Keddy, 1987), and they are applied in examples such as Pielou and Routledge (1976), Underwood (1978), Keddy (1983, 1984), and Shipley and Keddy (1987). The problem is that even if one can demonstrate significant departures from the null model it is not always possible to agree on the ecological interpretation (e.g., Connor and Simberloff, 1979; Grant and Abbott, 1980; Diamond and Gilpin, 1982; Wright and Biehl, 1982; Keddy, 1983, 1989a; Simberloff, 1983a,b; Shipley and Keddy, 1987); Thus although gradients are good systems for testing precise predictions about pattern, they are subject to the same problems of interpretation that have plagued other studies that try to infer pattern from process. If one is determined to explore questions about pattern, however, gradients and the above statistical tests are the obvious tools.

Given the difficulties inherent in inferring ecological processes from testing hypotheses about the pattern of species distributions, it may be better to chose a different set of dependent variables and to test hypotheses about state variables, which are more easily interpreted. They may include biomass and species richness (e.g., Al-Mufti et al., 1977; Silvertown, 1980; Moore et al., 1989) or competition intensity and biomass (e.g., Wilson and Keddy, 1986).

Incorporating Experiments

An experiment performed at one time at one site with one species has limited value, as illustrated by the difficulties in drawing generalizations about competition from collections of such studies (Connell, 1983; Scho-

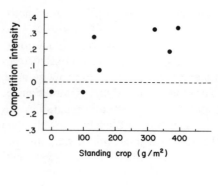

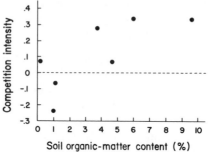

Figure 10.4. Competition intensity in lakeshore vegetation plotted against standing crop (top) and soil organic-matter content (bottom). By repeating the experiment at eight positions along the gradient we learn far more about competition than if we had simply asked if it was measurable at one selected site. (After Wilson and Keddy, 1986; Keddy, 1989a.)

ener, 1983; Ferson et al., 1986; Keddy, 1989a). In contrast, an experiment performed along a gradient may simultaneously indicate (1) the presence of a particular process, (2) trends in that process, and (3) constraints or boundary conditions upon it. For example, experimental work on barnacles (Connell, 1961), subalpine meadow plants (del Moral, 1985), grassland plants (Gurevitch, 1986), and lakeshore plants (Wilson and Keddy, 1986) indicate that interspecific competition is more intense in some sites than others. More importantly, they also provide clues as to which independent variables control competition, e.g., height on the shoreline (duration of inundation) for barnacles. Figure 10.4 gives an example of the information about competition intensity gained by using natural environmental gradients.

There is one area of ecological theory where experiments along gradients are particularly relevant: the body of ideas about resource partitioning and niche structure (e.g., MacArthur, 1972; Vandermeer, 1972; Pianka, 1973, 1981; Schoener, 1974; May, 1981; Giller, 1984). Many of the questions about niche structure in plant and animal communities can be tested explicitly using gradients. It is noteworthy that despite widespread evidence of inclusive niche structure (Colwell and Fuentes, 1975), and a growing body of literature on its connection with competitive hierarchies (Keddy, 1989) there is still the frequent assumption that species with different realized niches have different fundamental niches (e.g., Schoener 1986). Removal experiments along gradients are some of the best tests for inclusive niche structure (Colwell and Fuentes, 1975). Competitive hierarchies combined with inclusive niche structure can generate distribution patterns normally attributed to resource partitioning (Keddy, 1989a). Some examples of experiments using natural gradients to explore these ideas can be found in the work of Rabinowitz (1978), Sharitz and McCormick (1973), Mueller-Dombois and Ellenberg (1974), Grace and Wetzel (1981), and Snow and Vince (1984), but many opportunities remain to be explored.

Two Intersecting Gradients

If one gradient is useful, two can be more than doubly useful. It may be worthwhile to search for situations where two kinds of gradient interact. Examples include duration of inundation and disturbance by waves on rocky coastlines (e.g., Connell, 1972; Dayton, 1975; Dale, 1984; Underwood and Denley, 1984), moisture and elevation in forests on mountainsides (e.g., Whittaker, 1956), or slope and aspect in deserts (Yeaton and Cody, 1979). More generally, orthogonal stress and disturbance gradients would be excellent, as these gradients are prominent in recent theory (Grime, 1977; Southwood, 1988). Each gradient provides all the research opportunities described above but with the additional possibility of determining how the patterns and processes along one gradient are influenced by a second gradient.

To illustrate this point further, consider lakeshore wetlands. One can find an obvious water depth gradient that has been extensively described (Hutchinson, 1975; Spence, 1982). Orthogonal to this gradient is a gradient of exposure to waves that incorporates the direct effects of increasing disturbance with those of declining soil fertility (Hutchinson, 1975; Keddy, 1983). Because fertility and disturbance are essential elements of general ecological models (e.g., Grime, 1979; Southwood, 1988; Tilman, 1988), one can then ask how properties of the vertical gradient vary along the second gradient. Such properties can include the clustering of species distributional limits (Keddy, 1983), niche width (Keddy, 1983, 1984), distributions of rare species (Keddy, 1985), and the intensity of competition from shrubs (Keddy, 1989b). Figure 10.5 illustrates this approach by plotting niche width (measured along one gradient—elevation) against species richness (produced by a second gradient—exposure to waves) for lakeshore vegetation.

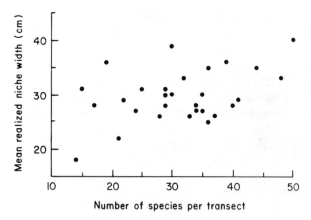

Figure 10.5. Use of two orthogonal gradients for testing ecological theory: mean niche width of shoreline plants plotted against species richness in 30 vegetation transects in a coastal plain lake in Nova Scotia, Canada. Niche width was measured along one gradient by measuring the mean elevational range occupied by plant species; the species richness gradient was obtained by arranging the transects along a second orthogonal environmental gradient, exposure to waves. Mean niche width is not correlated with species richness, suggesting that resource specialization is not the major factor controlling species richness in these lakes. (After Keddy, 1984)

Making Gradients When They Are Absent

Not all habitats provide the desired gradients. If they are not there, make them. The Park Grass experiments at the Rothamstead Experimental Station produced fertility and biomass gradients that have been extensively analyzed (Silvertown, 1980; Tilman, 1982; Digby and Kempton, 1987). More recently factorial disturbance and fertility gradients have been created in old field vegetation (e.g., Tilman 1987; Inouye and Tilman, 1988; Carson and Pickett, 1990).

Although such field experiments have their advantages, gradients can also be created in pots and microcosms. For example, Austin and Austin (1980) and Austin (1982) have created fertility gradients in pots and explored the responses of multispecies communities to them. Grime et al. (1987) have manipulated soil heterogeneity, grazing, and VA mycorrhizas in microcosms to explore their effects on diversity. Campbell (1988) has created orthogonal fertility and disturbance gradients in grassland microcosms. We learn much more from such studies than from those in which pairwise interactions are studied in a single environment.

Considering Scale

Allen and Starr (1982) have reminded ecologists that different phenomena occur at different physical and temporal scales. Building general theory requires that we explicitly state the scale at which the relations of interest operate. The same is true when comparing gradients. Ecological gradients and phenomena can range in scale from changes in geographic distributions with latitude (e.g., Simpson, 1964; Pianka, 1966; Pielou, 1977b; Currie and Paquin, 1987) to changes in parasite faunas along guts (e.g., Bush and Holmes, 1986; Stock and Holmes, 1987, 1988). If we are to compare patterns and processes along such different gradients, we need to carefully consider if and how different gradient lengths and positions are equivalent.

The following examples, dealing with the prediction of species richness in vascular plant communities, illustrate the importance of explicit consideration of scale when using gradients. At the continental scale, Currie and Paquin (1987) have shown that realized annual evapotranspiration statistically explains 76% of the variation in tree species richness. They proposed that this relation occurs because evapotranspiration is a measure of energy availability, and that the amount of energy available to be partitioned among species determines how many species can coexist. In contrast, at regional scales there is a strong relation between species richness and biomass, with species richness reaching a maximum at intermediate levels of biomass (e.g., Grime, 1973, 1979; Al-Mufti et al., 1977; Moore et al., 1989). The latter is attributed to interactions of stress, disturbance, and dominance. In situations with high biomass, it is assumed that only a few species can tolerate the effects of competitive dominants; at low biomass, only a few species can tolerate the effects of the disturbance or stress that has produced the low biomass. Therefore most species occur at intermediate levels of biomass, where neither dominance nor stress and disturbance predominate (Grime, 1973, 1979). At a lower scale yet, one can ask what predicts species richness at a local scale within a vegetation type. The regional model, with maximum richness at intermediate levels of biomass, is based on comparison of very different vegetation types; the first example compared vegetation types ranging from "woodlands" to "tall herbs" (Al-Mufti et al., 1977). Moore and Keddy (1989) have compared such richness-biomass patterns found at the regional scale (among vegetation types) with those at a lower scale still, the local scale (within vegetation types). The clear result was that this model applied at the regional scale (i.e., when vegetation types dominated by different species were included in the data set), but that it could not be detected within a single vegetation type. At the local scale, other factors such as differentiation of regeneration niches (Grubb, 1977) are presumably factors that control richness. Therefore as scale decreases from the continental to the local level, both the patterns of richness and the presumed mechanisms change.

A second problem with scale arises when considering the length of gradients necessary to test for specific relations. It is obvious from inspection of Figure 10.1 that if only low biomass regions occur in a habitat a positive relation is present (e.g., Wisheu and Keddy, 1989), and if only high biomass regions occur, a negative relation can be found (e.g., Silvertown,

1980). The full bitonic relation emerges only when a "sufficiently wide" range of biomass exists. The model is therefore not easily tested because all possible patterns are possible for "short" sections of the biomass continuum.

Gradients and Centrifugal Organization

The phenomenon of resource partitioning is so widespread (e.g., MacArthur, 1972; Schoener, 1974; Giller, 1984) that, whatever the underlying mechanisms, communities appear to be organized along resource gradients, or gradients that constrain access to resources. Descriptions of gradients may therefore indicate the principal axes along which communities are organized, and experiments along them may provide the tools for testing for general mechanisms. The phenomenon of "centrifugal organization" (Rosenzweig and Abramsky, 1986; Keddy, 1990a) emerges naturally from these two processes. In the case of desert rodents (Rosenzweig and Abramsky, 1986) there is a central, or core, habitat that is preferred by two species in the absence of competition but that is normally occupied by one competitive dominant. Coexistence occurs because each species has a secondary, peripheral habitat on which it is specialized. In the case of wetland plants (Keddy, 1990a), the secondary or peripheral habitats are replaced by gradients. The preferred habitats are fertile and undisturbed; they are dominated by large, leafy species with rhizomes, probably because light is the limiting resource. Different constraints such as disturbance and low productivity provide axes radiating out from the preferred central habitat (Figure 10.6). Different groups of species are arrayed along these gradients depending on the particular kind of stress or disturbance. For example, fluctuating water levels on sandy shores provide one kind of gradient, whereas ice scour on river banks provides another. Along any specific gradient there is evidence that species have inclusive niche structure (sensu Colwell and Fuentes, 1975) or at least are controlled by one-sided competition such that in the absence of neighbors they can move toward the preferred habitat but cannot move down it to sites with more disturbance or lower productivity.

It is not yet clear whether such organization occurs in other communities or is a general phenomenon in wetlands, but given the widespread occurrence of resource partitioning and constraints on resource acquisition, it seems likely. Certainly, it is possible to find similar gradient structure in other vegetation types. Keddy and MacLellan (1990) have shown that both the patterns and mechanisms of centrifugal organization occur in forests. In the Great Lakes–St. Lawrence forest region (Rowe, 1977), for example, fertile sites are dominated by *Acer saccharum* and *Fagus grandifolia* which would occupy the core habitat at the centre of Figure 10.6. As sites become warmer and drier *Quercus rubra* predominates, and eventual-

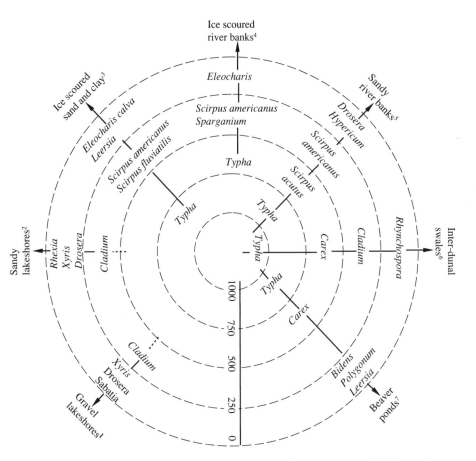

Figure 10.6. Centrifugal organization in wetlands. There is a central habitat with low adversity and low disturbance; this site is dominated by tall species with dense canopies and large rhizomes. Different gradients of adversity and disturbance radiate outward from this central habitat, providing a wider array of species and vegetation types with increasing severity of constraints. (After Keddy, 1990a).

ly it is replaced by such species as *Juniperus communis* (Gagnon and Bouchard, 1981) or *Pinus rigida* (Vander Kloet 1973). Colder and drier sites are dominated in turn by *Pinus strobus* and then *Pinus banksiana* (Gagnon and Bouchard, 1981). In contrast, as sites become wetter, *Ulmus americana* and *Fraxinus nigra* become more common; and eventually, in extreme cases, *Acer saccharinum* monocultures occur (e.g., Rowe, 1977). In soil depth gradients over limestone, *Thuja occidentalis* and *Juniperus communis* occupy sites with the least soil (Catling et al., 1975). The different forest site types in this area can therefore be organized along constraint gradients using the centrifugal model.

Conclusion

Gradients are a powerful research tool. They take the obstacles posed by heterogeneity and transform them to advantages. Pielou (1977a) observed more than a decade ago that gradients "are far more worthy of study than the elusive 'homogeneous environments' traditionally beloved of ecologists." The foregoing eight points illustrate some paths for making full use of such gradients in ecological research. Undoubtedly, more than eight points could be imagined, but this operator's manual is intended only to introduce the tool to potential users. As Pirsig (1974) observed, there are hundreds of ways to put a rotisserie together. "Technology presumes there's only one right way to do things, and there never is." As with rotisserie assembly, there is no single right way to work with gradients. Their potential for creative research remains largely unexplored.

Acknowledgments. I thank Chris Pielou for first introducing me to the use of environmental gradients. I thank also the people who have worked with me along natural wetland gradients and whose research stimulated the ideas presented here: Scott Wilson, Bill Shipley, Irene Wisheu, Dwayne Moore, Connie Gaudet. I also thank Irene Wisheu for her help in preparing this manuscript and Céline Boutin, David Currie, and Bill Shipley for their helpful comments on it. This work was supported by an operating grant from the Natural Sciences and Engineering Research Council of Canada.

References

Al-Farraj MM, Giller KE, Wheeler BD (1984) Phytometric estimation of fertility of waterlogged rich-fen peats using *Epilobium hirsutum*. Plant Soil 81:283–290

Allen TFH, Starr TB (1982) *Hierarchy*. University of Chicago Press, Chicago

Al-Mufti MM, Sydes CL, Furness SB, Grime JP, Band SR (1977) A quantitative analysis of shoot phenology and dominance in herbaceous vegetation. J Ecol 65:759–791

Antonvics J, Primack RB (1982) Experimental ecological genetics in *Plantago*. VI. The demography of seedling transplants of *P. lanceolata*. J Ecol 70:55–75

Austin MP (1982) Use of a relative physiological performance value in the prediction of performance in multispecies mixtures from monoculture performance. J Ecol 70:559–570

Austin MP, Austin BO (1980) Behaviour of experimental plant communities along a nutrient gradient. J Ecol 68:891–918

Bush AO, Holmes JC (1986) Intestinal helminths of leser scaup ducks: an interactive community. Can J Zool 64:142–152

Campbell BD (1988) Experimental tests of C-S-R theory. PhD thesis, University of Sheffield, Sheffield, UK

Carson WP, Pickett STA (1990) Role of resources and disturbance in the organization of an old-field plant community. Ecology 71:226–238.

Catling PM, Cruise JE, McIntosh KL, McKay SM (1975). Alvar vegetation in southern Ontario. Ontario Field Biol 29:1–25

Clements FE (1935) Experimental ecology in the public service. Ecology 16:342–363

Colinvaux P (1986) *Ecology*. Wiley, New York

Colwell RK, Fuentes ER (1975) Experimental studies of the niche. Annu Rev Ecol Syst 6:281–309

Connell JH (1961) The influence of interspecific competition and other factors on the distribution of the barnacle *Chthamalus stellatus*. Ecology 42:710–723

Connell JH (1972) Community interactions on marine rocky intertidal shores. Annu Rev Ecol Syst 3:169–192

Connell JH (1983) On the prevalence and relative importance of interspecific competition: evidence from field experiments. Am Nat 122:661–696

Connor EF, Simberloff D (1979) The assembly of species communities: chance or competition? Ecology 60:1132–1140

Currie DJ, Paquin V (1987) Large-scale biogeographical patterns of species richness of trees. Nature 329:326–327

Dale MRT (1984) The contiguity of upslope and downslope boundaries of species in a zoned community. Oikos 42:92–96

Dayton PK (1975) Experimental evaluation of ecological dominance in a rocky intertidal algal community. Ecol Monogr 45:137–159

Dayton PK (1979) Ecology: a science and a religion. In Livingston RJ (ed) *Ecological Processes in Coastal and Marine Systems*. Plenum Press, New York, pp 3–18

Del Moral R (1983a) Competition as a control mechanism in subalpine meadows. Am J Bot 70:232–245

Del Moral R (1983b) Vegetation ordination of subalpine meadows using adaptive strategies. Can J Bot 61:3117–3127

Del Moral R (1985) Competitive effects on the structure of subalpine meadow communities. Can J Bot 63:1444–1452

Diamond JM (1975) Assembly of species communities. In Cody ML, Diamond JM (eds) *Ecology and Evolution of Communities*. Harvard University Press, Cambridge, pp 342–444

Diamond JM, Gilpin M (1982) Examination of the "null" model of Connor and Simberloff for species co-occurrences on islands. Oecologia 52:64–74

Digby PGN, Kempton RA (1987) *Multivariate Analysis of Ecological Communities*. Chapman & Hall, London

Ferson S, Downey P, Klerks P, Weissburg M, Kroot I, Stewart S, Jacquez G, Ssemakula J, Malenky R, Anderson K (1986) Competing reviews, or why do Connell and Schoener disagree? Am Nat 127:571–576

Fonteyn PJ, Mahall BE (1981) An experimental analysis of structure in a desert plant community. J Ecol 69:883–896

Gagnon D, Bouchard A (1981) La vegetation de l'escarpement d'Eardley, parc de la Gatineau, Quebec. Can J Bot 59:2667–2691

Guach HG (1982) *Multivariate Analysis in Community Ecology*. Cambridge University Press, Cambridge

Gaudet CL, Keddy PA (1988) Predicting competitive ability from plant traits: a comparative approach. Nature 334:242–243

Giller PS (1984) *Community Structure and the Niche*. Chapman & Hall, London

Goldsmith FB (1973a) The vegetation of exposed sea cliffs at South Stack, Anglesey. I. The multivariate approach. J Ecol 61:787–818

Goldsmith FB (1973b) The vegetation of exposed sea cliffs at South Stack, Anglesey. II. Experimental studies. J Ecol 61:819–829

Goldsmith FB (1978) Interaction (competition) studies as a step towards the synthesis of seacliff vegetation. J Ecol 66:921–931

Grace JB, Wetzel RG (1981) Habitat partitioning and competitive displacement in Cattails (*Typha*): experimental field studies. Am Nat 118:463–474

Grant PR, Abbott I (1980) Interspecific competition, island biogeography and null hypotheses. Evolution 34:332–341

Grime JP (1973) Competitive exclusion in herbaceous vegetation. Nature 242:344–347

Grime JP (1974) Vegetation classification by reference to strategies. Nature 250:26–31

Grime JP (1977) Evidence for the existence of three primary strategies in plants and its relevance to ecological and evolutionary theory. Am Nat 111:1169–1194

Grime JP (1979) *Plant Strategies and Vegetation Processes*. Wiley, Chichester

Grime JP, Hunt R (1975) Relative growth rate: its range and adaptive significance in a local flora. J Ecol 63:393–422

Grime JP, Mackey JML, Hillier SH, Read DJ. (1987) Floristic diversity in a model system using experimental microcosms. Nature 328:420–422

Grime JP, Mason G, Curtis AV, Rodman J, Band SR, Mowforth MAG, Neal AM, Shaw S (1981) A comparative study of germination characteristics in a local flora. J Ecol 69:1017–1059

Grime JP, Hodgson JG, Hunt R (1986) *Comparative Plant Ecology: A Functional Approach to Common British Plants and Communities*. Allen & Unwin, London

Grubb PJ (1977) The maintenance of species richness in plant communities: the importance of the regeneration niche. Biol Rev 52:107–145

Gurevitch J (1986) Competition and the local distribution of the grass *Stipa neomexicana*. Ecology 67:46–57

Haefner JW (1978) Ecosystem assembly grammars: generative capacity and empirical adequacy. J Theor Biol 73:293–318

Haefner JW (1981) Avian community assembly rules: the foliage-gleaning guild. Oecologia 50:131–142

Harper JL (1977) *Population Biology of Plants*. Academic Press, London

Harper JL (1980) Plant demography and ecological theory. Oikos 35:244–253

Hickman JC (1975) Environmental unpredictability and plastic energy allocation strategies in the annual *Polygonum cascadense* (Polygonaceae). J Ecol 63:689–701

Hutchinson GE (1975) *A Treatise on Limnology*. Vol 3. Wiley, New York

Inouye RS, Tilman D (1988) Convergence and divergence of old-field plant communities along experimental nitrogen gradients. Ecology 69:995–1004

Jackson ST, Futyma RP, Wilcox DA (1988) A paleoecological test of a classical hydrosere in the Lake Michigan dunes. Ecology 69:928–936

Keddy PA (1983) Shoreline vegetation in Axe Lake, Ontario: effects of exposure on zonation patterns. Ecology 64:331–344

Keddy PA (1984) Plant zonation on lakeshores in Nova Scotia: a test of the resource specialization hypothesis. J Ecol 72:797–808

Keddy PA (1985) Wave disturbance on lakeshores and the within-lake distribution on Ontario's Atlantic coastal plain flora. Can J Bot 63:656–660

Keddy PA (1987) Beyond reductionism and scholasticism in plant community ecology. Vegetatio 69:209–211

Keddy PA (1989a) *Competition*. Chapman & Hall, London

Keddy PA (1989b) Effects of competition from shrubs on herbaceous wetland plants: a four year field experiment. Can J Bot 67:708–716

Keddy PA (1990a) Competitive hierarchies and centrifugal organization in plant communities. In Grace J, Tilman D (eds) *Perspectives on Plant Competition*. Academic Press, Orlando

Keddy PA (1990b) The use of functional as opposed to phylogenetic systematics: a first step in predictive community ecology. In Kawano S (ed) *Biological*

Approaches and Evolutionary Trends in Plants, Academic Press, Orlando, pp 387–406

Keddy PA, MacLellan P (1990) Centrifugal organization in forests. Oikos 59: 75–84

Leary RA (1985) A framework for assessing and rewarding a scientist's research productivity. Scientometrics 7:29–38

Legendre L, Legendre P (1983) *Numerical Ecology*. Elsevier, Amsterdam

Levitt J (1980) *Responses of Plants to Environmental Stresses*. 2nd ed. Academic Press, Orlando

Lewontin RC (1974) *The Genetic Basis of Evolutionary Change*. Columbia University Press, New York

Loehle C (1987) Hypothesis testing in ecology: psychological aspects and the importance of theory maturation. Rev Biol 62:397–409

MacArthur RH (1972) *Geographical Ecology*. Harper & Row, New York

May RM (1981) Patterns in multi-species communities. In May RM (ed) *Theoretical Ecology*. 2nd Blackwell, Oxford, pp 197–227

McCanny SJ, Keddy PA, Arnason TJ, Gaudet CL, Moore DRJ and Shipley B. (in press) Fertility and the food quality of wetland plants: a test of the resource availability hyopthesis. Oikos 58:000–000

Mitchley J, Grubb PJ (1986) Control of relative abundance of perennials in chalk grassland in southern England. I. Constancy of rank order and results of pot- and field-experiments on the role of interference. J Ecol 74:1139–1166

Moore DRJ, Keddy PA (1989) The relationship between species richness and standing crop in wetlands: the importance of scale. Vegetatio 79:99–106

Moore DRJ, Keddy PA, Gaudet CL, Wisheu IC (1989) Conservation of wetlands: do infertile wetlands deserve a higher priority? Biol Conserv 47:203–217

Morrison RG, Yarranton GA (1973) Diversity, richness, and evenness during a primary sand dune succession at Grand Bend, Ontario. Can J Bot 51:2401–2411

Morrison RG, Yarranton GA (1974) Vegetational heterogeneity during a primary sand dune succession. Can J Bot 52:397–410

Mueller-Dombois D, Ellenberg H (1974) *Aims and Methods of Vegetation Ecology*. Wiley, New York

Nilsson C, Keddy PA (1989) Predictability of change in shoreline vegetation in a hydroelectric reservoir, northern Sweden. Can J Fish Aquat Sci 45:1896–1904

Noble IR, Slatyer RO (1980) The use of vital attributes to predict successional changes in plant communities subject to recurrent disturbances. Vegetatio 43: 5–21

Orloci L (1978) *Multivariate Analysis in Vegetation Research*. 2nd Junk, The Hague

Peters RH (1980a) Useful concepts for predictive ecology. Synthese 43:257–269

Peters RH (1980b) From natural history to ecology. Perspectives in Biology and Medicine 23:191–203

Pianka ER (1966) Latitudinal gradients in species diversity: a review of the concepts. Am Nat 100:33–46

Pianka ER (1973) The structure of lizard communities. Annu Rev Ecol Syst 4:53–74

Pianka ER (1981) Competition and niche theory. In May RM (ed) *Theoretical Ecology*. 2nd ed. Blackwell, Oxford

Pielou EC (1977a) *Mathematical Ecology*. Wiley, New York

Pielou EC (1977b) The latitudinal spans of seaweed species and their patterns of overlap. J Biogeogr 4:299–311

Pielou EC (1984) *The Interpretation of Ecological Data*. Wiley, New York

Pielou EC, Routledge RD (1976) Salt marsh vegetation: latitudinal gradients in the zonation patterns. Oecologia 24:311–321

Pirsig RM (1974) Zen and the Art of Motorcycle Maintenance. Morrow, New York

Platt JR (1964) Strong inference. Science 146:347–353

Puerto A, Rico M, Matias MD and Garcia JA (1990) Variation in structure and diversity of Mediterranean grasslands related to trophic status and grazing intensity. Journal of Vegetation Science 1:445–452

Quinn JF, Dunham AE (1983) On hypothesis testing in ecology and evolution. Am Nat 122:602–617

Rabinowitz D (1978) Early growth of mangrove seedlings in Panama, and an hypothesis concerning the relationship of dispersal and zonation. J Biogeogr 5:111–1331

Rigler FH (1982) Recognition of the possible: an advantage of empiricism in ecology. Can J Fish Aquat Sci 39:1323–1331

Rosenzweig ML, Abramsky Z (1986) Centrifugal community organization. Oikos 46:339–348

Rowe JS (1977) Forest regions of Canada. Department of Fisheries and the Environment, Canadian Forestry Service Publication No. 1300

Saarinen E (ed.) (1982) *Conceptual Issues in Ecology*. D. Reidel, Dordrecht

Schoener TW (1974) Resource partitioning in ecological communities. Science 185:27–39

Schoener TW (1983) Field experiments on interspecific competition. Am Nat 122:240–285

Schoener TW (1986) Resource partitioning. In Kikkawa J and Anderson DJ (eds) *Community Ecology: Pattern and Process*, Blackwell Scientific Publications, Melbourne, pp 91–126

Shafi MI, Yarranton GA (1973) Vegetational heterogeneity during a secondary (postfire) succession. Can J Bot 51:73–90

Shafi MI, Yarranton GA (1973) Diversity, floristic richness, and species evenness during a secondary (post-fire) succession. Ecology 54:897–902

Sharitz RR, McCormick JF (1973) Population dynamics of two competing annual plant species. Ecology 54:723–740

Shipley B (1987) Pattern and mechanism in the emergent macrophyte communities along the Ottawa River (Canada). Ph.D. Thesis, University of Ottawa, Ottawa, Canada

Shipley B, Keddy PA (1987) The individualistic and community-unit concepts as falsifiable hypotheses. Vegetatio 69:47–55

Shipley B, Keddy PA (1988) The relationship between relative growth rate and sensitivity to stress in 28 emergent macrophytes. J Ecol 76:1101–1110

Shipley B, Keddy PA, Moore DRJ, Lemky K (1989) Regeneration and establishment strategies of emergent macrophytes. J Ecol 77:1093–1110

Siegel S (1956) Nonparametric statistics for the behavioral sciences. McGraw-Hill, New York

Silvertown J (1980) The dynamics of a grassland ecosystem: botanical equilibrium in the park grass experiment. J Appl Ecol 17:491–504

Simberloff D (1983a) Competition theory, hypothesis testing, and other community ecological buzzwords. Am Nat 122:626–635

Simberloff D (1983b) Sizes of coexisting species. In Futuyma DJ, Slatkin M (eds) *Coevolution*, Sinauer, Sunderland, pp 404–430

Simpson GG (1964) Species density of North American recent mammals. Syst Zool 13:57–73

Snow AA, Vince SW (1984) Plant zonation in an Alaskan salt marsh. II. An experimental study of the role of edaphic conditions. J Ecol 72:669–684

Southwood TRE (1988) Tactics, strategies and templets. Oikos 52:3–18

Spence DHN (1982) The zonation of plants in freshwater lakes. In Macfadyen A, Ford ED (eds) *Advances in Ecological Research* 12:37–125

Stock TM, Holmes JC (1987) Dioecocestus asper (Cestoda: Dioecocestidae): an interference competitor in an enteric helminth community. J Parasitol 73:1116–1123

Stock TM, Holmes JC (1988) Functional relationships and microhabitat distributions of enteric helminths of grebes (Podicepedidae): the evidence for interactive communities. J Parasitol 74:214–227

Tansley AG (1914) Presidential Address. J Ecol 2:194–203

Terborgh J (1971) Distribution on environmental gradients: theory and a preliminary interpretation of distributional patterns in the avifauna of the Cordillera Vilcabamba, Peru. Ecology 52:23–40

Tilman D (1977) Secondary succession and the pattern of plant dominance along experimental nitrogen gradients. Ecol Monogr 57:189–214

Tilman D (1982) Resource Competition and Community Structure. Princeton University Press, Princeton

Tilman D (1988) Plant Strategies and the Structure and Dynamics of Plant Communities. Princeton University Press, Princeton

Underwood AJ (1978) The detection of non-random patterns of distribution of species along an environmental gradient. Oecologia 36:317–326

Underwood AJ, Denley EJ (1984) Paradigms, explanations and generalizations in models for the structure of intertidal communities on rocky shores. In Strong DR, Simberloff D, Abele LG, Thistle AB (eds) Ecological Communities: Conceptual Issues and the Evidence. Princeton University Press, Princeton, pp 151–180

Vander Kloet SP (1973) The biological status of Pitch Pine, Pinus rigida Miller, in Ontario and New York. The Canadian Field-Naturalist 87:249–253

Vandermeer JH (1972) Niche theory. Annu Rev Ecol Syst 3:107–132

Van der Valk AG (1981) Succession in wetlands: a Gleasonian approach. Ecology 62:688–696

Weaver JE, Clements FE (1929) Plant Ecology. McGraw-Hill, New York

Weldon CW, Slauson WL (1986) The intensity of competition versus its importance: an overlooked distinction and some implications. Q Rev Biol 61:23–44

Wheeler BD, Giller KE (1982) Species richness of herbaceous fen vegetation in Broadland, Norfolk in relation to the quantity of above-ground plant material. J Ecol 70:179–200

Whittaker RH (1956) Vegetation of the Great Smoky Mountains. Ecol Monogr 26:1–80

Whittaker RH (1967) Gradient analysis of vegetation. Biol Rev 42:207–264

Whittaker RH (1975) Communities and Ecosystems. 2nd ed. Macmillan, London

Wiens JA (1983) Avian community ecology: an iconoclastic view. In Brush AH, Clark GA (eds) Perspectives in Ornithology. Cambridge University Press, Cambridge, pp 355–410

Wilson SD, Keddy PA (1985) Plant zonation on a shoreline gradient: physiological response curves of component species. J Ecol 73:851–860

Wilson SD, Keddy PA (1986) Measuring diffuse competition along an environmental gradient: results from a shoreline plant community. Am Nat 127:862–869

Wisheu IC, Keddy PA (1989) Species richness-standing crop relationships along four lakeshore gradients: constraints on the general model. Can J Bot 67:1609–1617

Wright SJ, Biehl CC (1982) Island biogeographic distributions testing for random, regular, and aggregated, patterns of species occurrence. Am Nat 119:345–357

Yeaton RI, Cody ML (1979) The distribution of cacti along environmental gradients in the Sonoran and Mohave deserts. J Ecol 67:529–541

11. Relations Among Spatiotemporal Heterogeneity, Population Abundance, and Variability in a Desert

Moshe Shachak and Sol Brand

There are at least two features of natural systems that ecologists agree on: (1) they are abiotically heterogeneous in time and space; and (2) they are variable in organism abundance (e.g., Sousa, 1984). In the literature there are discrepancies as to the usage of the terms heterogeneity and variability (see Chapter 1); therefore we define our usage of the terms. *Heterogeneity* refers to composition of parts of different kinds, and *variability* indicates changes in values of one kind (*Random House Dictionary*).

A number of questions arise as to the relation between environmental heterogeneity and population variability.

1. Under what conditions are heterogeneity and variability positively or negatively related to each other?
2. How do spatial and temporal heterogeneity affect population variability?
3. Is the relation between heterogeneity and abundance similar to that between heterogeneity and variability?

To approach these questions we suggest a way of finding the relation between heterogeneity and variability of a system. The relation can be explored by determining the common processes that integrate heterogeneity and variability. By definition, heterogeneity and variability relate differently to the processes. Environmental heterogeneity controls nonuniformity in resource distribution, whereas population variability is the product of the ecological processes initiated by resource distribution. A common approach for relating biological processes to environmental heterogeneity is to correlate spatiotemporal heterogeneity with population variables (e.g., Hallett and Pimm, 1979; Kotler and Brown, 1988). Kratz et al. (1987) have shown that the reverse is also possible. Abiotic processes operating within a system can be inferred by analyzing biological variability.

Our assumption is that the environmental heterogeneity and population variability relation can be assessed if the processes controlled by heterogeneity and the processes inferred from variability coincide. To test the assumption, we used data from our 15-year study on two populations of the desert isopod. *Hemilepistus reaumuri* in the Negev Desert. We compared the long-term abundance and variability of populations of *H. reaumuri* in two distinctly different landscape units: a rocky slope and a loessial valley. The rocky slope is representative of a spatially heterogeneous area, whereas the valley is relatively homogeneous.

The aim of the study was to integrate heterogeneity and variability by determining common processes. We inferred the processes from *H. reaumuri* population variability and water resource distribution on the two landscapes. Our specific objectives were to determine whether (1) *H. reaumuri* abundance and variability are dependent on heterogeneity; (2) *H. reaumuri* survivorship is heterogeneity-dependent; (3) there are common processes that integrate environmental heterogeneity with *H. reaumuri* population dynamics; and (4) it is possible to infer general relations among heterogeneity, abundance, and variability from our *H. reaumuri* case study.

Study Area and Methods

Life Cycle and Biology

Our long-term study (1973–1987) on the population dynamics of the desert isopod *Hemilepistus reaumuri* can be used to demonstrate the relations among spatiotemporal population abundance, variability, and environmental heterogeneity: *H. reaumuri* is suitable for the following reasons. (1) It is an annual animal with a simple life cycle (Linsenmair, 1972). (2) It inhabits most of the habitats in the Negev Desert (except for sand dunes). Its abundance varies considerably in time and space (Shachak and Yair, 1984). (3) It is relatively easy to locate its burrows. Each individual completes its life cycle at two sites: the site at which it hatched and matured and the site at which it mated and reproduced (Shachak and Brand, 1988). (4) It is possible to track its population dynamics with great accuracy. One can identify, in the field, almost all settling sites and determine if in the sites selected the animals successfully completed their life cycles (Shachak and Yair, 1984). Successful sites are defined as *safe sites* (sensu Harper, 1961). (5) The main factor in site suitability is soil moisture. If the site at a depth of 40 to 60 cm does not maintain at least 6 to 10% soil moisture all summer, the inhabitants perish (Shachak, 1980; Coenen-Stass, 1981).

A detailed description of *H. reaumuri*'s life cycle and biology has been published elsewhere (Shachak et al., 1979). Here we present information relevant to this study. *H. reaumuri* lives in monogamous family units in a single burrow dug by the family members. Each unit is comprised of parents that live together with the young until the parents' death at about September. During September and October the siblings continue to live in the same family unit, spending most of their time in the burrow and only about 1 to 2 hours each day outside foraging on organic matter and soil crust. From November to February the family is in a sedentary state and stays underground in its burrow, which is about 50 cm deep. Each burrow is occupied by about 30 to 50 siblings. During February the isopods vacate the sites in which they hatched and matured and search for new settling sites. The animals searching for new sites are referred to as *potential settlers*. Sites for burrowing are selected after a relatively long searching time, about 4 hours per day for a period up to a month. At the selected site the female digs a burrow (about 5 cm) in which pairs are formed. This process of burrowing and pairing during February to March is defined as *settling*. In April the females are pregnant, and in May the young hatch. Parents and young form a family unit, living in, excavating, and protecting their burrow together. During the postsettling period from April to February, the parents die, the offspring mature, and the isopods deepen their burrow to a depth of about 50 cm in quest of relatively high soil moisture.

Site quality assessment before settling is the most critical decision for family survivorship after settling. If the site selected for burrowing during February does not provide the soil moisture necessary for the family members all year, the inhabitants die. *H. reaumuri* feed on almost all the annual and perennial vegetation in the area. They also feed on soil crust, and most of their feces consist of soil. Food does not seem to a limiting factor in their spatial abundance among habitats. In addition, predation does not seem to be a major factor in the survival of the family after settling (Shachak et al., 1976; Shachak and Brand, 1988).

Study Areas

The two study areas are located in the northern Negev Desert, Israel, about 40 km south of Beer Sheva (Fig. 11.1) at an altitude of about 500 m. Rainfall is limited to the winter months (October to April). Average annual rainfall recorded over 30 years is 92 mm, with extreme values of 34 mm and 167 mm. The number of rain days varies from 15 to 42 per year, and only a few rains yield more than 20 mm per day (Yair and Shachak, 1987). Rainfall is low and Irregular, with a quotient of variation of ≥ 5.0 (Evenari et al., 1983). Mean monthly temperatures vary from 9°C in January to 25°C in August. Average daily relative humidity attains 60 to 70% in

A

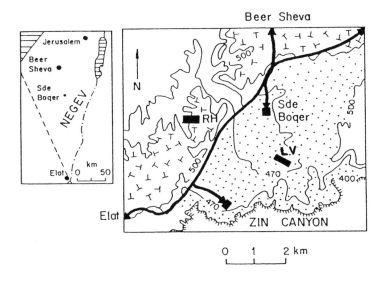

B

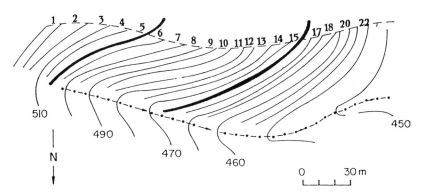

Figure 11.1. Study areas. (A) Location in the Negev Desert highlands.
▒▒▒= loessial valley; ⊥ ⊣= rocky slope; ■ RH = rocky study area; ■ LV = loessial
study area. Heavy lines indicate paved road. Numbers indicate the elevation
(meters). (B) Topography, habitats, and rows (1 to 22) of the rocky slope. (- - - - -)
divide line; (- · - · -) channel. Elevation ranged from 450 to 510 meters. Heavy
lines indicate the boundaries between habitats, and the other lines represent the
boundaries between rows.

winter and 40 to 50% in summer. For our study we chose two landscape units under the same climatic regime about 4 km from each other, a rocky slope and a loessial valley, which differ in their degree of heterogeneity (Fig. 11.1).

Rocky Slope

The rocky slope study area (11,325 m^2) was the whole north-facing slope of a small watershed. Owing to the geological structure, the hillside is sub-divided into distinct habitats based on surface properties and plant community composition. The lower part of the slope is covered by extensive stony colluvial soil that is more than 250 cm thick at the slope base. Vegetation covers 5 to 10% of the area and is uniformly spaced. The characteristic plant association is that of *Artemesia herba alba–Gymnocarpos decander* (Danin, 1972).

The midslope habitat is of massive crystalline limestone and bedrock that is exposed on 60 to 80% of the surface. Soil is found in soil strips located at the base of bedrock steps, joints, and bedding planes of the superficial rock strata. The vegetation cover (5 to 10%) is concentrated along the soil strips and soil-filled joints. The characteristic plant association is *Vartemia iphionides–Origanum dayi* (Danin, 1972).

The upper rocky unit is composed of thinly bedded limestone with flint concretions. The soil is shallow and appears in numerous small patches. Vegetation cover is 3 to 8%. The characteristic plant association is *Hammadetum scopariae lepidosum* (Yair and Danin, 1980).

Loessial Valley

The loessial valley study area (8000 m^2) is located in the lowest part of a 15 km^2 loessial plain. The area is covered by thick loessial serozem soil. The plant association was classified by Danin (1970) as *Hammadetum scopariae lepidosum*. It is composed of *Hammadetum scopariae lepidosum*, *Artemesia herba alba*, and *Reaumuria palestina* shrubs. The vegetation cover is 10 to 15%.

Methods

The purpose of this study was to determine the relation among environmental heterogeneity and *H. reaumuri* population abundance and variability. We did it by using two heterogeneity and two population variables.

Heterogeneity

The heterogeneity variables were ratios of rock/soil and runoff/rain. Rock/soil was chosen because the spatial heterogeneity in the rocky Negev Desert is determined mainly by a mosaic of rock and soil. The runoff/rain ratio is a measure of temporal heterogeneity of the water resource: The rainfall–

Table 11.1 Rocky Slope Spatial
Heterogeneity as Defined by the
Rock/Soil Surface Cover Ratio

Row[a]	Rock/Soil Cover
1	1.86
2	0.69
3	1.70
4	1.27
5	1.78
6	0.64
7	0.92
8	0.85
9	0.69
10	1.24
11	0.30
12	3.00
13	2.03
14	2.33
15	0.85
16	0.37
17	0.45
18	0.33
19	0.17
20	0.09
21	0.05
22	0.04

[a] Rows are in ascending order from
water divide to channel.

runoff relation controls the soil moisture regime, which is the single most
important factor for biological activity in the desert (Yair and Danin, 1980;
Olsvig-Whittaker et al., 1983). The rainfall–runoff relation in the area has
been studied by Yair and his colleagues (Yair and Lavee, 1982, 1985).
Information cited herein pertaining to rainfall and runoff was gained from
their studies.

The rocky slope heterogeneity was determined by subdividing the slope
into 22 parallel rows (Fig. 11.1). For each row we determined the rock/soil
ratio (Table 11.1). The rock/soil ratio of the loessial valley is 0; therefore,
according to our definition of heterogeneity, the area is spatially homogeneous.

Variability

To determine the abundance and variability of *H. reaumuri* populations we
measured the densities of settlers and successful families in the two areas
for 15 generations (1973–1987). Settlers were defined as the number of
new families formed during February. Successful families were the number

of families that survived after settling until the next settling season. The ratio of successful families to settlers was the survivorship.

The two variables were determined using permanent grid systems consisting of rows 5 m wide. The rocky slope grid consisted of 22 rows (60 to 160 m long) (Fig. 11.1), and the loessial area grid was a rectangle of 16 rows (50 m long). This row system made it possible to determine the number of *H. reaumuri* settlers and successful families in each area.

We found that the condition of the feces pile surrounding the isopod burrow during September provided the necessary information on settling and survivorship (Shachak, 1980). Burrows surrounded by large piles of feces (150 to 350 g) were safe sites inhabited by successful families. Burrows surrounded by small piles of feces during September (10 to 30 g) indicate unsuccessful families that did not survive to the next season. Settlers were the sum of successful and unsuccessful families. On the rocky area we determined isopod abundance in the total area and for soil cover only.

We used the coefficient of variation (CV) as our measure of variability because it is independent of the mean (Sokal and Rohlf, 1981). Use of the CV as a measure of variability is generally justified when the mean and the standard deviation are strongly correlated (Real and Rathcke, 1988) as they are in our areas.

Results

Abundance Patterns

Spatiotemporal Patterns

In the loessial valley there was no trend among rows, indicating that, on the average, settlers and successful families were uniformly distributed. In contrast, on the rocky slope a spatial pattern emerged for both settlers and successful families. There was an increase of *H. reaumuri* abundance from upper slope to midslope (row 12) and then a decrease toward the slope base (Fig 11.2). Row quality for the success of isopod families in the loessial valley was similar (maximum 21.6, minimum 19.2 families/100 m²). On the rocky slope, row quality was nonuniform (maximum 10.4, minimum 2.6 families/100 m²).

Although there was a spatial pattern in population abundances, we did not perceive any apparent temporal pattern (Fig. 11.3). Populations on both landscapes exhibited fluctuations in mean abundance. The loessial

---

Figure 11.3. Temporal abundance of *H. reaumuri* in the loessial valley (a) and rocky slope (b).

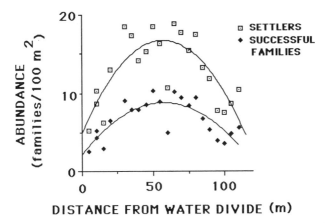

Figure 11.2. Spatial abundance pattern of *H. reaumuri* along the rocky slope. Each point represents a 15-year mean (1973–1987) for each row. Both curves are significant ($R = 0.8$ in both cases; $p < 0.01$).

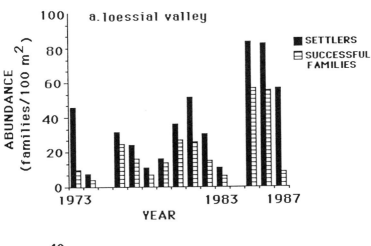

a

b

M. Shachak and S. Brand

A

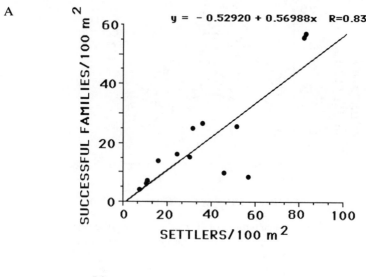

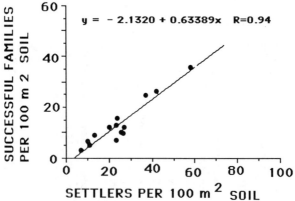

Figure 11.4. Temporal survivorship (successful families/settlers) of *H. reaumuri* in the loessial valley (A) and rocky slope (B). Each point represents a year.

valley populations fluctuated from 4.1 to 56.8 families/100 m². At the same time, rocky slope populations fluctuated from 1.7 to 18.9 families/100 m².

Settler and Successful Family Relations

We found high correlations between successful families and settlers for each generation (Fig. 11.4). On the average, about 58% (±21%) of the settlers in the loessial valley and about 54% (±13%) on the rocky slope survived. These differences are insignificant ($p > 0.5$, paired t test).

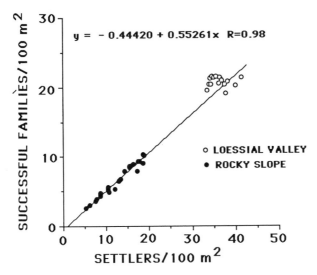

Figure 11.5. Spatial survivorship (successful families/settlers) of *H. reaumuri* in the loessial valley and rocky slope. Each point represents a row. The equation refers to the rocky slope.

The similarity and stability of the annual successful family/settler ratio (*survivorship*) indicates that relative survivorship of each generation was area- and density-independent. The survivorship was similar in the loessial valley and the rocky slope despite the differences in abundance. In both areas, survivorship was independent of year quality. During years with high and low densities of settlers, survivorship was similar (Fig. 11.4).

When we compared the survivorship among rows for 15 years, we found that the mean row-specific survivorship in the loessial area was 57% (±4%), which is significantly higher than the survivorship on the rocky slope (51 ± 4%) ($p < 0.001$). This finding implies that assessment of site quality for settling is better in the loessial valley than on the rocky slope. In addition, we found a row-specific survivorship pattern on the rocky slope with no apparent pattern for the loessial valley (Fig. 11.5). The pattern on the rocky slope is a positive relation between successful families and settlers ($r = 0.98$), which implies that on the rocky slope, despite the high variability in *H. reaumuri* abundance among rows, survivorship was similar. However, in the loessial valley, owing to homogeneity in abundance, we did not find an abundance–survivorship relation.

In summary, survivorship (successful family/settler ratio) is a relatively stable characteristic of *H. reaumuri* population dynamics in both time and space. This pattern contrasts with abundance, which is variable.

M. Shachak and S. Brand

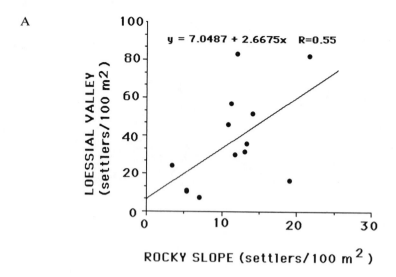

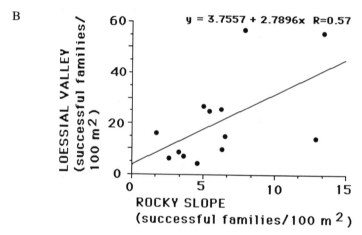

Figure 11.6. Abundance correlations between settlers (A) and successful families (B) of *H. reaumuri* in the loessial valley and rocky slope. Each point represents a year.

Loessial Valley and Rocky Slope Correlations

To determine if there is a relation between population densities of settlers and successful families on the rocky slope and loessial valley, we plotted the mean yearly density of the loessial area with that of the rocky area. The results (Fig. 11.6) depict the following trends.

1. There were significant correlations between year-specific settlers and successful families in each of the two landscape units ($p < 0.05$ in both cases). However, only about 32% of the variation was explained by year-specific conditions. This finding indicates that populations in both areas were controlled by the coarse-scale year-specific, and local, area-specific conditions.
2. The average Loessial valley/rocky slope abundance ratios were 2.7 for settlers and 2.8 for successful families. The ratios were much lower if densities on the rocky slope were calculated for soil cover: 1.4 and 1.5, respectively, for settlers and successful families. Even when densities per unit soil cover were compared, the densities in the loessial area were significantly higher ($p < 0.05$ in both cases, paired t tests) than in the rocky area for both settlers (37.4 ± 25.4 versus 22.3 ± 10.3 families/100 m^2 soil cover) and successful families (20.8 ± 17.4 versus 11.9 ± 6.9 families/100 m^2 soil cover).

Rock, Soil, and Abundance

We tested whether variability in isopod abundance was related to row heterogeneity in terms of the rock/soil ratio on the rocky slope (Table 11.1, Fig. 11.7). We found a positive relation between rock/soil ratio and abundance of setters ($R = 0.72$; $p < 0.01$) and successful families ($R = 0.67$; $p < 0.01$), which suggests that the rock outcrop near the soil increased soil quality in terms of isopod abundance.

Variability Patterns

We looked for variability patterns for two reasons: (1) to determine the heterogeneity–variability relation; and (2) to test the hypothesis that system and population processes can be inferred from variability (Connell and Sousa, 1983; Kratz et al., 1987).

Year-Specific Variability

Year-specific variabilities (CVs) among rows for settlers and successful families of the loessial area were significantly lower than on the rocky slope ($p < 0.001$ in both cases) (Table 11.2), which implies that *H. reaumuri* perceived the loessial area as relatively more homogeneous than the rocky slope during a specific year. There was an insignificant difference ($p > 0.05$) in the variability between successful families and settlers in the

A

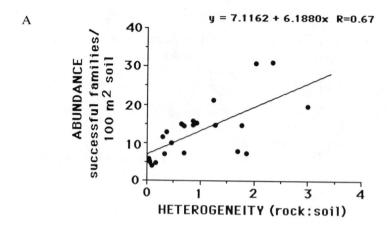

B

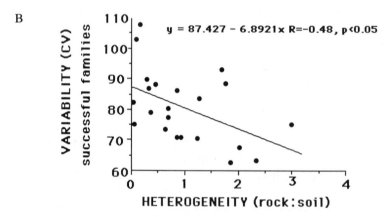

Figure 11.7. Relation between spatial heterogeneity (rock/soil) and (A) abundance and (B) variability of *H. reaumuri* on the rocky slope. Each point represents a row.

rocky area, whereas successful family variability was significantly higher ($p < 0.001$) than for settlers in the loessial area. This finding indicates that it was more difficult for the isopods to locate safe sites in the relatively homogeneous loessial area than on the heterogeneous rocky slope.

In addition, we found that relative variability in the loessial valley was significantly inversely related to settler density ($R = 0.69$; $p < 0.05$) for both settler and successful families, whereas there was no significant relation on the rocky slope. During years with high settler density, relative variability among rows in the loessial area is lower than during years with low settling density. If we assume that the density of settlers during a given

Table 11.2. Year-Specific Variability (CV) of *H. reaumuri* Abundance in Two Landscapes in the Negev Desert, 1973–1987

	Loessial Valley		Rocky Slope	
Year	Settlers	Successful Families	Settlers	Successful Families
1973	24.4	42.4	55.6	86.5
1974	70.2	85.9	60.7	68.0
1975	—	—	55.3	48.4
1976	9.8	24.7	67.0	78.3
1977	30.8	35.6	65.4	67.1
1978	31.9	33.4	52.1	62.8
1979	24.8	27.9	61.6	61.1
1980	21.6	24.2	62.4	51.7
1981	20.1	34.0	50.9	56.3
1982	27.6	39.0	67.3	73.0
1983	36.5	40.8	79.7	77.1
1984	—	—	—	69.3
1985	8.7	10.8	63.4	60.6
1986	13.8	23.0	60.0	64.7
1987	16.9	37.8	68.8	84.9

—, no data available.

year is related to year quality, from the *H. reaumuri* perspective the rows in the loessial valley were perceived as more heterogeneous during low quality years. On the rocky slope the CVs were similar ($p > 0.05$) for high and low quality years, which means that the isopods perceived the heterogeneity of the area in high and low quality years similarly on the rocky slope only.

Row-Specific Variability

Row-specific relative variabilities for settlers and successful families in the loessial area and rocky slope showed that the mean temporal variability (CV) of *H. reaumuri* abundance for 15 years in the loessial area was significantly ($p < 0.01$) higher (settlers 71.9, successful families 88.5) than on the rocky slope (settlers 62.2, successful families 78.5) (Table 11.3). This result suggests that the quality of the rows for settling in the loessial valley through time were relatively more variable than on the rocky slope.

We tested the assumption that row-specific variability in the rocky area was related to row heterogeneity in terms of the rock/soil ratio (Fig. 11.7). We found a significant inverse relation between the rock/soil ratio and row-specific variability ($R = 0.48$; $p < 0.05$).

To summarize our results, three trends emerged on the relation of population abundance and variability:

Table 11.3. Row-Specific Variability (CV) of *H. reaumuri* Abundance in Two Landscapes in the Negev Desert, 1973–1987

Row	Loessial Valley		Rocky Slope	
	Settlers	Successful Families	Settlers	Successful Families
1	64.1	69.4	50.0	62.7
2	64.6	73.4	75.6	77.3
3	69.4	76.9	67.7	92.9
4	63.5	80.1	73.8	83.5
5	63.1	81.1	60.2	88.5
6	71.3	89.2	61.1	73.3
7	61.8	81.7	48.6	70.8
8	67.3	93.4	62.7	70.8
9	74.1	86.4	59.5	80.1
10	66.6	88.9	61.6	70.3
11	68.9	78.0	67.5	89.9
12	77.4	93.9	63.6	75.0
13	83.6	104.5	52.7	67.6
14	85.5	98.9	59.3	63.4
15	83.8	113.1	58.4	86.2
16	85.4	106.6	56.3	79.0
17			65.4	88.2
18			63.9	86.8
19			77.6	77.8
20			66.7	82.2
21			51.7	75.0
22			63.8	82.2

1. The row-specific mean and variability of *H. reaumuri* abundance through time was higher in the loessial valley than on the rocky slope.
2. The year-specific mean *H. reaumuri* abundance was higher in the loessial valley than on the rocky slope. However, the variability was lower.
3. There was, a direct relation between the rock/soil ratio and *H. reaumuri* row-specific abundance and an inverse relation between the rock/soil ratio and variability.

Discussion

Hemilepistus reaumuri populations responded to spatiotemporal heterogeneity with high stability in survivorship despite high variability in abundance. Our objective is to suggest a unifying process that explains the stability in survivorship, variability in abundance, and environmental heterogeneity for *H. reaumuri*. In addition, based on our case study, we suggest some general conclusions on the relation between environmental heterogeneity and population variability.

Heterogeneity and Survivorship

We found two similar spatiotemporal trends for settler and successful family abundances: (1) On the rocky hill there was relatively high abundance of *H. reaumuri* at midslope that decreased toward the upper and lower parts of the slope (Fig. 11.2). (2) There were relatively higher abundances of isopods in the loessial valley than on the rocky hill (Fig. 11.6).

Despite the apparent similarity between settler and successful family trends, the two population variables are controlled by different processes. Settler abundance is the consequence of the individual isopod's decision where to settle after a site selection process (Shachak and Brand, 1988). The abundance represents isopod assessment of area quality for settling. High settling density implies that individuals assessed high numbers of safe sites in the area. Site selection is characterized by high mortality, as only about 12% of the potential settlers select sites. We do not know the environmental cues used by *H. reaumuri* for deciding where to settle; however, their densities are controlled by their own decisions (Shachak and Brand, 1988).

Abundance of successful families represents the relation between isopod assessment of the area and its real quality. Isopods that make good safe site assessments complete their life cycles. The number of safe sites reflects area quality. Of the animals that settled, about 57% in the loessial valley and 51% on the rocky slope properly assessed site quality and completed their life cycles.

Site quality assessment by the individual is a precise (Shachak and Brand, 1988) and stable (Fig. 11.6) process. This process controlled the similarity in the observed pattern of settlers and successful families on the rocky slope (Fig. 11.2). Stability in survivorship is an indication of the degree to which the animals recognize environmental heterogeneity. We found survivorship on the rocky slope to be more stable than in the loessial valley, which indicates that the ability to assess the area is greater in the more heterogeneous landscape.

At settling, the isopods have the same spatial distribution pattern as the successful families on the rocky slope. These patterns are related to safe site densities (Shachak and Brand, 1988). A safe site is a location that maintains at least 6 to 10% soil moisture at a depth of about 50 cm during the dry season (Shachak, 1980; Coenen-Stass, 1981). Safe site density is therefore related to the soil moisture regime. The better the soil moisture regime of an area, the higher is the probability of providing a large number of safe sites.

We tested for the temporal relation between the water regime, as defined by annual rainfall amount, and *H. reaumuri* abundances for the rocky hill and the loessial valley. The correlations in both cases were insignificant. Therefore we tried to determine other factors that could explain isopod abundance and variability in relation to soil moisture regime.

Hydrological studies on the rocky slope showed that soil moisture regime is related to rainfall redistribution due to runoff as well as rainfall quantity and intensity (Yair and Lavee, 1982, 1985). Soil moisture regime at a given site is controlled by the runoff water volume generated above the site that infiltrates into it. On the slope, processes of runoff generation and infiltration create a typical spatial pattern of soil moisture distribution along the slope. The pattern is relatively high water concentration at mid-slope, decreasing toward the upper and lower slopes (Yair and Lavee, 1982).

The patterns of settlers and successful families along the slope (Fig. 11.2) are similar to the soil moisture distribution pattern, indicating a relation between water redistribution and population processes. Therefore we suggest the following hypothesis as an explanation for *H. reaumuri* spatiotemporal abundance and variability. Year-specific rainfall and area-specific runoff water, i.e., spatiotemporal heterogeneity, control the distribution and variability of safe sites. Assessment of safe site distribution by settlers determines successful family abundance and variability.

Heterogeneity and Variability

A question arises as to the possibility of explaining two apparently contradictory variability patterns by the above hypothesis. The two patterns are as follows: (1) Year-specific variability is relatively higher on the rocky slope than in the loessial valley; and (2) row-specific variability is relatively lower on the rocky slope than in the loessial area. To explain area- and year-specific variability of successful family abundances, we use two terms, frequency (f) and magnitude (m), to describe runoff properties.

The loessial valley is part of a large basin that experiences flooding infrequently. There may even be years with no flooding. However, when the area is flooded it may remain submerged for a relatively long period. Thus the area experiences low frequency in runoff but a high magnitude of resource input when water is available.

The rocky hill exhibits different runoff characteristics. Soil underlying rocky outcrops receives runoff water in addition to rain after most rain events (Yair and Lavee, 1982). Soil closer to the rocky outcrops absorbs runoff water more frequently and in greater magnitude than soil farther away from the rock strata. However, the quantity of runoff water generated on one slope does not reach the magnitude of runoff generated in a large basin such as the loessial valley.

We suggest that *H. reaumuri* spatiotemporal abundance and variability is related to the frequency and magnitude of runoff infiltration into the soil. The year-specific number of sites is controlled by the number of runoff events and water volume generated during each event. We assume a direct relation between the number of events, the water volume produced during each event, and the probability of a soil patch becoming a safe site for settlers. Furthermore, we assume the following relation among frequency and magnitude of runoff and the abundance and variability of *H. reaumuri*: direct between the product of frequency and magnitude ($f*m$) and abundance, and inverse between frequency (f) and variability. To test the assumption, we relate variability on the rocky slope and differences in spatiotemporal variability in the loessial valley and on the rocky slope to site-specific runoff properties.

Variability on the Rocky Slope

Studies on the rocky slope revealed that the magnitude and frequency of runoff infiltration into a soil patch is positively related to the size of the rocky area that generates most of the runoff and negatively related to the soil volume into which it infiltrates (Yair and Lavee, 1982). It is not possible to measure frequency and magnitude of runoff generation at each potential site for isopod settling. Therefore the rock/soil ratio, which is our heterogeneity index, is also a valuable index for site quality estimation in terms of runoff.

The results (Fig. 11.7) supported our assumption on the relations among frequency, magnitude, abundance, and variability. The mean of *H. reaumuri* abundance is directly related to the rock/soil ratio, and the variability is inversely related to the ratio. This finding implies a direct relation between abundance and the frequency magnitude product ($f*m$) and an inverse relation between variability and frequency (f) of runoff. It suggests that the relations among heterogeneity, abundance, and variability are controlled by the rock/soil ratio via the rainfall–runoff relation.

Loessial Valley and Rocky Slope Temporal Variability

Evenari et al. (1983) found that runoff frequency and magnitude is scale-dependent. Small watershed runoff generation and infiltration are characterized by relatively small magnitude but high frequency in comparison to large watersheds. We therefore, suggest that on the rocky slope *H. reaumuri* abundance and variability are responses to small-scale watershed processes, whereas in the loessial valley they are responses to large-scale drainage basin processes that collect water from many small watersheds.

The relatively higher product of frequency and magnitude ($f*m$) in the loessial valley is mainly due to a high magnitude and explains the higher abundance of isopods. The lower frequency explains the higher variability.

The population properties on the rocky slope are responses to small-scale heterogeneity, i.e., rock–soil spatial distribution. Population dynamics in the loessial valley are controlled by water accumulation in the soil from many hill slopes. It is a large-scale landscape response to heterogeneity.

Loessial Valley and Rocky Slope Spatial Variability

The frequency and magnitude assumptions can be used also to interpret our findings of relatively higher year-specific spatial variability on the rocky slope than in the loessial valley. The loessial valley is relatively homogeneous in terms of topography and soil cover. Thus a large amount of runoff causes relatively uniform water distribution among rows, which reduces year-specific spatial variability in *H. reaumuri* abundance. The uniformity is emphasized when large quantities of water are collected in the valley.

On the other hand, the rocky slope may be viewed as encompassing a diversity of frequencies and magnitudes of runoff events. It is due to the heterogeneity in rock–soil spatial distributions. Such heterogeneity creates the observed spatial variability.

Conclusions

Isopod abundance and variability are perceived as interactions among site selection, surface heterogeneity, and spatiotemporal soil moisture distribution. We can summarize the relation between water resource distribution processes and *H. reaumuri* population dynamics into a general model (Fig. 11.8). The model provides insight into the central questions of the heterogeneity–population dynamics relation. We stress the following points.

1. Environmental heterogeneity and population variability are positively related. Spatial variability is higher in the more spatially heterogeneous area (rocky slope), whereas temporal variability is higher in the more temporally heterogeneous area of the water regime, the loessial valley.
2. Spatial and temporal heterogeneity are interrelated. Their combined effect may increase or decrease variability. On the rocky slope the high spatial heterogeneity, due to rock and soil distribution, decreases the temporal population variability, whereas the spatial location of the loessial valley increases temporal population variability.
3. The abundance–variability relations are area-dependent. In the loessial valley abundance and variability are positively related, whereas on the rocky slope they are negatively related.
4. Site selection is the organism's response to environmental heterogeneity. Survivorship stability depends on the degree of site selection.

If population abundance and variability are mainly controlled by spatiotemporal resource availability, which is in turn controlled by environmental heterogeneity, the implications of our study are as follows.

1. Redistribution of a limiting resource, such as water redistribution in a desert, could be of primary importance in controlling spatiotemporal population abundance and variability. Spatial and temporal population

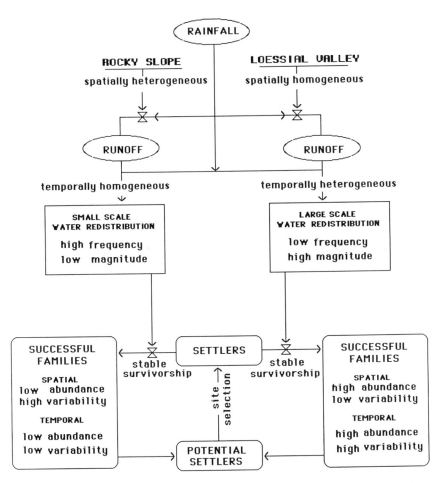

Figure 11.8. Relations among heterogeneity, scale, and population dynamics of *H. reaumuri*. Variables and processes at the center of the diagram (rainfall, settlers, potential settlers, site selection) are common to both landscapes. Variables and processes at the left (rocky slope) and right (loessial valley) are area-specific. "High" and "Low" are relative terms used for comparison of the loessial valley and rocky slope.

variabilities are expected to be positively related to the heterogeneity of resource redistribution in space and time.

2. Where population variability is concerned, both spatial and temporal heterogeneity and their relations should be considered because both phenomena control distribution of the same resource. However, the combined effect may increase or decrease the effect of each independently.

3. When site quality is measured in terms of population responses, both

abundance and variability should be considered because each factor may be positively or negatively related to the other. The abundance and variability relation depends on the spatiotemporal frequency and magnitude of available resources.

4. Variability in survivorship in a heterogeneous environment is inversely related to site selection. High ability to assess site quality results in low variability in survivorship.

In summary, we suggest that there is no general predictive relation among environmental heterogeneity, population abundance, and variability. Each relation depends on the specific processes that integrate environmental heterogeneity, resource distribution, population abundance, and variability.

Acknowledgments. We thank Sonia Rozin and Zvi Ofer of The Blaustein Institute for Desert Research and students of the geography department of the Hebrew University of Jerusalem for their assistance in the field work. This chapter is publication #112 from The Mitrani Center for Desert Ecology, Blaustein Institute for Desert Research, Ben Gurion University, Sede Boqer Campus, Israel.

References

Coenen-Stass D (1981) Some aspects of the water balance of two desert woodlice, *Hemilepistus aphganicus* and *Hemilepistus reaumuri* (Crustacea, Isopoda, Oniscoidea). Comp Biochem Physiol 70A:405–419

Connell JH, Sousa WP (1983) On the evidence needed to judge ecological stability or persistence. Am Nat 121:789–823

Danin A (1970) A phytosociological ecological study on the northern Negev of Israel. PhD thesis, Hebrew University, Jerusalem

Danin A (1972) Mediterranean elements in rocks of the Negev and Sinai deserts. Notes R Bot Gardens Edinb 31:29–49

Evenari M, Shanan L, Tadmor N (1983) *The Negev: The Challenge of a Desert.* Oxford University Press, London

Hallet JG, Pimm SL (1979) Direct estimation of competition. Am Nat 113:593–600

Harper JL (1961) Approaches to the study of plant competition. In Malthorpe FL (ed) *Mechanisms in Biological Competition.* Symposium Society for Experimental Biology 15. Cambridge University Press, Cambridge, pp 1–39

Kotler BP, Brown JS (1988) Environmental heterogeneity and the coexistence of desert rodents. Annu Rev Ecol Syst 19:281–307

Kratz TK, Frost TM, Magnuson JJ (1987) Inferences from spatial and temporal variability in ecosystems: long term zooplankton data from lakes. Am Nat 129:830–846

Linsenmair KE (1972) Die Bedeutung familienspezifischer "Abzeichen" für den Familienzusammenhalt bei der sozialen Wustenassel *Hemilepistus reaumuri* (Crustacea, Isopoda, Oniscoidea). Z Tierpsychol 31:131–161

Olsvig-Whittaker L, Shachak M, Yair A (1983) Vegetation patterns related to environmental factors in a Negev Desert watershed. Vegetatio 54:153–165

Random House Dictionary of the English Language. Random House, New York

Real L, Rathcke BJ (1988) Patterns of individual variability in floral resources. Ecology 69:728–735

Shachak M (1980) Energy allocation and life history strategy of the desert isopod, *Hemilepistus reaumuri*. Oecologia 45:404–413

Shachak M, Brand S (1988) Relationship among settling, demography and habitat selection: an approach and a case study. Oecologia 76:620–627

Shachak M, Yair A (1984) Population dynamics and the role of *Hemilepistus reaumuri* in a desert ecosystem. In Sutton SL, Holdich DM (eds) *The Biology of Terrestrial Isopods*. Symposium of the Zoological Society of London 53. Oxford Science Publication, London, pp 295–314

Shachak M, Chapman EA, Steinberger Y (1976) Feeding, energy flow and soil turnover in the desert isopod, *Hemilepistus reaumuri*. Oecologia 24:57–69

Shachak M, Steinberger Y, Orr Y (1979) Phenology, activity and regulation of radiation load in the desert isopod, *Hemilepistus reaumuri*. Oecologia 40:133–140

Sokal RR, Rohlf FJ (1981) *Biometry*. 2nd ed. WH Freeman, San Francisco

Sousa WP (1984) The role of disturbance in natural communities. Annu Rev Ecol Syst 15:353–391

Yair A, Danin A (1980) Spatial variation in vegetation species as related to the soil moisture regime over an arid limestone hillside, northern Negev, Israel. Oecologia 47:83–88

Yair A, Lavee, H (1982) Factors affecting the spatial variability of runoff generation over arid hillsides, southern Israel. Isr J Earth Sci 31:133–143

Yair A, Lavee H (1985) Runoff generation in arid and semiarid zones. In Anderson MG, Burt TP (eds) *Hydrological Forecasting*. Wiley, London, pp 183–220

Yair A, Shachak M (1987) Studies in watershed ecology. In Berkofsky L, Wurtele MG (eds) *Progress in Desert Research*. Rowman and Littlefield Publishers, Totowa, New Jersey, pp 145–193

12. Ecological Consequences of Heterogeneity of Consumable Resources

Shahid Naeem and Robert K. Colwell

The business of consuming resources is of paramount importance to the biology of all species. Whether consuming inorganic resources, e.g., photons or sulfur, or consuming organic resources, e.g., other organisms or their products, virtually every aspect of the biology of a species is affected by the quantity, quality, and availability of consumable resources. In the simplest systems—such as a uniformly distributed population under homogeneous environmental conditions consuming resources that renew at constant rates (as in a chemostat)—population growth is generally a straightforward outcome of consumption and the metabolic processes that govern the production of new individuals. What happens to populations, however, when systems are complex? That is, what happens when resources vary in time or space or when resources are consumed by other species in the community? We address this question by examining the ecological consequences of heterogeneity of consumable resources in multi-species systems.

We define consumable resources as food, nutrients, vitamins, minerals, oxygen, and carbon dioxide but not resources such as mates, space, nest sites, or time. To simplify the text, we refer to consumable resources simply as "resources." When ambiguity might arise from this usage, we are more explicit and include the adjective "consumable."

Resources and Environmental Heterogeneity

Laboratory studies can approach environmental constancy through manipulation of environmental factors, and theoretical studies can eliminate troublesome environmental variation in models by declaring conditions to be constant. Field studies, however, must contend with the vagaries of real environments. This difference between laboratory/theoretical and field studies potentially accounts for the differences frequently observed between results of laboratory/theoretical and field studies. For this reason, ecologists have devoted considerable effort to studying the ways in which environmental heterogeneity affects biological communities. Several authors have reviewed these efforts (Giesel, 1976; Levin, 1976; Wiens, 1976, 1977; Connell, 1978; papers in Steele, 1978; Strong, 1983; Pickett and White, 1985; Chesson, 1986; Chesson and Case, 1986).

Until recently, greater effort has focused on the consequences of heterogeneity of physical aspects of the environment with less attention to heterogeneity of consumable resources (Tilman, 1987). Given that consuming resources or being consumed as a resource are critical activities in the biology of all species, what accounts for this bias? Community ecology has been traditionally a study of densities, or the distribution and abundance and dynamics of species, often without quantitative assessments of resources. The reason for this failing is simple: Resources are difficult to study.

If the resource is a simple abiotic nutrient, e.g., water or inorganic nitrogen, identifying and manipulating the resource may be possible. Such situations are frequently limited, however, to studies using artificial microcosms (see examples below). For most organisms, use of laboratory microcosms is not a feasible method. Biotic resources are more difficult to work with. As MacArthur (1972) noted (see below), a biotic resource may consist of several species or only part of a species, which makes experimental work difficult. Although not impossible, one cannot easily quantify or manipulate phloem for a polyphagous plant sucker, plankton for guilds of oceanic filter feeders, availability of insect prey for guilds of insectivorous vertebrates, host tissues for a guild of parasites, or fruits for migrating flocks of tropical frugivorous birds.

Another difficulty with studying consumable resources is that they may not be discrete. That is, two or more resources may be perfectly or imperfectly substitutable or perfectly or imperfectly complementary to one another (MacArthur, 1972; Leon and Tumpson, 1975, Tilman, 1980, 1988). The degree of substitutability or complementarity of resources is important because it affects the mechanism by which species interact (Leon and Tumpson, 1975; Tilman, 1980).

Despite these difficulties, a theoretical and experimental literature on the ecological consequences of heterogeneity of consumable resources is growing rapidly. In this survey we examine this literature in an effort to delineate some of the special consequences of heterogeneity of consumable resources in contrast to the broader, more general ecological consequences of environmental heterogeneity dealt with elsewhere in this book.

General Consequences of Resource Heterogeneity

Investigation of the effects of resource heterogeneity on community structure imposes considerable demands for both researchers and their study systems. Examining these demands by hypothetical example can serve as a way of outlining some of the ecological consequences of resource heterogeneity with which empirical and theoretical studies must contend.

Figure 12.1 shows a simple system consisting of three patches at three different times. Each time period represents a "season" within a "year" of three time periods. The level of resources in each patch is indicated by the shaded area in the circle. This figure examines the consequences of an intermediate degree of constancy and contingency of resources among patches over space and time, i.e., intermediate predictability (sensu Col-

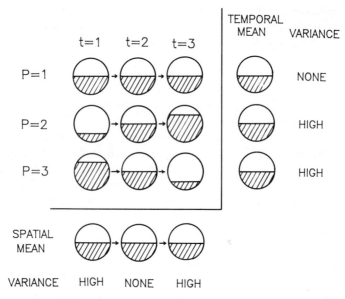

Figure 12.1 Resource heterogeneity in a simple three-patch system. Shown are three patches (P) represented by circles, each with its level of resources indicated by a cross-hatched area. The patches are shown at three time periods (t). The mean and variance of the resource level for each patch is indicated by the circles in the margins: spatial at the bottom, temporal on the right. Note that all patches have the same mean resource level. Furthermore, the spatial mean resource level is the same at any given time. The patches, however, vary greatly with respect to individual variation in resource level.

well, 1974). We treat these three patches as the regional system; that is, this set of patches is isolated from other sets of patches in other regions in a global habitat. Movement between regions can occur but is rare.

Even in this simple system we can identify several types of heterogeneity or sources of variation in resources: (1) temporal (among column) variability within patches (rows), among patches, and among regions (sets of patches); (2) spatial (among row) variability within seasons (column), among seasons, and among years; and (3) variability among patches and seasons considered jointly.

Note that the mean patch resource level in this figure is the same for all three patches at any season (column means) or for any patch over all three seasons (row means). This apparent homogeneity of resources over space and time, when only means are considered, hides the considerable variability within seasons and within patches, as well as variation in the *degree* of variability.

In what ways might such variability affect species? Because species differ in their movement patterns among patches, their consumption rates, and their efficiencies of conversion of resources to new individuals, the sources of resource variation outlined in this system have different consequences for the growth, distribution, and abundance of species. Figure 12.2 shows how individuals of different species may experience different levels of resources when species differ in their mobility and capacity for habitat selection. That is, dispersal characteristics and life history strategies determine how "grainy" resources appear from the point of view of the organism (MacArthur and Levins, 1967; Levins, 1968; Vandermeer, 1970). The least mobile species, which are confined to a single patch (species A and B in Fig. 12.2), experience the heterogeneity of resources in a coarse-grained fashion. Mobile species, in contrast, may experience different degrees of resource heterogeneity in a fine-grained fashion (species C and D in Fig. 12.2). The mean and variance of the resource level experienced by a species is therefore partly a function of its capacity for movement and habitat selection. Thus even among species that are identical in all other aspects of their biology, differences in this capacity may result in differences in population growth.

What would an experiment require to examine thoroughly the consequences of all aspects of resource heterogeneity? Essentially, the experiment would follow a repeated measures multivariate analysis of variance (MANOVA) design with replication. [See Morin (1986) for discussion on the use of this method for community-level studies and Morrison (1976) for a technical discussion]. Two types of manipulation are necessary—manipulations of densities (Bender et al., 1984) and manipulations of resources (Tilman, 1987)—to distinguish between density effects, resource effects, and interaction between the two.

To examine resource effects in an idealized case, experimental manipulation of resources, for the purposes of MANOVA, might require at a

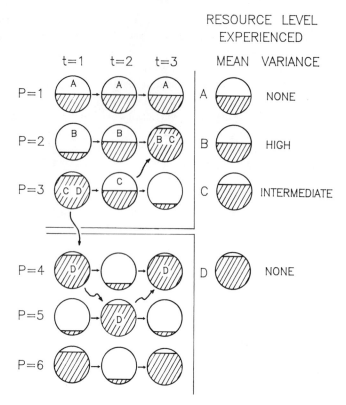

Figure 12.2. Consequences of resource heterogeneity with respect to different movement patterns of species among patches in a simple system. See Figure 12.1 for an explanation of this system. Letters represent species that experience the heterogeneity of resources differently. Species A and B are sedentary. Species C and D can move from patch to patch and select the best patch available during a given time interval. Species D is the most mobile and can move to a different region, as illustrated by the second set of patches below. As in Figure 12.1, the temporal means for all patches are the same, and spatial means are the same for all time periods. Here, however, species experience different resource levels and different variation in resource levels, as indicated in the right-hand columns of circles. These differences can ultimately lead to differences in growth of populations and affect interactions among species.

minimum three sets of three patches at three initially different resource levels (e.g., low, intermediate, and high). Resources are manipulated to provide five temporal patterns of change in resources (five treatments): (1) high to low; (2) low to high; (3) constant at low levels; (4) constant at intermediate levels; and (5) constant at high levels. These manipulations test for the effects of temporal variability in resources as well as the effects of variability in initial conditions. In summary, the requirements are three replicates of five treatments, requiring 15 patches, each censused three times. The choice of three as a minimum allows evaluation of the linearity of interactions. In actuality, the appropriate number of replicates is a function of the desired level of type II error and requires specific knowledge of the variance of the experimental system (Winer, 1971; Hurlbert, 1984). The actual number of replicates required is likely to be higher than three.

To examine density effects in an idealized case, several experimental manipulations of densities are required. Consider a study community consisting of two species, A and B. The experimental manipulations require use of monospecific (A or B) treatments at different initial densities (e.g., high, intermediate, and low). Additionally, use of mixed-species treatments (A and B) is necessary to test for interactions between A and B. The mixed-species treatments consist of combinations of A and B in densities representing all pairwise combinations of three densities. For this simplest of idealized cases, in summary, density manipulations require six monospecific treatments and nine multispecies treatments, each replicated three times.

Thus the investigation of a simple three-patch, two-species system, using no more than three levels for each factor, requires 225 patches in which a total of 675 censuses must be made over the course of the experiment. Finally, among-year and among-region variability could also be studied. To examine these sources of variability would require repeating the experiment (sticking to our convention of threes) in three geographically distinct regions over 3 years, bringing our total number of patches to 675 and number of censuses to 2025. It is a formidable set of requirements, even for laboratory-microcosm studies. For field work that frequently entails working on fragile, intricate habitats, with species sensitive to being manipulated, and with consumable resources that are not easily identified or manipulated, such an experiment might well be impossible to conduct. Clearly, heterogeneity of resources complicates life for study organisms and researchers alike.

Theoretical Studies

Ecological theory has compiled a rough inventory of the many consequences of environmental heterogeneity, of which heterogeneity of consumable resources is a special case. In this section we survey some of the theory that touches on the issues of resource heterogeneity. In some cases we include models that broadly examine environmental heterogeneity but are nevertheless readily interpretable in terms of the more specific phenomenon of resource heterogeneity.

Most classic models of population growth do not include parameters specifically for resources (Leon and Tumpson, 1975; Lynch, 1978; Tilman,

1980). That is, models have intrinsic rates of population increase (r_i) and carrying capacities (K_i) that, although implicitly related to resources, are not resource parameters, nor are they solely determined by resources. Classic competition models include terms for the interaction (alphas); but, again, although the interaction is resource-mediated, the role of resources is rarely implicit. [MacArthur (1972, pp. 51–58), however, did analyze a two-species competition model with explicit renewal equations for resources.] Classic predator–prey models include parameters such as an attack rate and a coefficient of conversion of prey biomass into predator biomass. These parameters, however, are assumed constant and reflect neither the potential variability in a resource (prey) species' ability to avoid or escape consumers (predators) nor variability in resource quality. Furthermore, parameters for the resources that prey consume are not included in the models; they too are assumed constant.

In classic treatments of biotic interactions, the effect of one species on the growth of another is usually modeled as the product of the *densities* or frequency of co-occurrence of interacting species, without terms for resources. The assumption that these coefficients are fixed implies an assumption that the consequences of species interacting with one another are directly related to densities of the interacting species—hence the focus on density manipulations in empirical work.

Two general, strongly influential ecological principles have emerged from this body of theory: the principle of competitive exclusion (Hardin, 1960) and, by extension, the corollary that n limiting resources support n or fewer species of consumers (Levin, 1970; MacArthur, 1972). These principles are derived from assumptions that resources and interaction coefficients are constant and linear (Ayala et al., 1973; Armstrong and McGehee, 1976a, 1980; Kerfoot et al., 1985; Smith et al., 1988). When the assumptions of constancy are relaxed, these principles do not hold.

The classic Lotka-Volterra type models have been modified in three principal ways to examine the consequences of resource heterogeneity. One way is by giving resources a "life of their own" so that they grow and are affected by the consumption of two or more consumer species; that is, resources are governed by intrinsic biotic processes (e.g., MacArthur, 1972; Koch, 1974b; Armstrong and McGehee, 1976a; Levine, 1976; Hsu et al., 1978; Hsu and Hubbell, 1979; Abrams, 1980, 1988; Hsu, 1981; Vance, 1984). A second way is to allow resources to vary but under the control of extrinsic environmental processes, such as seasonal effects (e.g., Hutchinson, 1961; Richerson et al., 1970; Grenney et al., 1973; Koch, 1974a; Levins, 1979, Cushing, 1980; Butler and Waltman, 1981). A third way is by introducing explicit spatial and temporal patchiness to the resources (e.g., Cohen, 1970; Levins and Culver, 1971; Horn and MacArthur, 1972; Levin, 1974, 1978; Slatkin, 1974; Caswell, 1978; Shorrocks et al., 1979; Atkinson and Shorrocks, 1981; Hanski, 1981, 1982a,b; Gotelli and Simberloff, 1988). We review each of these approaches below.

Life of Their Own

Giving resources a life of their own consists in considering two species of consumers, most simply two species of predators competing for a single prey species (resource). These models frequently consist of three or more Lotka-Volterra population growth models in which growth of both predators and the prey are functions of each other's densities as well as their own intrinsic growth parameters. Armstrong and McGehee (1980) referred to this class of models as "biotic resource models." In a survey of this class of models, Caswell (1978) noted that most fell into a class he called "closed-equilibrium" models: "closed" because they permit no immigration or emigration and "equilibrium" because they include no parameters for stochastic processes, which means the models tend toward equilibrium values given certain initial conditions. A system of three equations behaves in a much more complex way, even under equilibrium conditions with no emigration or immigration, than systems of two equations. These more complex models provide more opportunities for coexistence than simple models.

Extrinsically Generated Fluctuations in Resources

Extrinsically generated fluctuations in resources can also permit coexistence of competing species (Hutchinson, 1961, Richerson et al., 1970; Grenney et al., 1973, Koch, 1974a; Cushing, 1980). Armstrong and McGehee (1980) called these models "abiotic resource models," and Caswell classified them as "closed-nonequilibrium" models. The models frequently invoke differential performance among consumers at different levels of resources. For example, one species might be a better competitor at high resource levels, whereas the inferior species may be reproductively superior at low resource levels. So long as the environment fluctuates between low and high resource conditions, the density of the superior competitor fluctuates. The competitively inferior species can recover from a population decline caused by competition by recruiting rapidly when the density of the superior competitor is low. Vance (1984) also demonstrated that coexistence could be obtained in a seasonal environment if the hierarchy of competition changes. His model achieves this result with a trade-off, one species being a better exploitative competitor in one environment whereas the other is a better interference competitor in another environment.

In a different manner, Levins (1979) also demonstrated that coexistence is possible if resources are variable. If species growth is a nonlinear function of resource density, and resources vary, then $\text{Var}(R)$, where $R =$ the density of the resource, functions as a second resource. This model offers a different view of consumers, where some "consume" the mean of the resource, and others "consume" components of the total variance of the resource (Chesson, 1986).

Spatial or Temporal Heterogeneity

Spatial or temporal heterogeneity, or open patchiness, when included in models also modifies the predicted Lotka-Volterra outcomes of species interactions, where "open" means that immigration and emigration occur among patches. These models, which fall into Caswell's (1978) scheme as "open-equilibrium" or "open-nonequilibrium" models, examine the effects of separating groups of interacting species into discrete habitat patches. The patches consist of two or more types, each type optimal for only one species or a fraction of the species in the community. Levin (1974) noted that simple models of this kind (e.g., Cohen, 1970; Levins and Culver, 1971; Horn and MacArthur 1972) ignore density- or frequency-dependent dynamics. Instead, they examine the frequency of co-occurring species on patches; and in the case of competition, the probability of patch extinction is higher on those patches where competitors co-occur. More complex models (e.g., Levin, 1974; Shorrocks et al., 1979; Atkinson and Shorrocks, 1981) allow individual patches to follow Lotka-Volterra dynamics locally, but densities are also affected by immigration, emigration, or periodic local extinction due to abiotic factors. Both simple and more complex models achieve the same results (Caswell, 1978). In some models movement is modeled as random among patches, whereas in others movement follows diffusion gradients (Levin, 1974).

These models cover physical aspects of habitat patchiness as well as resource patchiness, but their conclusions are relevant to resource heterogeneity. The models demonstrate that patchiness operates to increase the diversity of coexisting species in four ways: (1) it provides spatial or temporal refuge for prey or competitively inferior species; (2) it provides opportunities for priority effects to allow competing species to persist regionally; (3) if patches differ, it permits competing species to coexist on their respective optimal patch types; and (4) it can create nonequilibrium conditions that persist for so long that equilibrium effects of Lotka-Volterra dynamics are effectively irrelevant (e.g., Sale, 1977; Caswell, 1978; Connell, 1978; Powell and Richerson, 1985). With condition (1), "refuge" often means physical shelter, in the case of heterogeneity of the physical environment; but in the case of heterogeneity of consumable resources, a "refuge" may be, instead, a time or place where an organism has access to the resource without interruption or interference. Ecological consequences of such effects have been examined for interspecific competition (Schoener, 1974; Crowell and Pimm, 1976; Hallet and Pimm, 1979; Rosenzweig et al., 1984, 1985), but other interactions have been less well studied.

Regardless of the model type, resource patchiness permits more possibilities for coexistence of n species on fewer than n resources than is permissible with homogeneous resource distributions.

Empirical Studies

A full review of the empirical literature, which has grown rapidly over the years, is beyond the scope of this chapter. What follows is a *survey* of work that we believe demonstrates some of the consequences of resource heterogeneity as discussed in the previous sections of this chapter. We draw information from microbial, animal, and terrestrial plant studies to show that empirical results confirm many of the predictions of theory but consistently reveal unanticipated complexities.

At the end of the summary of each paper, or set of papers, we state a "lesson" that the empirical study provides with respect to heterogeneity of resources. We also list, in parentheses, where lessons of a given study overlap or are similar to the lessons derived from other studies.

Microbes: Bacteria

Experiments that are possible to conduct with microbes are impossible to carry out at the same scale using less dynamic, longer-lived, numerically less abundant plants or animals, especially given the current patterns of academic research and funding. With respect to the study of resource heterogeneity, microbes have an additional advantage over higher organisms. The advantage is that in many instances several species of microbes in an assemblage may differ only in their ability to use a single, simple, clearly defined, chemical or biochemical resource, which makes experimental manipulations of resources relatively straightforward. Despite these advantages, however, microbial ecology has been and continues to be predominantly a laboratory science. In situ studies are rare in microbial ecology (Brock, 1987). Thus most of our examples are laboratory studies.

Microbial ecology has long dealt with the importance of resources and resource heterogeneity in population dynamics (Fredrickson and Stephanopoulus, 1981; Veldkamp et al., 1984). As early as 1949, S. Winogradski (cf. Veldkamp et al., 1984) recognized that nutrient conditions strongly affected the structure of bacterial communities. He noted that under conditions of low nutrients "autochthonous" bacterial species capable of using nutrients in low concentration dominated the bacterial communities but were virtually absent under high nutrient conditions. This observation began a tradition of studying bacterial dynamics under differing nutrient conditions. The tool microbiologists developed for these studies is the *chemostat*, a laboratory container for bacterial growth in which all environmental variables and all resource inputs and outputs are under the experimenters' control (for review see Veldkamp, 1977). This method is a highly artificial means of examining microbes, but it has been useful for understanding the mechanisms of resource utilization, the dynamics of bacterial populations, and the patterns observed in natural systems.

Because the mouth is considered a natural chemostat (Rogers et al., 1987) results of chemostat studies on dental plaque communities may be applicable to other real-world situations. Rogers et al. (1987) examined growth of *Streptococcus milleri* and *Streptococcus mutans*, two bacteria found in dental plaque, and their biotic interactions in a laboratory chemostat. Under limiting arginine and glucose conditions, Rogers et al. obtained a stable coexistence of both species, with *S. mutans* being the numerically dominant species. When additional arginine was added, however, *S. milleri* rapidly outcompeted *S. mutans* until the added arginine was consumed, at which point *S. milleri* returned to its previous low population size. A similar addition (pulse) of glucose caused *S. mutans* to increase, but the effect of the pulse on *S. milleri* was negligible. These results indicated that when the first limiting resource (arginine) of the superior competitor (*S. milleri*) was no longer limiting, the organism could outcompete the inferior competitor (*S. mutans*) for the second limiting resource (glucose).

In a second experiment, glucose was added while arginine remained low. The additional glucose could not be used by *S. milleri* because the low levels of arginine held its population growth in check. *S. mutans*, however, could take advantage of the resource. In fact, when Rogers et al. added glucose and arginine simultaneously, *S. milleri* increased rapidly after the pulse of arginine, as before, but this time the additional glucose permitted *S. mutans* to persist, unlike the earlier treatment.

These results fit the theory of *n* limiting resources necessary to support *n* species. One complication, however, is that, although one or the other species was reduced to very low densities, in no case did competitive exclusion occur; instead, competitive equilibria were obtained. One other complication occurred. Changing the pH caused *S. mutans* and *S. milleri* to change places in numerical dominance. Physical aspects of the environment affect the outcome. Given that the dental plaque community is a complex community that experiences environmental and resource heterogeneity every time a meal is consumed and ecological disturbance when teeth are brushed, and that space, substrate properties, and other factors also affect these communities (Rogers et al., 1987), it is difficult to determine how applicable results from chemostats are to real mouths. Nevertheless, though dental plaque may not be as glamorous as the rocky intertidal zone or a tropical rainforest, the results provide an interesting lesson.

Lesson 1: *The principle that* n *species requires* n *or more limiting resources can apply to some simple systems for specific sets of environmental conditions.*

More complex, multispecies, multinutrient studies are also common in bacterial ecology (for review see Harder and Dijkhuizen, 1982). A particularly interesting example (Gottschal et al., 1979) involved diauxic species (species that can substitute one resource for another) in a three-species,

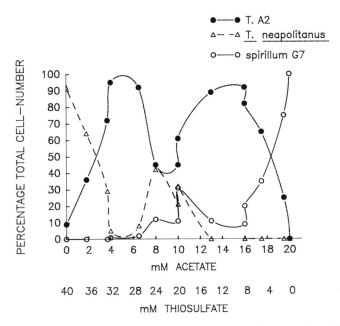

Figure 12.3. Effect of concentration of two resources on three species of bacteria (*Thiobacillus neapolitanus*, *Thiobacillus* A2, and spirillum G7) when grown together in a chemostat with thiosulfate and acetate as growth-limiting substrates. Relative abundances of the three species are plotted as percentage of total cells found in the chemostat after species densities and concentrations of substrates had reached a steady state. See text for further details. (From Gottschal et al., 1979.)

two-resource study. In a nutrient replacement series, where nutrients ranged from high thiosulfate/low acetate to low thiosulfate/high acetate conditions, Gottschal et al. (1979) monitored the growth of mixed assemblages of *Thiobacillus* A2, *Thiobacillus neapolitanus*, and *Spirillum* G7. *T. A2* is a facultative chemolithotroph that can use thiosulfate or acetate, *T. neapolitanus* is an obligate lithotroph requiring thiosulfate, and *Spirillum* G7 is a heterotroph requiring acetate. The results are shown in Figure 12.3. Note that when the resources are composed primarily of thiosulfate or primarily of acetate, the respective "specialist" outcompetes the other specialist, leaving, as classic theory predicts, two species utilizing two resources. A range of intermediate resource combinations, however, yield coexistence among all three species on two resources, and this result is contrary to the predictions of classic theory. Such a competitive equilibrium, however, is an unstable situation because it requires that resource levels remain constant. In nature, consumption is likely to change resource levels, and different resources are not likely to renew themselves at the same rate. Nutrients would therefore fluctuate regularly in the environment between states that favor different competitive equilibria and states that lead to competitive exclusion. It would not be surprising that assemblages of all three species would persist on two resources in a heterogeneous environment. This system seems ideal for testing ideas concerning substitutable and complementary resources (Leon and Tumpson, 1975; Tilman, 1980).

Lesson 2: *Even in simple systems, when resources fluctuate, more than n species can persist on fewer than n resources if species differ in their ability to use substitutable resources.*

Microbes: Mixed Assemblages

In the laboratory, Rashit and Bazin (1987) manipulated the level of environmental heterogeneity experienced by natural assemblages of microbes (bacteria and protozoa). They conducted their experiments, however, under both high resource and low resource conditions. The environmental heterogeneity consisted of periodic removal of portions of the fluid, including microbes, and replacement with fresh medium. Their results showed that this type of environmental fluctuation leads to a reduction of densities of all species, but whether biotic diversity decreases or increases depends on the level of resources. At low resource levels, environmental fluctuation takes its toll on all species and biotic diversity is reduced. When resources are high, however, those species that can recruit quickly, even if they are competitively inferior species or vulnerable prey species, can exploit the transient favorable conditions and grow rapidly while competition or predation pressure is low.

Rashit and Bazin (1987) offered the model illustrated in Figure 12.4. Their model assumes that diversity isoclines are a function of resource productivity and severity of environmental fluctuations, or

$$\frac{\delta D}{\delta P} = \frac{\delta D}{\delta F} = 0$$

where D = a measure of species diversity, P = a measure of resource productivity, and F = a measure of severity of environmental fluctuations. The isoclines differ for different sets of species that have different intrinsic rates of population increase (r). In Rashit and Bazin's (1987) study, for example, the sets of species were bacteria, a high-r set; flagellates, an intermediate-r set; and protozoan predators, a low-r set. Depending on the intersection points and shapes of the diversity isoclines, different levels of resource productivity support different communities at the same levels of environmental heterogeneity. Thus even if all patches were exposed to identical conditions of environmental heterogeneity, differences in resource productivity among the patches would result in differences in community composition among patches. Some support for these ideas can be found in the work of

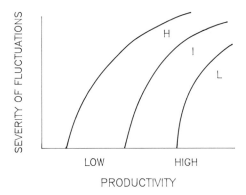

Figure 12.4. Model outlining the relations among biotic diversity, resource productivity, and disturbance caused by environmental fluctuations. Fluctuations consist in environmental effects that uniformly reduce populations of most species in the habitat, e.g., a severe disturbance. Productivity represents rate of resource renewal or input into the habitat. Curves represent diversity isoclines for various taxonomic groups (indicated by H, I, and L) given different combinations of degree of severity of environmental fluctuation and resource productivity. The assumption is that greater resource productivity in a habitat can compensate for severe environmental fluctuations, but the degree of compensation decreases with increasing severity of disturbance. Isoclines from left to right indicate species or taxonomic groups of species, with high (H), intermediate (I), and low (L) intrinsic rates of population growth (e.g., bacteria, flagellates, and protozoan predators, respectively). See text for further detail. (From Rashit and Bazin, 1987).

Jackson and Berger (1984), who examined differences in survival among protozoan species at different levels of resources.

Interestingly, similar results were obtained earlier by Cairns et al. (1971) by a simpler experiment. These authors looked at protozoa colonizing a spongy polyurethane substrate that could be periodically squeezed, eliminating portions of the protozoan assemblage, much as the removal regime in Rashit and Bazin's work. The manipulation resulted in greater protozoan diversity than with the controls (unsqueezed substrate).

Lesson 3: *The ecological consequences of physical environmental heterogeneity, such as that produced by disturbance, depend on resource levels.* (See also lessons 5 and 14.)

Microbes: Phytoplankton

Like bacteria, phytoplankton offers advantages for the study of ecological consequences of spatial and temporal resource heterogeneity (for review see Tilman et al., 1982; Powell and Richerson, 1985). From the vast phytoplankton literature, space allows us to consider only three studies that we believe demonstrate the importance of resource heterogeneity.

In laboratory microcosms, Turpin and Harrison (1979) examined the effects of temporal heterogeneity of nitrogen on mixed assemblages of phytoplankton. Each treatment received the same total daily nitrogen input but at different flow rates. One treatment received its nitrogen steadily throughout the entire day, a second received it in eight pulses, and a third received it in one pulse. The entire assemblage was affected, but two numerically dominant species were most strongly affected, *Skeletonema costatum* and *Chaetoceros simplex*. Under continuous addition of nitrogen, *C. simplex* dominated, but when the addition of nitrogen was discontinuous *S. costatum* dominated.

Similarly, in a series of laboratory microcosm experiments, Grover (1988) found that phosphorus nutrient pulses, but not steady-state nutrient input, facilitated coexistence of two competitors, *Fragilaria crotonensis* and a species of *Synedra*. Such nutrient fluxes are known to occur in nature (Tilman et al., 1982). Therefore this type of resource-mediated coexistence may be a common feature of phytoplankton communities.

Lesson 4: *Even when the long-term mean abundance of a limiting resource is kept constant, patterns of short-term heterogeneity can affect the distribution and abundance of species.*

Stross and Pemrick (1974) documented temporal variation among phytoplankton in nutrient uptake (which they called "clock control"). They described two idealized classes: (1) small cells that maintained, at high metabolic costs, the enzymatic machinery necessary for rapid nutrient uptake with short notice of a change in concentration (fast clock); and (2) large cells that responded to nutrient fluctuations more slowly (slow clock) because they produced enzymes necessary for nutrient uptake only after detecting nutrient increases. The latter class could outcompete the former under steady-state conditions, but their slow response permitted the former class to take advantage of nutrient fluxes before competitive effects took their toll. "Clock control" is a biological example of Levins' (1979) theory that $Var(R)$ can function as if it were a resource. In this case, large cells cannot use the pulses in the resources as effectively as can small cells.

Lesson 5: *Resource heterogeneity facilitates coexistence among competing species when species differ in their ability to use the variance of the resource.* (See also lesson 14.)

Terrestrial Plants

Bazzaz and colleagues (e.g., Bazzaz and Carlson, 1984; Zangrel and Bazzaz, 1984, Bazzaz and Garbutt, 1988) have shown that CO_2, water level, nutrient levels, and community composition affect plant community dynamics in complex ways. This work was necessarily conducted in environmental chambers where CO_2 levels could be reliably manipulated. All these studies demonstrated that the effects of increased CO_2 is dependent on plant metabolic type (i.e., C3 or C4) and that CO_2-level effects were seldom linear, the strongest effects with often occurring at intermediate CO_2 levels. Bazzaz and Carlson (1984) demonstrated that the effect of the CO_2 level depends on the water level. Zangrel and Bazzaz (1984) demonstrated that the effect of the CO_2 level depends on light levels and, to a lesser degree, on nutrient (nitrogen, phosphorus, and potassium) levels, especially when nutrients are low. Finally, Bazzaz and Garbutt (1988) demonstrated that species composition of the plant community affected species-level and community-level responses to CO_2, suggesting that the complex nonlinear interactions among CO_2, plant metabolic type, water, and nutrient factors are further complicated by effects related to community composition.

Lesson 6: *The effects of heterogeneity of one resource may depend on, or be a function of, the level of another resource.*

Inouye and Tilman (1988) examined plant community dynamics in old fields in Minnesota for a period of 4 years. They examined three factors; (1) historical factors (age of field since abandonment); (2) disturbance effects (one-half of the experimental plots were disturbed by tilling); and (3) the effects of variation in nutrients (nine treatments of twice-yearly additions of nitrogen fertilizer). They measured relative abundance (in biomass) of the plant species present and percent similarity (PS) among sets of replicate plots blocked to test for single factor effects and interactions among the three factors. The results were complex. All factors had significant effects. In general, however, the clearest consistent effect was that, over time, plant assemblages receiving different amounts of nitrogen diverged (among-treatment PS decreased). Year-to-year differences among replicate plots *within* treatments, however, showed either increasing PS (convergence), decreasing PS (divergence), or no clear pattern, depending on the history of the field or the presence or absence of disturbance at the onset of the experiment.

Lesson 7: *Ecological consequences of historical factors depend on past and current resource levels.* (See also lesson 3.)

Many terrestrial plants invade new habitats (resource patches) via clonal growth. Two types of strategy occur. Clonal ramets can "habitat-select" (e.g., Salzman, 1985, Eissenstat and Caldwell, 1988), or ramets can move indiscriminately into adjacent habitats (e.g., Hartnett and Bazzaz, 1983; Salzman and Parker, 1985). Thus in a community some plants may grow selectively into patches that have the optimal or nearest to optimal nutrients for the forward-moving ramets, and some plants may grow into suboptimal habitat by reallocating photosynthate and nutrients via rhizomes from ramets in more optimal habitats.

Lesson 8: *The effects of resource heterogeneity on the distribution and abundance of species may depend on the ability of colonizing species to habitat select.*

Animals: Arthropods: Herbivores

Organic nitrogen varies within plants, among plants, among regions, and on other scales (reviewed in Mattson, 1980). Because nitrogen is frequently limiting to animals, variation in nitrogen frequently affects the growth and reproduction of herbivores (Mattson, 1980). Nitrogen availability can also affect a plant's ability to produce plant secondary compounds (Mattson, 1980). Because plant susceptibility to herbivory is affected by the type and quantity of secondary compounds, which in turn affect the dynamics of herbivorous insects (e.g, Feeney, 1976; Rhoades and Cates, 1976), one would expect plant, herbivore, and herbivore consumer assemblages to be affected by variation in nitrogen (e.g., Price et al., 1980).

Fertilizer treatment provides a ready means by which nutrients, e.g., nitrogen, can be experimentally manipulated. Several multitrophic level studies have used this approach (e.g., Hurd et al., 1971; Hurd and Wolf, 1974; Valiela et al., 1975; Onuf et al., 1977; Vince et al., 1981; Strauss, 1987; see also references to the Park Grass Experiments in Tilman, 1982, and Digby and Kempton, 1987).

Fertilizer treatment can have dramatic effects. Hurd and Wolf (1974) described fertilizer effects that could be detected not only in plants but in their herbivores and in the herbivore's predators as well, including spiders, reduviids (Hemiptera), and at least one parasitic chalcid (Hymenoptera). Thus assemblages at three trophic levels may be affected by variation in nitrogen and other nutrients.

Lesson 9: *Heterogeneity in resource level causes effects that can be carried through the entire food chain.*

Strauss (1987) demonstrated the potentially complex results that variability in nitrogen level may create in plant–herbivore communities. Strauss added nitrogen fertilizer to experimental plots in an old field in Minnesota. Focusing on abundance of herbivorous insects on *Artemesia ludoviciana*, she found that most insects showed density responses to fertilizer treatments. Some species, however, increased in abundance whereas others decreased. Further analyses showed that the increase in sap-sucking insect abundance (membracids and aphids) led to an increase in the presence of ants that tended membracids and aphids, especially in plots where ant nests were present. This increase in ants seemed to be responsible for the decrease in chewing insects, such as beetles. Choice tests confirmed that beetles preferred the fertilized plants over control plants, but in the field where ants were present beetles decreased in abundance.

Lesson 10: *Heterogeneity of resource level can cause complex indirect as well as direct effects on food webs.* (See also lessons 9 and 16.)

Vince et al. (1981) obtained similar results with nitrogen fertilizer additions to salt marsh plots. They too obtained increases in insect herbivores; but, unlike the results of Strauss (1987), both chewers and suckers increased. Furthermore, predators (in this case spiders) showed an increase in density with nitrogen fertilization; and where spiders were abundant, the plant hopper prey (*Prokelisia marginata*) density was low. However, Vince et al. noted one further complication. The response of the insect assemblage differed between low and high marsh areas.

Lesson 11: *Effects of resource heterogeneity may vary even for slightly different habitats.* (See also lessons 9 and 16.)

Fox and Macauley (1977) examined general phenolic, tannin, and foliar nitrogen of several species of *Eucalyptus* in an Australian forest. Feeding rates of herbivorous insects were remarkably constant despite high species to species variability in general phenolic, tannin, or nitrogen content. In contrast, the growth rate of the same insects showed positive correlations with nitrogen content but not with other variables.

Lesson 12: *Variation In Plant-nutritive resources may be more important for herbivores than antiherbivore compounds in plants.*

Auerbach and Strong (1981), on the other hand, observed different responses for different types of herbivores to variation in nitrogen content of *Heliconia* plants (Musaceae). Foliar nitrogen was increased in plants by use of nitrogen fertilizer. This treatment affected growth of six generalist species of herbivorous caterpillars but not the growth of two specialist chrysomelid beetle larvae.

Lesson 13: *The effect of variation in resource quality may depend on whether the species are specialists or generalists.* (See also lesson 2.)

Animals: Arthropods: Decomposers

Naeem (1988b) demonstrated that resource heterogeneity facilitates coexistence between slime mites (*Sarraceniopus darlingtoniae*) and larvae of the midge (*Metriocnemus edwardsi*) that inhabit pitchers of the California pitcher plant (*Darlingtonia californica*) and compete for food resources (decomposing insects). Variation in height among unmanipulated pitchers results in variation in the quantity of resources found in different pitchers. Where resources were abundant, mites could recruit but midge larvae could not. This difference in life history between mites and midges permitted mites to exploit the variation in resources among patches (pitchers). That is, pitchers with unusually high resource levels for the size of the pitcher provide a refuge for the mites.

Similarly, Naeem (1988a) demonstrated in two other studies that the distribution and abundance of fauna inhabiting the water-filled bracts of *Heliconia* plants are affected by resource heterogeneity. In *Heliconia* the resources consisted of decomposing corollas of flowers that accumulate in the bracts and form the primary organic input into these microaquatic systems. Again, differences in life history among the species permitted differential exploitation of variation in the resources among patches (bracts).

Further support for this effect of resource heterogeneity was suggested in Vegter's (1987) study of seasonal variability in abundance of fungal mycilia. Variability in this resource is potentially responsible for the co-existence of competing species of collembola with different life history strategies. Citing Levins' 1979 model, Vegter suggested that some collembola life history strategies track long-term trends in resource abundance, whereas other collembola have life histories that allow them to exploit short-term fluctuations in resource availability.

Heterogeneity of resources in decomposer communities may not always affect community structure. For example, studies of assemblages inhabiting carrion of different sizes show no detectable associations between the distribution and abundance of species in the decomposer community and the size of the carrion being consumed (Beaver, 1972, 1977; Hanski and Kuusela, 1977; Kuusela and Hanski, 1982). These carrion studies, however, examined only fly species, and it is possible that the life histories of these flies were too similar to permit differential exploitation of variation in resources among patches. Different results might be obtained were other taxa included in the analyses.

Lesson 14: *Ecological consequences of resource heterogeneity depend on the life history strategies of the species in the assemblage.* (See also lesson 5).

Animals: Zooplankton

DeMott and Kerfoot (1982) monitored consumption of bacteria and phytoplankton by zooplankters (*Daphnia rosea, Bosmina longirostris,* and *Diaphanosoma brachyurum*) in field and laboratory experiments. The authors enriched nutrient levels (using phosphate-nitrate fertilizer) in large enclosures suspended in Lake Mitchell, Vermont. They also examined the importance of fish predation by establishing some unenriched enclosures with and without predatory trout. The enrichment increased microbial resources in the enclosures. This increase in abundance of resources led to increased *Bosmina* and *Daphnia* densities when compared to controls. Densities of *Diaphanosoma*, however, decreased, presumably because of the increased densities of the competitively superior *Daphnia*. In unenriched enclosures where fish were present, fish preyed primarily on *Daphnia*. This predation reduced *Daphnia* densities, which led to a dramatic increase in the competitively inferior *Diaphanosoma*. *Bosmina* densities, however, were not significantly affected by this treatment when compared with controls. Thus *Daphnia* and *Diaphanosoma* coexistence in the lake appears to be predator-mediated. Coexistence of *Bosmina* with *Daphnia* and *Diaphanosoma*, however, appears to be due to *Bosmina*'s dietary flexibility. Though the diets of all three overlap, *Bosmina* differs from *Daphnia* and *Diaphanosoma* because it can feed effectively on small prey when large prey are rare. Thus coexistence of these three species is the complex result of several factors including predation, differences in the feeding capabilities of the species, and variation in resource abundance and composition.

Similarly, DeMott (1983) observed that coexistence between *Daphnia rosea* and *Daphnia pulicaria*, which compete for algal cells, is due to differences in feeding abilities of the two species. The competitive interaction between *D. rosea* and *D. pulicaria* varied as a function of resource composition (percent "resistant" or difficult-to-digest algal cells). Thus even if the total abundance of algae were to remain constant, variation in species composition of the algal resources would affect the distribution and abundance of the consumers.

Lesson 15: *Heterogeneity of resource composition or quality may foster coexistence among competing species even if the absolute resource level remains constant.* (See also lessons 2, 5, and 16.)

Neill and Peacock (1980), Neill (1981, 1984), and Peacock (1982) examined the effects of manipulating resources, predator densities, and competitor densities in freshwater zooplankton communities in two oligotrophic, fishless lakes in British Columbia, Canada. Experimental evidence suggested that the effects on community structure due to predation by the larval predatory insect *Chaoborus* were not as great as the effects of variation in resource levels (Neill, 1981). This phenomenon was examined in greater detail by Neill and Peacock (1980). They manipulated nutrients by addition of phosphate-nitrate fertilizer to experimental enclosures held within the lake and compared effects of this treatment with or without *Chaoborus*. The results were complex. In enclosures without *Chaoborus*, densities of several zooplankton species increased, though the relative abundance of species differed little from those enclosures with *Chaoborus*. This result supported the hypothesis that *Chaoborus* predation does not significantly alter zooplankton diversity in natural lakes. Community composition, however, was strongly affected by nutrient level. At low and moderate nutrient levels, rare zooplankton species increased in abundance even in the presence of *Chaoborus*. At high nutrient concentrations *Chaoborus* survival increased dramatically, and the resulting increase in *Chaoborus* predation ultimately led to the extinction of several zooplankton species.

Using similar methods, Neill (1984) examined rotifer density responses to predator (*Chaoborus*), competitor (crustacea, especially *Daphnia* spp.), and nutrient manipulations (fertilizer). Increasing nutrients had only short-term positive effects on rotifers because the increase in nutrients also increased densities of competitors. Removal of predators allowed competitor (nonrotifer) populations to increase, which also resulted in depressed rotifer populations. Removal of competitors (*Daphnia*) allowed rotifers to increase. The most dramatic increase in rotifers occurred when competitors were removed and nutrients increased.

Vanni (1987) similarly examined the effects of experimental manipulations of nutrients and species densities in large suspended enclosures in Dynamite Lake, Illinois. This extensive study included examinations of rotifers, crustacea, and predatory bluegill sunfish. Fish predation primarily altered size–class frequencies and life histories (e.g., change in age of first reproduction, volume of eggs) of zooplankton populations. Fish predation, however, had little or no significant effect on the densities of zooplankton in enclosures. Manipulations of nutrients, however, produced much more dramatic changes in the zooplankton.

The results from these studies are contrary to the view sometimes held that predation is the major ecological factor structuring aquatic communities (e.g., Hrbacek, 1962; Brooks and Dodson, 1965; Zaret, 1980, papers in Kerfoot and Sih, 1986). In some circumstances, alteration of nutrients and resources can have more dramatic and far-reaching effects on zooplankton community structure than predation alone.

Lesson 16: *Resource heterogeneity can affect predator-mediated effects on community structure.* (See also lessons 9 and 10.)

Vertebrates

Schmitt and Holbrook (1986; Holbrook and Schmitt, 1988) demonstrated that competition between striped (*Embiotoca lateralis*) and black (*E. jacksoni*) surfperch appears to be important only when invertebrate prey densities, the shared common resource, are low.

Schluter (1982) measured insect and seed (resource) abundance, finch diets, and finch densities across an altitudinal transect for 5 months on Isla Pinta. Of three hypotheses considered, the data supported the hypothesis that heterogeneity in the distribution and abundance of resources were the dominant factors determining the distribution and abundance of finches.

Further evidence of the importance of resources in the Galapagos finches comes from Grant and Boag (1980; Boag and Grant, 1984; Grant, 1986 and Price (1985). These studies argued, from data on seed abundances, finch abundances, climate, and year-to-year variation, that competition for resources occurs primarily during drought years. During periods of normal rainfall, diets are sufficiently different and resources sufficiently abundant that interspecific competition for resources is not an important factor.

Lesson 17: *When resources are heterogeneous, competition may be important only at times of low resources.*

Pimm and Pimm (1982) examined the response of a guild of Hawaiian honeycreepers (Drepanididae), nectarivorous birds, to seasonally varying resources (flowers with nectar). They found that total densities of honeycreepers (consumers) did not change, but the distribution and abundance within local areas did change because of temporal heterogeneity of resources. That is, the exact pattern of distribution and abundance is complex and depends on the resource preferences and competitive status of the birds.

Lesson 18: *Resource heterogeneity can affect local or short-term variation* *in the distribution and abundance of species while having no effect on re-gional or long-term variation.*

Among the more complicated community-level vertebrate examples of the potential effects of resource heterogeneity are the studies of nutrients, plants, lemmings (Schultz 1964, 1969; Batzli et al., 1981; Batzli 1983), and lemming predators: snowy owls, skuas, glaucous gulls, weasels, ermines, and arctic foxes (Batzli et al., 1980). Schultz (1969) manipulated nutrients by fertilizing plots. These plots produced significantly higher densities of lemming nests than did untreated plots. Because it is known that microtine fluctuations significantly affect predator population dynamics (e.g., Batzli et al., 1980), such nutrient-induced fluctuations in lemming densities are likely to also affect predator dynamics.

The causes of microtine cycles are complex and almost certainly involve intrinsic as well as extrinsic factors (see synthesis by Begon et al., 1986, pp. 563–571). More complete studies are needed, but clearly temporal fluctuations in nutrients may affect plant abundance and quality, which consequently affect lemming densities. In turn, lemming and lemming–predator densities mutually affect each other. Coexistence of so many predators using one resource is undoubtedly facilitated by the temporal heterogeneity of lemming resources and lemmings as resources.

Lesson 19: *Temporal heterogeneity of resources may contribute to cyclic fluctuations in food webs.*

Discussion

Our examples come from three, fields: microbial ecology, terrestrial plant ecology, and animal ecology. All three, however, have been influenced by the same ideas concerning the ecological relations between resources and consumer. We offer Figure 12.5 as evidence of this commonality. Figure 12.5 shows three portrayals of the theory that n limiting resources supports n or fewer species of consumers (see above), what Powell and Richerson (1985) called "a hoary ecological tenet." Each example in Figure 12.5 has been the motivation of much empirical work concerning resource heterogeneity. MacArthur's (1972) formulation was inspired predominantly by the Pearl/Verhulst and Lotka/Volterra models for population growth. Taylor and Williams (1975), however, were influenced by Monod's microbial growth model and Michaelis-Menten enzyme kinetic equations, and they appear to have come to their strikingly similar model independently. This convergence is not surprising given that Monod/Michaelis-Menten models are formally analogous to Lotka-Volterra models (Smouse, 1980). Tilman (1982), basing his model on the two approaches, with additional influence from Leon and Tumpson (1975), produced a hybrid model. All three models have incorporated the same approach. Note that

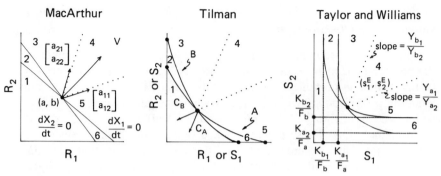

Figure 12.5. Three models demonstrating the principle that n species require n or more limiting resources. The models are from MacArthur (1972), Tilman (1982), and Taylor and Williams (1975). We have mostly used the authors' original notations but have made some changes for the purposes of simplification. Solid lines represent zero population growth isoclines. Dotted lines are variously defined but generally represent abundance of both resources as functions of consumption by one consumer species when the other is absent. Numbers 1 through 6 represent regions where none, one, or both species may persist (see text for further detail). *MacArthur's model:* a_{ij} = probability that during a unit of time an individual of species i encounters and eats an individual of resource i. Short arrows represent the vectors indicated in brackets; they show direction of change of combined abundance of both resources if only consumer i is present. (a,b) = stable equilibrium point. R_i = population size of resource species i. v = the vector representing the sum of the vectors indicated in brackets; it shows the direction of change of combined abundance of both resources in the absence of consumers. X_1 = the population size of consumer species i. *Tilman's model:* A and B represent two consumer species. C_A and C_B are consumption vectors for consumer species A and B, respectively; they show direction of change of combined resources consumed when only one consumer species, indicated by subscript, is present. The slope of the vector is the ratio of per-capita consumption rates of both resources for the species indicated by the subscript. S_1 = concentration of substrate i; R_1 is as in MacArthur's model. *Taylor and Williams' model:* F_a and F_b are maximal growth rates under optimal conditions for consumer species a and b, respectively, divided by the dilution rate of the chemostat. K_{a_1} and K_{b_1} are substrate affinity constants for consumer species a and b, respectively, or one-half concentration of substrate i at maximum population growth. S_1 is as in Tilman's model, is the concentration of substrate i at equilibrium, Y_{a_1} and Y_{b_1} are the yield coefficients (constants) for species a or b, respectively, when grown in isolation on substrate i. For further treatment of these models readers should refer to the original papers.

MacArthur's model has been most influential among terrestrial plant and animal ecologists, Tilman's model among phytoplankton and terrestrial plant ecologists, and Taylor and William's model among phytoplankton and microbial ecologists. Taylor and Williams produced their model specifically for continuous-flow chemostats.

The models have several things in common. Resources are represented by a phase plane defined by axes that measure constant levels of resource (R_i) abundance or substrate abundance (S_i). The plane is subsequently divided into regions by zero-population-growth isoclines (solid lines) for two competing species and by two lines representing vectors of resource use by the two competing species (dotted lines). All authors incorporated stability analyses of the six regions to describe combinations of both resources that are either too low to permit either species to persist (region 1), sufficient for only one of two species to survive (regions 2 and 6), sufficient for either species to persist individually but not together (regions 3 and 5), and sufficient to support both species indefinitely (region 4). By extension of this type of analysis to more than two species and more than two resources, all authors concluded that stable equilibria cannot be found when the number of limiting resources is lower than the number of competing species.

We refer the reader to the original sources for more thorough treatments of the models. It should be apparent from even this limited treatment, however, that the models are the same. This similarity of models from three fields shows the prevalance of the notion that resources, when limiting, constrain biotic diversity.

The real world is vastly more complex than anything Figure 12.5 can represent. Our simple hypothetical example (Figs. 12.1 and 12.2) revealed some of the complexity resource heterogeneity adds to (real) biological communities when compared to communities in (hypothetical) homogeneous environments. Resource heterogeneity creates opportunities for differences in the biology of species (e.g., their patterns of movement, life histories, or abilities to use resources) to translate into differences in the way populations grow. Such differences can ultimately affect the way species interact with one another, which in turn affects the distribution and abundance of species in habitats. When results of theoretical and empirical work conflicts with predictions of classic models, the conflict is often the product of differences in distribution and abundance of species that resource heterogeneity generates.

The remarkable variety of mathematical approaches authors have used demonstrates considerable industry and inspiration on the part of modelers, but the conclusions of most of these models vary little: Heterogeneity of resources theoretically permits coexistence of more than *n* competing species on fewer than *n* resources.

The variety of theoretical approaches exceeds the variety of empirical approaches. Of the four model types (classic, giving-resources-a-life-of-their-own, extrinsically generated heterogeneity, and open-patch models), 72% of the empirical work we reviewed (33 of 46 studies) relates to models that incorporate extrinsically generated sources of heterogeneity of re-sources. We do not imply that this bias in the literature (or our reading of it) suggests that one type of model is more relevant than other types to the workings of the real world. This bias probably reflects the difficulty of empirically testing other classes of models using multispecies systems and may indeed reflect sampling error in this survey. The survey, nevertheless, covered a wide variety of studies, ecosystems, and organisms. Even with a major emphasis on extrinsically generated resource heterogeneity, the large number of lessons obtained in our survey attests to a greater richness of behavior in natural systems than many models would lead us to believe.

As with the predictions of theory, however, with few exceptions the major conclusions of empirical studies are predominantly the same: Heter-ogeneity of resources permits coexistence of more than n competing species on fewer than n resources.

Lest the contrast between older models and more recent work be mis-construed as evidence that the authors of older models had a narrow or limited appreciation for the complexity of the real world, we offer this 1972 quotation of MacArthur.

Generalizing [from preceding theoretical considerations], no more than n species can coexist if they are limited by only n resources. This interesting statement has two weaknesses. First, if other limiting factors enter—in other words if the equa-tions of growth of the consumers involve their own populations or predator popula-tions as well as the resource levels—it no longer holds; we could then have more species than resources. Second, no one knows precisely what a "resource" is. It is not a species, because the fruit and bark of a tree can be harvested almost indepen-dently by two kinds of consumer insects and hence probably constitute two re-sources. And the various grass species eaten indiscriminately by some grazers may only constitute one resource. To the extent that we have to tailor our definition of resource to match the species present, we have made no progress.

MacArthur presciently described two of the principal problems with the theory that the variety of theoretical and empirical investigations would address since the publication of his book.

In conclusion, heterogeneity of consumable resources is important be-cause it is a common feature of ecosystems and modifies predicted outcom-es of biotic interactions based on simple models of community dynamics. Resource heterogeneity can both constrain and facilitate biotic diversity. These effects make the interpretation of empirical work difficult, especially descriptive field data. Descriptive work profits from documentation of re-source levels and models from incorporation of parameters for resource heterogeneity. Experiments are more meaningful when conducted at mean, high, and low resource levels (as determined from field data). These aug-mentations of current ecological practices will lead ultimately to a better understanding of what is occurring in heterogeneous ecosystems.

Summary

Virtually every aspect of an organism's biology is affected by the quality and quantity of resources available for consumption. Spatial or temporal variation in resources therefore contributes to variability in growth, development, and performance of organisms. This variability affects interactions among species that share common resources and can affect patterns in the distribution and abundance of species in communities. We have examined the ecological consequences of heterogeneity of consumable resources on the structure of biological communities. Our examination includes a survey of theoretical and empirical literature in ecology, and we present 19 "lessons" culled from the literature that summarize some of the complex effects heterogeneity of resources creates.

1. The principle that n species requires n or more limiting resources may apply to ecologically simple systems.
2. More than n species may persist on fewer than n resources when resources fluctuate and if consumer species differ in their ability to use resources.
3. The ecological consequences of physical environmental heterogeneity, such as disturbance, is dependent on resource levels.
4. Even when mean resource levels are constant over the long term, short-term resource heterogeneity can affect community structure.
5. More than n species may persist on fewer than n resources if consumer species differ in their ability to use the *variance* of the resource.
6. The effects of heterogeneity of one resource may be dependent on, or a function of, the level of another resource.
7. Ecological consequences of historical factors are dependent on past and current resource levels.
8. The effects of resource heterogeneity on the distribution and abundance of species may depend on the ability of colonizing species to select a habitat.
9. The effects of resource heterogeneity on one trophic level can carry through to another level in a food chain.
10. Heterogeneity of resource level can cause complex indirect as well as direct effects on food webs.
11. Effects of resource heterogeneity may vary even for slightly different habitats.
12. Variation in plant-nutritive resources for herbivores may be more important than antiherbivore compounds in plants.
13. The effect of variation in resource quality may depend on whether the species are specialists or generalists.
14. Ecological consequences of resource heterogeneity depend on the life history strategies of the species in the assemblage.
15. Heterogeneity of resource composition or quality may foster coexis-

tence among competing species even if the absolute resource level remains constant.

16. Resource heterogeneity can affect predator-mediated effects on community structure.
17. When resources are heterogeneous, competition may be important only at times of low resources.
18. Resource heterogeneity can affect local or short-term variation in the distribution and abundance of species while having no effect on regional or long-term variation.
19. Temporal heterogeneity of resources may contribute to cyclic fluctuations in food webs.

Acknowledgment. We thank D. Carey and M. Power for critical readings of the manuscript. Preparation of the manuscript was supported by the US National Science Foundation (BSR86-04929).

References

Abrams PA (1980) Consumer functional response and competition in consumer-resource systems. Theor Pop Biol 17:80–102

Abrams PA (1988) Resource predictability-consumer species diversity: simple models of competition in spatially heterogeneous environments. Ecology 69:1418–1433

Armstrong RA, McGehee R (1976a) Coexistence of species competing for shared resources. Theor Pop Biol 9:317–328

Armstrong RA, McGehee R (1976b) Coexistence of two competitors on one resource. J Theor Biol 56:499–502

Armstrong RA, McGehee R (1980) Competitive exclusion. Am Nat 115:151–170

Atkinson D, Shorrocks B (1981) Competition on a divided and ephemeral resource: a simulation model. J Anim Ecol 50:461–471

Auerbach MJ, Strong DR Jr (1981) Nutritional ecology of *Heliconia* herbivores: experiments with plant fertilization and alternate hosts. Ecol Monogr 51:63–83

Ayala FJ, Gilpin MJ, Ehrenfeld EG (1973) Competition between species and theoretical models and experimental tests. Theor Pop Biol 4:331–356

Batzli GO (1983) Responses of Arctic rodent populations to nutritional factors. Oikos 40:396–406

Batzli GO, White RG, MacLean SF, Pitelka FA, Collier BD (1980) The herbivore based trophic system. In Brown J, Miller PC, Tieszen LL, FL Bunnell (eds) *An Arctic Ecosystem: The Coastal Tundra at Barrow, Alaska*. Dowden, Hutchinson & Ross, Stroudsburg

Batzli GO, Jung HG, Guntenspergen G (1981) Nutritional ecology of microtine rodents: linear-forage rate curves for brown lemmings. Oikos 37:112–116

Bazzaz FA, Carlson RW (1984) The response of plants to elevated CO_2. I. Competition among an assemblage of annuals at different levels of soil moisture. Oecologia 62:196–198

Bazzaz FA, Garbutt K (1988) The response of annuals in competitive neighborhoods: effects of elevated CO_2. Ecology 69:937–946

Beaver RA (1972) Ecological studies on Diptera breeding in dead snails. I. Biology of the species found in *Cepea nemoralis* (L.). Entomologist 105:41–52

Beaver RA (1977) Non-equilibrium "island" communities: Diptera breeding in dead snails. J Anim Ecol 46:783–798

Begon M, Harper JL, Townsend CR (1986) *Ecology: Individuals, Populations, and Communities.* Sinauer Associates, Sunderland

Bender EA, Case TJ, Gilpin ME (1984) Perturbation experiments in community ecology: theory and practice. Ecology 65:1–13

Boag PT, Grant PR (1984) Darwin's Finches (*Geospiza*) on Isla Daphne Major, Galapagos: breeding and feeding ecology in a climatically variable environment. Ecol Monogr 54:463–489

Brock TD (1987) The study of microorganisms in situ: progress and problems. In Fletcher MT, Gray RG, Jones JG (eds) *Ecology of Microbial Communities.* Cambridge University Press, New York

Brooks JL, Dodson SI (1965) Predation, body size and composition of the plankton. Science 150:28–35

Butler GJ, Waltman P (1981) Bifurcation from a limit cycle in a two predator-one prey ecosystem modeled on a chemostat. J Math Biol 12:295–310

Cairns J Jr, Dickson KL, Yongue WH Jr (1971) The consequences of nonselective periodic removal of portions of fresh water protozoan communities. Trans Am Microsc Soc 70:71–80

Caswell H (1978) Predator-mediated coexistence: a nonequilibrium model. Am Nat 112:127–154

Chesson PL (1986) Environmental variation and the coexistence of species. In Diamond J, Case TJ (eds) *Community Ecology.* Harper & Row, New York, pp 240–256

Chesson PL, Case TJ (1986) Nonequilibrium community theories: chance, variability, history, and coexistence. In Diamond J, Case TJ (eds) *Community Ecology.* Harper & Row, New York, pp 229–239

Cohen JE (1970) A Markov contingency table model for replicated Lotka-Voltera systems near equilibrium. Am Nat 104:547–559

Colwell RK (1974) Predictability, constancy, and contingency of periodic phenomena. Ecology 55:1148–1153

Connell JH (1978) Diversity in tropical rain-forests and coral reefs. Science 199:1302–1310

Crowell KL, Pimm SL (1976) Competition and niche shifts of mice introduced onto islands. Oikos 27:251–258

Cushing JM (1980) Two species competition in a periodic environment. J Math Biol 10:485–400

DeMott WR (1983) Seasonal succession in a natural *Daphnia* assemblage. Ecol Monogr 53: 321–340

DeMott WR, Kerfoot WC (1982) Competition among cladocerans: nature of the interaction between *Bosmina* and *Daphnia*. Ecology 63:1949–1966

Digby PGN, Kempton RA (1987) *Multivariate Analysis of Ecological Communities.* Chapman & Hall, London

Eissenstat DM, Caldwell MM (1988) Seasonal timing of root growth in favorable microsites. Ecology 69:870–873

Feeny PP (1976) Plant apparency and chemical defense. In Wallace WJ, RL Mansell (eds) *Recent Advances in Phytochemistry. Vol 10. Biochemical Interactions Between Plants and Insects.* Plenum Press, New York, pp 1–40

Fox LR, Macauley BJ (1977) Insect grazing in *Eucalyptus* in response to variation in leaf tannins and nitrogen. Oecologia 29:145–162

Fredrickson AG, Stephanopoulos G (1981) Microbial competition. Science 213: 972–979

Giesel, JT (1976) Reproductive strategies as adaptations to life in temporally heterogeneous environments. Annu Rev Ecol Syst 7:57–79

Gotelli NJ, Simberloff D (1988) The distribution and abundance of tall grass prairie plants: a test of the core-satellite hypothesis. Am Nat 130:18–35

Gottschal JC, Devries SC, JG Kuenen (1979) Competition between the facultative-ly chemolithotrophic *Thiobacillus* A2, an obligately chemolithotrophic *Thiobacillus* and a heterotrophic *Spirillum* for inorganic and organic substrates. Arch Microbiol 121:241–249

Grant PR (1986) *Ecology and Evolution of Darwin's Finches.* Princeton University Press, Princeton

Grant PR, Boag PT (1980) Rainfall on the Galapagos and the demography of Darwin's finches. Auk 97:227–244

Grenney WJ, Bella DA, Curl HC Jr (1973) A theoretical approach to interspecific competition in phytoplankton communities. Am Nat 107:405–425

Grover JP (1988) Dynamics of competition in a variable environment: experiments with two diatom species. Ecology 69:408–417

Hallet JG, Pimm SL (1979) Direct estimation of competition. Am Nat 113:593–600

Hanski I (1981) Coexistence of competitors in patchy environment with and without predation. Oikos 37:306–312

Hanski I (1982a) Communities of bumblebees: testing the core-satellite species hypothesis. Ann Zool Fenn 19:65–73

Hanski I (1982b) Dynamics of regional distribution: the core and satellite species hypothesis. Oikos 38:210–221

Hanski I, Kuusela S (1977) An experiment on competition and diversity in the carrion fly community. Ann Entomol Fenn 43:108–115

Harder W, Dijkhuizen L (1982) Strategies of mixed substrate utilization in micro-organisms. Philos Trans Ry Soc Lond [Biol] 297:459–480

Hardin G (1960) The competitive exclusion principle. Science 131:1292–1297

Hartnett DC, Bazzaz FA (1983) Physiological interaction among intraclonal ramets in *Solidago canadensis*. Ecology 64:799–797

Holbrook SJ, Schmitt RJ (1988) The combined effects of predation risk and food reward on patch selection. Ecology 69:125–134

Horn HS, MacArthur RH (1972) Competition among fugitive species in a harlequin environment. Ecology 53:749–752

Hrbacek J (1962) Species composition and the amount of zooplankton in relation to fish stock. Rozpravy Cesk Akad Ved, Rada Matemat Prirodnich Ved 72:1–116

Hsu SB (1981) On a resource based ecological competition model with interference. J Math Biol 12:45–52

Hsu SB, Hubbell SP (1979) Two predators competing for two prey species and analysis of MacArthur's model. Math Biosci 47:143–171

Hsu SB, Hubbell SP, Waltman P (1978) A contribution to the theory of competing predators. Ecol Monogr 48:337–349

Hurd LE, LL Wolf (1974) Stability in relation to nutrient enrichment in arthropod consumers of old-field successional systems. Ecol Monogr 44:465–482

Hurd LE, Mellinger MV, Wolf LL, McNaughton SJ (1971) Stability and diversity at three trophic levels in terrestrial ecosystems. Science 173:1134–1136

Hurlbert SH (1984) Pseudoreplication and the design of ecological field experiments. Ecol Monogr 54:187–211

Hutchinson GE (1961) The paradox of the plankton. Am Nat 95:137–145

Inouye RS, Tilman D (1988) Convergence and divergence of old-field plant communities along experimental nitrogen gradients. Ecology 69:995–1004

Jackson KM, Berger J (1984) Survival of ciliate protozoa under starvation conditions and at low bacterial levels. Microb Ecol 10:47–59

Kerfoot WC, Sih A (eds) (1986) *Predation: Direct and Indirect Impacts on Aquatic Communities.* University Press of New England, Hanover

Kerfoot WC, DeMott WR, Levitan C (1985) Non-linearities in competition interactions: component variables or system response? Ecology 66:959–965

Koch AL (1974a) Coexistence resulting from an alternation of density dependent and density independent growth. J Theor Biol 44:373–386

Koch AL (1974b) Competitive coexistence of two predators utilizing the same prey under constant environmental conditions. J Theor Biol 44:387–395

Kuusela S, Hanski I (1982) The structure of carrion fly communities: the size and the type of carrion. Holarct Ecol 5:337–348

Leon J, Tumpson D (1975) Competition between two species for complementary or substitutable resources. J Theor Biol 50:185–201

Levin SA (1970) Community equilibria and stability, and an extension of the competitive exclusion principle. Am Nat 104:413–423

Levin SA (1974) Dispersion and population interaction. Am Nat 108:207–228

Levin SA (1976) Population dynamic models in heterogenous environments. Annu Rev Ecol Syst 7:287–310

Levin SA (1978) Pattern formation in ecological communities. In Steele JH (ed) *Spatial Pattern in Plankton Communities*. Plenum Press, New York, pp 433–467

Levine SH (1976) Competitive interactions in ecosystems. Am Nat 110:903–910

Levins R (1968) *Evolution in Changing Environments*. Princeton University Press, Princeton

Levins R (1979) Coexistence in a variable environment. Am Nat 114:765–783

Levins R, Culver D (1971) Regional coexistence of species and competition between rare species. Proc Natl Acad Sci USA 68:246–248

Lynch M (1978) Complex interactions between natural coexploiters—*Daphnia* and *Ceriodaphnia*. Ecology 59:552–564

MacArthur RH (1972) *Geographical Ecology*. Princeton University Press, Princeton

MacArthur RH, Levins R (1967) Competition, habitat selection, and character displacement in a patchy environment. Proc Natl Acad Sci USA 51:1207–1210

Mattson WJ Jr (1980) Herbivory in relation to plant nitrogen content. Annu Rev Ecol Sys 11:119–162

Morin PJ (1986) Salamander predation, prey facilitation, and seasonal succession in microcrustacean communities. In Kerfoot CW, Sih A (eds) *Predation: Direct and Indirect Impacts on Aquatic Communities*. University Press of New England, Hanover, pp 174–187

Morrison DF (1976) *Multivariate Statistical Methods*. McGraw-Hill, New York

Naeem S (1988a) Ecological consequences of temporal heterogeneity on aquatic communities of microbes and metazoa. PhD dissertation, University of California, Berkeley

Naeem S (1988b) Resource heterogeneity fosters the coexistence of a mite and a midge in a pitcher plant. Ecol Monogr 58:215–227

Neill WE (1981) Importance of *Chaoborus* predation upon the structure and dynamics of a crustacean zooplankton community. Oecologia 48:164–177

Neill WE (1984) Regulation of rotifer densities by crustacean zooplankton in an oligotrophic montane lake in British Columbia. Oecologia 61:175–181

Neill WE, Peacock A (1980) Breaking the bottleneck: invertebrate predators and nutrients in oligotrophic lakes. In Kerfoot WC (ed) *Evolution and Ecology of Zooplankton Communities*. University Press of New England, Hanover, pp 715–724

Onuf CP, Teal JM, Valiela I (1977) Interactions of nutrients, plant growth and herbivory in a mangrove ecosystem. Ecology 58:514–526

Peacock A (1982) Responses of *Cyclops bicuspidatus thomasi* to alterations in food and predators. Can J Zool 60:1446–1462

Pickett STA, White PS (eds) (1985) *The Ecology of Natural Disturbance and Patch Dynamics*. Academic Press, Orlando

Pimm SL, Pimm JW (1982) Resource use, competition, and resource availability in Hawaiian honeycreepers. Ecology 63:1468–1480

Powell T, Richerson PJ (1985) Temporal variation, spatial heterogeneity, and competition for resources in plankton systems: a theoretical model. Am Nat 125:431–464

Price PW, Boutan CE, Gross P, McPheron BA, Thomson JN, Weins AE (1980) Interactions among three trophic levels: influence of plants on interactions between insect herbivores and natural enemies. Annu Rev Ecol Sys 11:41–65

Price TD (1985) Reproductive responses to varying food supply in a population of Darwin's finches: clutch size, growth rates and hatching synchrony. Oecologia 66:411–416

Rashit E, Bazin M (1987) Environmental fluctuations, productivity, and species diversity: an experimental study. Microb Ecol 14:101–112

Rhoades DF, Cates RG (1976) Toward a general theory of plant antiherbivore chemistry. In Wallace WJ, Mansell RL (eds) *Recent Advances in Phytochemistry. Vol 10. Biochemical Interactions Between Plants and Insects*. Plenum Press, New York, pp 168–213

Richerson P, Armstrong R, Goldman CR (1970) Contemporaneous disequilibrium, a new hypothesis to explain the paradox of the plankton. Proc Natl Acad Sci USA 67:1710–1714

Rogers AH, Zilm PS, Gully NJ (1987) Influence of argenine on the coexistence of *Streptococcus mutans* and *S. milleri* in glucose-limited mixed continuous culture. Microb Ecol 14:193–202

Rosenzweig ML, Abramsky Z, Brand S (1984) Estimating species interactions in heterogeneous environments. Oikos 43:329–340

Rosenzweig ML, Abramsky Z, Kotler B, Mitchell W (1985) Can interaction coefficients be determined from census data? Oecologia 66:194–198

Sale PF (1977) Maintenance of high diversity in coral reef fishes. Am Nat 111:337–359

Salzman AG (1985) Habitat selection in a clonal plant. Science 228:603–604

Salzman AG, MA Parker (1985) Neighbors ameliorate local salinity stress for a rhizomatous plant in a heterogeneous environment. Oecologia 65:273–277

Schluter D (1982) Distributions of Galapagos ground finches along an altitudinal gradient: the importance of food supply. Ecology 63:1504–1517

Schmitt RJ, Holbrook SJ (1986) Seasonally fluctuating resources and temporal variability of interspecific competition. Oecologia 69:1–11

Schoener TW (1974) Competition and the form of habitat shift. Theor Pop Biol 6:265–307

Schultz AM (1964) The nutrient recovery hypothesis for Arctic microtine cycles. II. Ecosystem variables in relation to Arctic microtine cycles. In Crisp DJ (ed) *Grazing in Terrestrial and Marine Environments*. Blackwell, Oxford, pp 57–68

Schultz AM (1969) A study of an ecosystem: the arctic tundra. In van Dyne GM (ed) *The Ecosystem Concept in Natural Resource Management*. Academic Press, Orlando, pp 77–93

Shorrocks B, Atkinson W, Charlesworth P (1979) Competition on a divided and ephemeral resource. J Anim Ecol 48:899–908

Slatkin M (1974) Competition and regional coexistence. Ecology 55:128–134

Smith DW, Cooper SD, Sarnelle O (1988) Curvilinear density dependence and the design of field experiments on zooplankton competition. Ecology 69:868–870

Smouse PE (1980) Mathematical models for continuous culture growth dynamics of mixed populations subsisting on a heterogenous resource base. I. Simple competition. Theor Pop Biol 17:16–36

Steele JH (ed) (1978) *Spatial Pattern in Plankton Communities*. Plenum Press, New York

Strauss SY (1987) Direct and indirect effects of host-plant fertilization on insect community. Ecology 68:670–1678

Strong DR Jr (1983) Natural variability and the manifold mechanisms of ecological communities. Am Nat 122:636–660

Stross RG, Pemrick SM (1974) Nutrient uptake kinetics in phytoplankton: a basis for niche separation. J Phycol 10:164–169

Taylor PA, Williams PJLeB (1975) The critical studies on the coexistence of competing species under controlled flow conditions. Can J Microb 21:90–98

Tilman D (1980) Resources: a graphical-mechanistic approach to competition and predation. Am Nat 116:362–393

Tilman D (1982) *Resource Competition and Community Structure*. Princeton University Press, Princeton

Tilman D (1987) The importance of the mechanisms of interspecific competition. Am Nat 129:769–774

Tilman D (1988) *Plant Strategies and the Dynamics and Structure of Plant Communities*. Princeton University Press, Princeton

Tilman D, Kilham SS, Kilham P (1982) Phytoplankton community ecology: the role of limiting nutrients. Annu Rev Ecol Syst 13:349–372

Turpin DH, Harrison PJ (1979) Limiting nutrient patchiness and its role in phytoplankton ecology. J Exp Mar Biol Ecol 39:151–166

Valiela I, Teal JM, Sass WJ (1975) Production and dynamics of salt marsh vegetation and the effects of experimental treatment with sewage sludge: biomass, production and species composition. J Appl Ecol 12:973–982

Vance RR (1984) Interference competition and the coexistence of two competitors on a single limiting resource. Ecology 65:1349–1357

Vandermeer JH (1970) The community matrix and the number of species in a community. Am Nat 104:73–83

Vanni MJ (1987) Effects of food availability and fish predation on a zooplankton community. Ecol Monogr 57:61–88

Vegter JJ (1987) Phenology and seasonal resource partitioning in forest floor *Collembola*. Oikos 48:175–185

Veldkamp H (1977) Ecological studies with the chemostat. Adv Microbiol Ecol 1:59–94

Veldkamp H, Van Gemerden H, Harder W, Laanbroek HJ (1984) Competition among bacteria: an overview. In Klug MJ, Reddy CA (eds) *Proceeding of the Third International Symposium on Microbial Ecology*. American Society Microbiologists, Washington, DC, pp 279–290

Vince SW, Valiela I, Teal JM (1981) An experimental study of the structure of herbivorous insect communities in a salt marsh. Ecology 62:1662–1678

Wiens JA (1976) Population responses to patchy environments. Annu Rev Ecol Syst 7:81–120

Wiens JA (1977) On competition in variable environments. Am Sci 65:590–597

Winer BJ (1971) *Statistical Principles in Experimental Design*. McGraw-Hill, New York

Zangrel AR, Bazzaz FA (1984) The response of plants to elevated CO_2. II. Competitive interactions between annual plants under varying light. Oecologia 62:412–417

Zaret TM (1980) *Predation and Freshwater Communities*. Yale University Press, New Haven

13. Spatial Heterogeneity During Succession: A Cyclic Model of Invasion and Exclusion

Juan J. Armesto, Steward T.A. Pickett, and
Mark J. McDonnell

The concept of heterogeneity is one of the most important and most widely applicable ideas in ecology. It is relevant at all scales and levels of organization and yet is expressed differently in each context (Hutchinson, 1953). One of the principal problems in understanding heterogeneity is to discover its patterns and, subsequently, its functional significance in its various manifestations. In this chapter we take heterogeneity to be the spatial pattern of elements of a biotic assemblage. Old fields, because of their commonness, variety, and manipulability, are ideal systems in which to investigate the pattern of heterogeneity and its ecological significance. Such studies can also contribute to understanding mechanisms of succession (Finegan, 1984; Pickett et al., 1987).

Successional studies have seldom addressed heterogeneity directly, perhaps in part because of the practical difficulties involved in measuring and interpreting heterogeneity in ecological systems. However, it may also reflect the slow demise of superorganismic ideas. Vegetational changes during succession have traditionally been examined at the whole community level, hence focusing on spatial and temporal scales associated with the change in species composition and abundance within entire fields. Our purpose in this chapter is to examine the causes and consequences of heterogeneity during succession. We provide evidence that some characteristic temporal patterns of heterogeneity exist in old-field succession. We show that these patterns are generalizable to all sorts of vegetation and propose a general model to account for these patterns.

Causes of Heterogeneity

Heterogeneity may be caused by a great variety of factors. Therefore generality may depend on the ability to understand how different interactions affect heterogeneity. Here we indicate some of the important biotic and habitat features that might generate heterogeneity and focus on old fields as a useful case. Soil patchiness is perhaps the most widely appreciated abiotic cause of heterogeneity in successional plant communities. The ridge and furrow microtopography originating from row cropping persists and influences pattern in successional communities for a time (Sterling et al., 1984). Old-field species can respond to patchiness in soil characteristics, e.g., fertility (Tilman, 1984) and moisture (Evans and Dahl, 1955; Bazzaz, 1969; Zedler and Zedler, 1969).

Biotic influences can modify or amplify the abiotic pattern through a variety of mechanisms. Differential seed dispersal is one of the most prominent patterns of biotic heterogeneity to overlay soil heterogeneity. As a result of the interaction of these two sources of heterogeneity, soil seed banks vary greatly from place to place in a community (Pickett and McDonnell, 1989). Different microsites may differentially accumulate dispersing seeds (Reader and Buck, 1986). Patterns of invasion in old fields may also reflect the location of propagule sources both outside and inside the field. Sites close to fruiting trees may receive more seed than open patches (McDonnell, 1988), and even sterile plants can act as perches and enhance local seed input (McDonnell and Stiles, 1983). Spatial distribution of species richness (Fig. 13.1) and woody species cover (Fig. 13.2) in two old fields at the Institute of Ecosystem Studies, New York, clearly indicate the impact of propagule sources adjacent to fields on species presence and performance in old fields. Trees along fences and secondary woodlots serve as sources. This pattern is rarely documented in successional studies, and the influence it may have on rate and course of succession is therefore unquantified.

Another major cause of heterogeneity is canopy effects on the light environment. Even in low-statured, dynamic communities such as heathlands and successional fields, patterns of light intensity determined by the canopy are important sources of heterogeneity (Barclay-Estrup, 1971). Canopy in old fields can limit light in the understory to less than 5% of ambient conditions (Armesto and Pickett, 1985). Such limitation is alleviated by moderate to severe disruption of the overstory. These disruptions, resulting in canopy gaps, may have an important role in vegetation dynamics, providing sites for the establishment of later successional species in cases where such recruitment is inhibited by shading (Armesto and Pickett, 1986).

Areas of soil disrupted by animals commonly support plant assemblages different from those in adjacent undisturbed areas (Ross et al., 1968; Platt, 1975). Nutrient availability can be markedly changed by burrowing animals (Inouye et al., 1987a). For instance, spatial heterogeneity of nutrient availability and vegetation resulted from pocket gopher activity in Minnesota

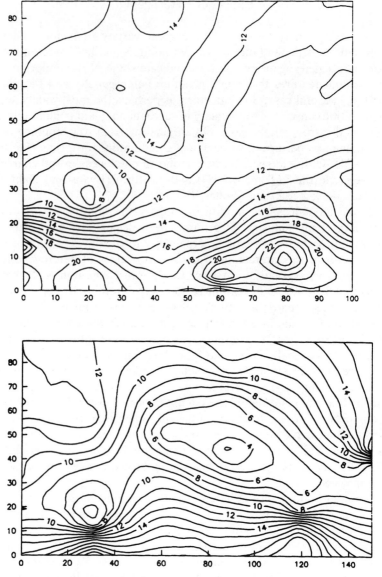

Figure 13.1. Plant species richness isolines for two old fields at the Cary Arboretum of the Institute of Ecosystem Studies (IES), New York, in relation to forest edges and hedgerows. Edges without labels represent continuous old field. Numbers along the axes represent linear distances in meters along forest and hedgerow borders. (Top) A continuously mown meadow (IES grid 177). (Bottom) A 40-year-old abandoned pasture (IES grid 130). We used a stratified random sampling based on five 2.0×0.5 m plots located along six transects from the edge to the middle of the field. Transect lines were separated by 20 m in field A and 30 m in field B. Equal richness contours were estimated by kriging.

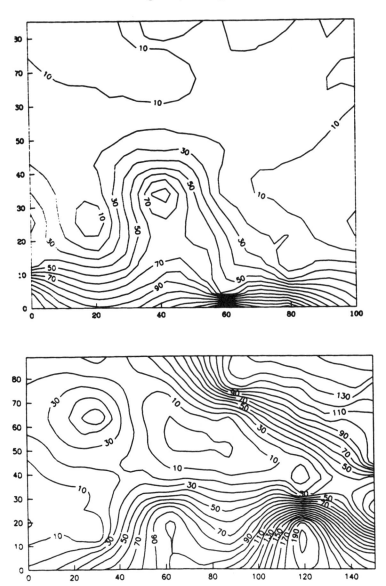

Figure 13.2. Percent cover isolines for woody species in two old fields at the Cary Arboretum of the Institute of Ecosystem Studies, New York, in relation to forest edges and hedgerows. Edges without labels represent continuous old field. Numbers along the axes represent linear distances in meters on field edges. (Top) A continuously mown meadow. (Bottom) A 40-year-old abandoned pasture. Methods as in Figure 13.1.

old fields (Inouye et al., 1987a). Animal diggings are frequently sites of emergence from the seed bank or are hospitable sites for seed rain.

Despite the large number of causes of heterogeneity and the variety of examples of its importance, the subject has rarely been the central focus of study in successional communities. Therefore we draw together the meager information available and propose generalizations and hypotheses that can orient future work.

Definition and Measurement of Heterogeneity

Techniques to measure heterogeneity vary from simple indices, such as diversity indices (Peet, 1974), to more sophisticated statistical (Palmer, 1987) or geostatistical analyses (Robertson, 1987). Such variation presents a problem in reviewing the literature on old-field succession. In addition, heterogeneity may be assessed for species composition versus species performance (e.g., cover, biomass, or reproduction). Different authors have used different approaches to assess heterogeneity, and these methods are rarely comparable. One relatively simple quantification of heterogeneity in a community is to determine the percent similarity in a given biotic or abiotic parameter between any two sampling units and to average the pairwise similarities across the entire community (e.g., Muraoka et al., 1984; Inouye et al., 1987b). Average similarity between pairs of plots should be lower as communities become internally more heterogeneous. This approach resembles the basic principle of geostatistical semivariograms (Robertson, 1987) in which dissimilarities are related graphically to the linear distance between sampling units. Semivariance analysis provides information on the spatial scale as well as the intensity of patchiness.

When analyzing our own data to search for temporal trends in heterogeneity in old-field succession, we use average percent similarity as an inverse index of heterogeneity (Muraoka et al., 1984). We do not use semivariograms because we are not concerned here with the spatial scale of patchiness, although this variable should ultimately be addressed through succession. In addition, few of the data available in the literature have been cast in terms of semivariance. We use floristic composition as the biotic parameter to be measured and estimate similarity between plots using Jaccard's index (Greig-Smith, 1983). This index is straightforward and readily interpreted ecologically (Muraoka et al., 1984). When assessing the patterns exhibited in other data sets, we must be content with the indices used in published accounts.

Patterns of Heterogeneity Through Time

A review of the literature strongly supports the proposition that secondary successions exhibit specific and repeatable patterns of heterogeneity, although the data are scattered and often not directly comparable. Heterogeneity in old-field succession has often been subsumed in studies of diversity employing information theoretical indices (Peet, 1974). For instance, over 11 year of rainforest succession in tropical Australia, heterogeneity measured via H' increased (Webb et al., 1972). In that study, however, sampling was intense for the first 7 years, but only one measurement was taken during the subsequent 4 years.

Mellinger and McNaughton (1975) showed a marked increase in diversity (and hence heterogeneity) through years 5, 6, 16, and 17 that stabilized roughly by year 36 of old-field succession in central New York. Some of their data resulted from consecutive observation of the same fields, although part was a chronosequence. The long gaps between sample dates and the confounding of richness and evenness by the diversity index used makes the trends difficult to interpret. The study by Bazzaz (1975) of a chronosequence in southern Illinois distinguished species evenness, which can be considered an index of heterogeneity in species composition within fields. Using this criterion, heterogeneity increased through ages 1, 4, 15, 25, and 40 but oscillated markedly during the first 15 years.

A still more intense study of heterogeneity in succession employed separate measures for richness, diversity, and redundancy in 51 field and secondary forest stands of various ages (Nicholson and Monk, 1974). Redundancy, which is inversely related to evenness and therefore to heterogeneity, decreased monotonically from early old-field succession to forests up to 200 years old. However, there is a large dispersion of values during the early stages of succession, and there were occasional decreases of heterogeneity in pine stands associated with understory dominance by the introduced *Lonicera japonica*. Again, the data suggest large fluctuations in heterogeneity in early successional old fields.

One permanent plot study of very early succession in northern Ohio assessed spatial heterogeneity as the coefficient of variation in H' based on abundance data in 36 plots. Heterogeneity decreased markedly from the first to the second year and steadily but less precipitously from the third to the fourth year (Tramer, 1975). The first 5 years of succession in Argentinean grassland exhibited two significant peaks of heterogeneity, as measured by an index similar to Jaccard's, designed to assess heterogeneity in species composition (Facelli et al., 1987). There was no net increase or decrease in heterogeneity over the 5-year period. Facelli et al. (1987) found low heterogeneity to be associated with coalescence of patches. One of the longest studies of secondary succession that specifically addressed spatial heterogeneity, in this case using correlations among species abundances as an index, found 2.5 irregular cycles of heterogeneity in 51 years of postfire succession (Shafi and Yarranton, 1973).

Within-field heterogeneity, measured as the average similarity in species composition between pairs of plots (à la Muraoka et al., 1984), increased with time for 22 old fields varying in age from 1 to 56 years (Inouye et al., 1987b). The coarse temporal resolution of the data set, however, may hide

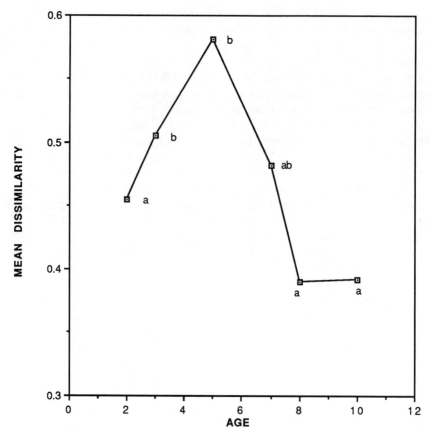

Figure 13.3. Changes in dissimilarity between plots during 10 years of old-field succession at the Hutcheson Memorial Forest Center in central New Jersey. Values are a direct measure of within-field heterogeneity assessed via Jaccard's index between all pairs of an array of 15 randomly located plots. Data points keyed with different letters are significantly different at the 5% level (Duncan's multiple range test).

significant short-term fluctuations in spatial heterogeneity. Some cycles of heterogeneity can actually be observed in the data and figure of Inouye et al. (1987b), although no test for their significance was presented. Overall, plots in younger fields tended to be more similar to one another than plots in older fields. The increase in heterogeneity in older fields was related to greater species richness.

A two-phase trend has been documented in a 10-year chronosequence in two old fields at the Hutcheson Memorial Forest Center (HMF). (Fig. 13.3). The trends were assessed using the Jaccard index, as noted above. An initial increase in average similarity between plots from years 2 to 5 was followed by a decrease from years 7 to 10. This short-term cycle in heterogeneity can be interpreted in relation to alternate periods of canopy integrity and disruption driven by the contrasting life histories of old-field species. Initially, all plots were dominated by the spring annual *Ambrosia artemisiifolia* (Armesto and Pickett, 1985). After its demise, a combination of annuals and biennials took over, thereby increasing heterogeneity, until *Solidago canadensis* clones became dominant in the entire field and decreased heterogeneity (Fig. 13.3).

The data presented and reviewed here consistently show that compositional heterogeneity of plots within fields changes during succession. General long-term trends may indicate net decreasing (Inouye et al., 1987b) or increasing heterogeneity over time (Bazzaz, 1975). Other studies show no overall upward or downward trends (Facelli et al., 1987). All studies reviewed, however, suggested that there are cycles of heterogeneity that tend to be more pronounced during early rather than late succession. This conclusion is admittedly tentative and is subject to testing against additional data sets or analyses as they become available. However, the apparent generality of the observation prompts us to erect a conceptual model to focus study of heterogeneity in succession. A mechanistic model can encourage the search for pattern, promote the testing of our proposition, and help specify the appropriate conditions for its test.

Model of Cycles of Heterogeneity During Succession

A model based on characteristics of the disturbance regime, the invasion process, and the life cycles of dominant species can explain the cycles of heterogeneity documented above. The particular elements and assumptions of the model are as follows.

1. *Large area, limited seed rain.* Initially, invasion in fields derives directly from the seed bank. Seed dispersal into fields, especially for woody plants, is limited by the distance to sources and the short dispersal capacities of many species. We assume here that the size of the disturbed area is large relative to, for instance, a treefall gap in a forest.

2. *Internal Source of colonists.* Species that are dominant the first year of old-field succession are commonly fecund herbs, which saturate the seed banks of open areas (Livingston and Allessio, 1968). Terms such as weed, colonist, ruderal, and r-strategist have been used to describe such species.

3. *Synchronous establishment.* These initial invaders are usually annuals (e.g., *Ambrosia artemisiifolia*), which monopolize the open space by synchronous germination early in the growing season and by overtopping later emerging herbs that otherwise might be superior competitors (Grime, 1979).

4. *Life history diversification.* First-year dominants are replaced by

biennial and perennial species that survive over winter and get an early start during the second year of succession (e.g., Raynal and Bazzaz, 1976). The life-spans and population size structures of this second group are diverse due to their autecologies and to asynchronous establishment.

5. *Coalescence of patches*. Patches of perennial species tend to coalesce during the following 5 to 10 year of succession, especially in the case of vegetatively expanding clonal species (e.g., *Solidago* spp.). These species may monopolize the space because they are capable of compensating for competitive disadvantage in unfavorable sites by reallocating photosynthate from ramets exploiting more favorable microsites (Hartnett and Bazzaz, 1985), or their phalanx spread competitively excludes other species.

6. *Synchronous demise of colonists*. The demise of the early successional dominants is synchronous relative to later successional species in the absence of disturbance. Upon the demise of the early dominants, a new cycle of establishment and dominance is initiated. If mortality of dominants is scattered through time because of asynchronous initial establishment or dominance of species with diverse life cycles, new species invade at different times, resulting in an age-structured population of dominants. Such asynchronous mortality and invasion would generate small-scale patchiness in establishment and regeneration.

The various steps outlined in the assumptions above can be represented schematically (Fig. 13.4). The model suggests several predictions about changes in heterogeneity for those cases in which succession moves from annuals to perennial herbs to woody species (Fig. 13.5). These predictions apply to large disturbed sites on relatively homogeneous soil.

The model predicts a net increase in heterogeneity superimposed on the cycles during succession (Fig. 13.5). The net increase appears because species capable of monopolizing the space, e.g., synchronously germinating annuals or rapidly invading clonal perennials, become less important later in succession. As succession proceeds, the dynamics change from a coarse scale to a finer spatial scale related to the patches formed by subsequent episodes of mortality and establishment. To test the latter prediction, we need techniques that permit assessment of the scale of patchiness (e.g., Robertson, 1987).

The number and amplitude of cycles of heterogeneity (Fig. 13.5) are related to the life cycles of the dominant species. Because annual species are frequently the first dominants, initial cycles would be short and of high amplitude. The height of the peak of heterogeneity following the death of the first-year dominant may be related to patchiness and size of the propagule pool. Heterogeneity again decreases as clonal perennials expand to occupy the site. It is likely that reduced heterogeneity lasts longer because the dominant is a longer-lived species. Scattered species establishment may occur throughout this period, producing age-structured populations of later successional dominants. We predict that this effect would reduce the

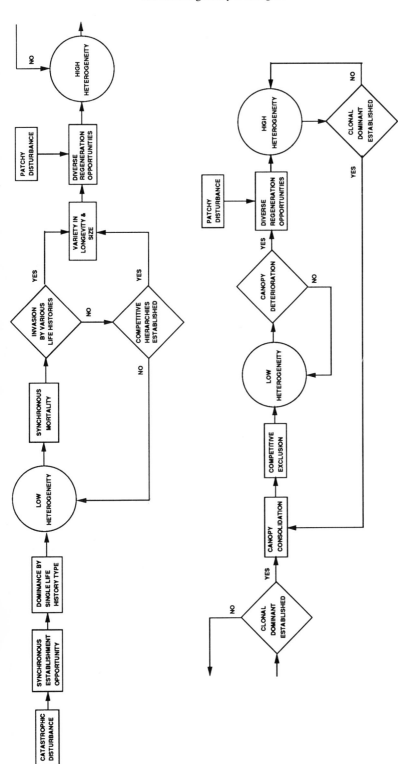

Figure 13.4. Mechanistic model for changes in heterogeneity during succession. The model is based on characteristics of disturbance, pattern of establishment, and life cycles of the dominant species. Low heterogeneity is associated with dominance by species that establish synchronously or with clonal species. High heterogeneity is associated with canopy deterioration and asynchronous establishment of species with various life histories. The system state of "high heterogeneity" at the right end of the top row of the model connects via the broken arrows to the decision point of "clonal dominant established" at the left end of the bottom row.

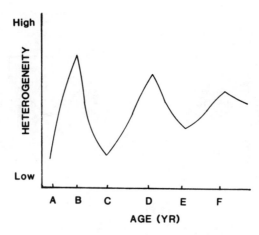

Figure 13.5. Changes in heterogeneity during secondary succession predicted for old fields in the eastern United States by the model of Figure 13.4. Letters along the axes represent known stages of succession in central New Jersey (Bard, 1952; Pickett, 1982; Armesto and Pickett, 1986). A = dominance by the annual *Ambrosia artemisiifolia*; B = dominance by a mixture of annuals and biennials; C = dominance by *Solidago* spp. clones; D = dominance by a mixture of shrubs and herbs; E = dominance by *Juniperus virginiana*; F = mixed forest.

amplitude and increase the wavelength of the heterogeneity cycles later in succession.

The predicted pattern can be altered in detail by specific conditions. Prior history or treatment of the site may alter the outcome of the model. Plowed fields, for instance, may be initially patchy because different species establish in ridges and furrows (Sterling et al., 1984). The dynamics predicted by the model should apply to each patch type if those patches are large enough. The presence of recruitment foci (McDonnell and Stiles, 1983) for avian-dispersed species would affect the predictions of our model and increase heterogeneity earlier in succession. If the seed rain is the dominant source of colonists relative to the seed bank immediately after the field is abandoned, initial heterogeneity would be higher (Fig. 13.5) and fluctuations would be greatly reduced.

Generality of Heterogeneity Cycles

We have seen that both site conditions and biotic interactions are capable of generating heterogeneity in successional communities. We have reviewed the small number of data available on heterogeneity through old-field succession to seek generality in temporal trends of spatial heterogeneity. To encourage further study and testing of these ideas, we have presented a conceptual model based on disturbance characteristics, life history, and autecology of species through succession. The structural essence of the model is that high and low periods of spatial heterogeneity alternate through succession (Figs. 13.4 and 13.5). The functional essence of the model is that periods of high heterogeneity represent periods of species invasion and establishment, whereas the alternating periods of low heterogeneity represent periods of species exclusion (Fig. 13.4). Similar alternating phases of invasibility and exclusion have been documented for forest and heathland successions (Watt, 1947; Bormann and Likens, 1979; Oliver, 1981; Christensen and Peet, 1984) and for arid ecosystems (Yeaton, 1978).

The model of alternating phases of exclusion and invasion is consistent with experiments on canopy disruption and species coexistence in early successional old fields (Allen and Forman, 1976; Armesto and Pickett, 1985, 1986). It also explains the temporal patterns of spatial heterogeneity inferred from the other studies of old-field successions in other regions in North America (Odum, 1960; Bazzaz, 1975; Tramer, 1975; Inouye et al., 1987b). Odum (1960) noted that during the first decade of succession ecosystem properties stabilized rapidly. He speculated that periods of stability would persist so long as the community was dominated by a particular life form. Such a suggestion is echoed in patterns discovered elsewhere. Greig-Smith (1961), in a study of *Ammophila*, found that initial patchiness disappeared in later stages of succession as patches coalesced. This idea has been extended by Brereton (1971) to include an initial random phase if the plants are present in low density. He suggested that each of the phases indicates alternating control by abiotic requirements of the species versus control by interactions among plants. A model of forest development in large disturbed areas on uniform soils (Armesto and Fuentes, 1988) conveys similar predictions: an initial phase of low heterogeneity followed by an increase in patchiness in old-growth forest due to gap-phase regeneration.

To adequately test the proposed model (Fig. 13.4), several conditions must be met: (1) Rigorous documentation of trends in spatial heterogeneity through time is needed. (2) A variety of successions must be compared. (3) Continuous samples through time are necessary. Finally, it is important to use indices and techniques that permit decoupling of spatial heterogeneity from other aspects of community structure and composition. The mechanistic predictions of the model, with their foundation on life history, autecology, and propagule supply, are then testable across a wide range of successions and environments.

Acknowledgments. J.J. Armesto acknowledges the support of the Cary Summer Fellowship of the Institute of Ecosystem Studies. This work is a contribution to the program of the Institute of Ecosystem Studies, with financial support from the National Science Foundation (BSR-8918551) and the Mary Flagler Cary Charitable Trust.

References

Allen EB, Forman RTT (1976) Plant species removals and old-field community structure and stability. Ecology 57:1233–1243

Armesto JJ, Fuentes ER (1988) Tree species regeneration in a mid-elevation, temperate rainforest in Isla de Chilo, Chile. Vegetatio 74:157–159

Armesto JJ, Pickett STA (1985) Experiments on disturbance in oldfield plant communities: impact on species richness and abundance. Ecology 66:230–240

Armesto JJ, Pickett STA (1986) Removal experiments to test mechanisms of plant succession in oldfields. Vegetatio 66:85–93

Barclay-Estrup P (1971) The description and interpretation of cyclical processes in a heath community. III. Micro-climate in relation to the *Calluna* cycle. J Ecol 59:143–166

Bard GE (1952) Secondary succession on the piedmont of New Jersey. Ecol Monogr 22:195–216

Bazzaz FA (1969) Succession and species distribution in relation to erosion in southern Illinois. Trans Ill State Acad Sci 62:430–435

Bazzaz FA (1975) Plant species diversity in old field successional ecosystems in southern Illinois. Ecology 56:485–488

Bormann FH, Likens GE (1979) *Pattern and Process in a Forested Ecosystem.* Springer-Verlag, New York

Brereton AJ (1971) The structure of the species populations in the initial stages of salt-marsh succession. J Ecol 59:321–338

Christensen NL, Peet RK (1984) Convergence during secondary forest succession. J Ecol 72:25–36

Evans FC, Dahl E (1955) The vegetational structure of an abandoned field in southeastern Michigan and its relation to environmental factors. Ecology 36:685–706

Facelli JM, D'Angela E, Leon RJC (1987) Diversity changes during pioneer stages in a subhumid pampean grassland succession. Am Midl Nat 117:17–24

Finegan B (1984) Forest succession. Nature 312:109–114

Greig-Smith P (1961) Data on pattern within plant communities. II. *Ammophila arenaria* (L.) Link. J Ecol 49:703–708

Greig-Smith P (1983) *Quantitative Plant Ecology.* 3rd ed. University of California Press, Berkeley

Grime JP (1979) *Plant Strategies and Vegetation Processes.* Wiley, New York

Hartnett DC, Bazzaz FA (1985) The integration of neighbourhood effects by clonal genetics in *Solidago canadensis.* J Ecol 73:415–427

Hutchinson GE (1953) The concept of pattern in ecology. Proc Acad Natl Sci Phila 105:1–12

Inouye RS, Huntly NJ, Tilman D, Tester JR (1987a) Pocket gophers, *Geomys bursarius*, vegetation and soil nitrogen along a successional sere in East Central Minnesota, USA. Oecologia 72:178–184

Inouye RS, Huntly NJ, Tilman D, Tester JR, Stillwell M, Zinnel KC (1987b) Old-field succession on a Minnesota sand plain. Ecology 68:12–26

Livingston RB, Allessio ML (1968) Buried viable seed in successional fields and forest stands, Harvard Forest, Massachusetts. Bull Torrey Bot Club 95:58–69

McDonnell MJ (1988) Landscapes, birds, and plants: dispersal patterns and vegetation change. In Downhower JF (ed) *The Biogeography of the Island Region of Western Lake Erie.* Ohio State University Press, Columbus, pp 214–220

McDonnell MJ, Stiles EW (1983) The structural complexity of old field vegetation and the recruitment of bird-dispersed plant species. Oecologia 56:109–116

Mellinger M, McNaughton SJ (1975) Structure and function of successional vascular plant communities in central New York. Ecol Monogr 34:161–182

Muraoka J, Armesto JJ, Pickett STA (1984) Comparisons of spatial heterogeneity in fields of different successional ages. Bull Ecol Soc Am 65:66 (abstract)

Nicholson SA, Monk CD (1974) Plant species diversity in old-field succession on the Georgia Piedmont. Ecology 55:1075–1085

Odum EP (1960) Organic production and turnover in old field succession. Ecology 41:34–49

Oliver CD (1981) Forest development in North America following major disturbances. Forest Ecol Manag 3:153–168

Palmer MW (1987) Variability in species richness within Minnesota oldfields: a use of the variance test. Vegetatio 70:61–64

Peet RH (1974) The measurement of species diversity. Annu Rev Ecol Syst 5:285–307

Pickett STA (1982) Population patterns through 20 years of oldfield succession. Vegetatio 49:45–59

Pickett STA, McDonnell MJ (1989) Seed bank dynamics in temperate deciduous forest. In Leck MA, Parker VT, Simpson RL (eds) *Ecology of Soil Seed Banks.* Springer-Verlag, New York, pp 123–147

Pickett STA, Collins SL, Armesto JJ (1987) Models, mechanisms and pathways of succession. Bot Rev 53:281–360

Platt WJ (1975) The colonization and formation of equilibrium plant species associations on badger disturbances in a tall grass prairie. Ecol Monogr 45:285–305

Raynal DJ, Bazzaz FA (1976) Interference of winter annuals with *Ambrosia artemissifolia* in early successional fields. Ecology 56:35–49

Reader RJ, Buck J (1986) Topographic variation in the abundance of *Hieracium floribundum*: relative importance of differential seed dispersal, seedling establishment, plant survival, and reproduction. J Ecol 74:815–822

Robertson GP (1987) Geostatistics in ecology: interpolating with known variance. Ecology 68:744–748

Ross BA, Tester JR, Breckenridge WJ (1968) Ecology of mima-type mounds in northwestern Minnesota. Ecology 49:172–177

Shafi MI, Yarranton GA (1973) Vegetation heterogeneity during a secondary (postfire) succession. Can J Bot 51:73–90

Sterling A, Peco B, Casado MS, Galiano EF, Pineda FD (1984) Influence of microtopography on floristic variation in the ecological succession in grassland. Oikos 42:334–342

Tilman D (1984) Plant dominance along an experimental nutrient gradient. Am Nat 125:1445–1453

Tramer EJ (1975) Regulation of plant species diversity on an early successional old-field. Ecology 56:905–914

Watt AS (1947) Pattern and process in the plant community. J Ecol 35:1–22

Webb LJ, Tracey JG, Williams WT (1972) Regeneration and pattern in the subtropical rainforest. J Ecol 60:675–696

Yeaton RI (1978) A cyclic relationship between *Larrea tridentata* and *Opuntia leptocaulis* in the northern Chihuahua desert. J Ecol 66:651–656

Zedler JB, Zedler PH (1969) Association of species and their relationship to microtopography within old fields. Ecology 50:432–442

14. Physical Heterogeneity and the Organization of Marine Communities

James P. Barry and Paul K. Dayton

Benthic and pelagic marine habitats, like terrestrial landscapes, can be viewed as mosaics of environmental quality produced by spatial and temporal variation in the physical and biological constraints encountered by populations. In marine systems hydrodynamic processes (water column stability, temperature and nutrient gradients, turbulent oceanographic features, storm disturbances) and biological processes (competition, grazing, predation) affect the recruitment and survival of populations. The emerging view of the dynamics of communities and populations as nonequilibrium systems (Chesson and Case, 1986; DeAngelis and Waterhouse, 1987) increases the interest in processes that generate or perpetuate heterogeneity. That variable environmental parameters cause gradients in community structure or the distribution of species is by no means a novel observation. Merriam's (1898) concept of life zones relating plant distributions to changes in air temperature along an elevation gradient was an early explanation of environmentally controlled community patterns. Terrestrial biomes (Clements and Shelford, 1939) are defined by large-scale changes in physical and biological characteristics of the landscape. In some marine habitats, however, these divisions may be less apparent, but they are nevertheless similarly heterogeneous over many scales in space and time.

In this chapter we explore the fundamental, but frequently obscure, role of environmental heterogeneity in controlling the distribution and abundance of species in marine communities. Physical features and processes that affect the distribution of species in planktonic marine communities have been reviewed by several authors (Bainbridge, 1957; Fasham, 1977, 1978; Haury et al., 1978; Smith 1978a; Angel 1984; Denman and Powell, 1984; Horne and Platt, 1984; Legendre and Demers, 1984; Tett and Edwards, 1984; Mackas et al., 1985; Harris, 1987; Smith, 1987; Haury and Pieper, 1988). These papers dealt mostly with the relation between oceanographic structure and the distribution of biomass for plankton populations. Legendre and Demers (1984) pointed out the greater awareness of physical-biological coupling in marine systems, which they termed "hydrodynamic biological oceanography." They emphasized, as we reiterate here, that "the various physical, chemical, and biological factors of the environment are considered as the proximal agents through which hydrodynamic variability is transmitted to living organisms." This perspective, they believed, requires a comprehensive sampling program to include consideration of the pertinent spatial and temporal scales of physical as well as biological variables. We agree that the role of physical processes in generating and maintaining patterns of biotic abundance and distribution has been underemphasized.

This view, of course, does not negate or diminish the importance of the many biotic factors known to structure communities. Rather than downplay the relevance of biotic interactions, we intend to underscore the value of a perspective that considers these interactions in the context of physically variable environments. Because the scale of observation often can determine the perceived importance of any factor, biotic or abiotic, it is important to consider the generality of our interpretation on other scales. Most biotic and abiotic processes are scale-dependent. Our intent here is to motivate ecologists to consider physical as well as biological patterns because they constitute the fundamental constraints to individuals, populations, and species. In particular, we argue that the outcome of many biotic interactions are influenced or mediated by abiotic environmental patterns. As such, physical heterogeneity plays a subtle but fundamental role in structuring communities, even in the context of species interactions.

We review some of the major sources of variability in the oceans that affect the structure of planktonic and benthic communities and highlight the physical features that are most important to benthic species in coastal communities. We then discuss some implications of this viewpoint for ecologists in terms of research approaches or perspectives. How can ecologists deal with habitat heterogeneity? What spatial or temporal scale(s) do we use?

Fundamental Processes Leading to Heterogeneity in Marine Ecosystems

On the scale of an ocean basin (thousands of kilometers), zonal winds, driven by the uneven heating of the earth's surface, cause circulation patterns in the ocean that divide it into distinct hydrographic regions that differ markedly in horizontal and vertical motion, nutrient fluxes, and seasonality (Reid et al., 1978). Fractionation of the kinetic energy from large-scale oceanic gyres and energy derived from baroclinic instabilities (gradients in potential energy) lead to the formation of several types of mesoscale (tens to hundreds of kilometers) features (Horne and Platt, 1984; Mackas et al., 1985). Oceanic circulation on this spatial scale is dominated by the earth's rotational forces, often leading to the formation of eddies. Several factors contribute to this fractionation process, particularly along coastlines. Energy from mesoscale features, tidal and regional wind forcing, and interactions of eddies or other sources of motion with bottom topography or continental margins can generate upwelling, fronts, eddies, continental shelf waves, filaments, internal waves, and other phenomena, all of which can affect the productivity and distribution of pelagic and coastal organisms. Although this energy originated from global-scale processes, it is fractionated in a turbulent cascade of energy to smaller and smaller scales until it is ultimately dissipated through molecular processes (Horne and Platt, 1984; Mackas et al., 1985).

There is a close correspondence between the temporal and spatial scales of these various physical phenomena and aspects of the biotic heterogeneity associated with them. Haury et al. (1978) illustrated this concept using a "Stommell diagram" of time and space scales for zooplankton biomass (Fig. 14.1); variability on small spatial scales occur over short time scales, whereas the largest variations (e.g., biogeographical provinces) vary over thousands of years. One exception to this general pattern is diel vertical migration, which ranges over a wide variety of spatial scales. The Stommell diagram, although it is based on limited information, clearly illustrates the general correspondence of time and space scales in the oceans and identifies several scales of oceanographic features relevant to patterns in pelagic marine communities. We return to this diagram later when faced with the question of how ecologists can integrate consideration of scale into sampling programs.

Heterogeneity can arise in marine communities from the nonuniform dispersion of individuals by physical processes, by spatially varying patterns of population growth, or both. In planktonic communities, small species with limited mobility (e.g., phytoplankton) are generally aggregated or dispersed by advective processes. Convergent circulation patterns coupled with limited, but directional, mobility (e.g., phototaxis) can cause the physical accumulation of individuals in particular areas. If the well mixed surface layer of the ocean is relatively shallow (thin) and phytoplankton are maintained in the euphotic zone, population growth can further increase their density. Larger organisms are less directly affected by advective processes but aggregate by behavioral patterns, usually in response to patterns of food availability or reproduction.

In general, benthic and pelagic patterns are linked to physical processes that regulate the stability of the water column and the flux of nutrients into

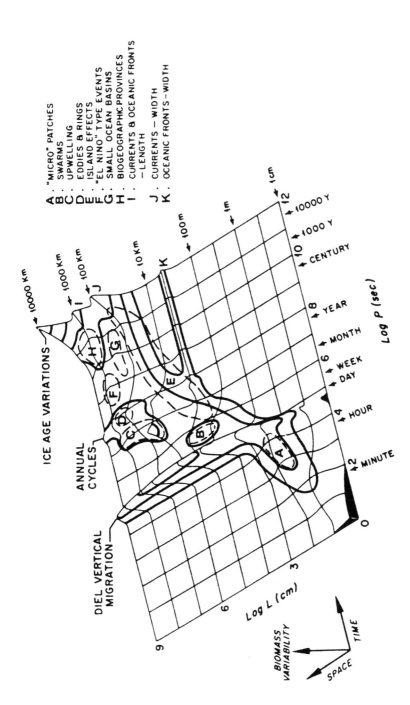

Figure 14.1. Stommell diagram of space and time scales of zooplankton abundance in the oceans. This conceptual model indicates the ranges of biomass variability for zooplankton populations. (From Haury et al., 1978.)

the euphotic zone; these processes control primary productivity in the surface waters over much of the ocean. Processes that increase stability and
nutrient levels in surface waters often generate the highest rates of primary
production. These conditions are, however, paradoxical; increases in water
column stability inhibit (vertical) advection and vice versa (Legendre,
1981). In other words, if the water column is strongly stratified (stable),
nutrients that are abundant below the euphotic zone are not mixed into the
surface waters where they are needed by phytoplankton. The other extreme, low stability and a deep mixed layer, also leads to low primary
production. In this case, however, nutrients are present in sufficient concentration owing to the more rapid and deep mixing of the surface waters,
but phytoplankton are mixed to depths where light levels are too low.
Areas with typically high rates of primary production tend to vary between
these extremes. For example, the temporal pattern of generation and relaxation of coastal upwelling events accomplishes both stability and vertical
mixing via pulses of nutrients into the surface waters (upwelling) followed
by increases in water column stability (relaxation) and high primary productivity, a coupling of temporal and spatial patterns on the proper scale
for phytoplankton growth. Elevated standing stocks of phytoplankton,
even for short periods, have cascading consequences throughout the food
chain; upper trophic levels respond with increased feeding rates, leading to
higher growth and reproduction. In addition, the high sedimentation rates
of organic matter in productive regions similarly increase benthic productivity. Pearson and Rosenberg (1987) ranked those processes that structure
marine infaunal communities in a hierarchical scheme derived principally
from the distribution of organic carbon. Because they both are important
for determining the amount and temporal patchiness of organic fluxes to
the sea floor, depth and latitude are primary influences on the structure of
marine sedimentary communities. Local or in situ processes such as water
motion, stochastic events, and biotic interactions are subordinate.

Biologically Relevant Hydrodynamic Features

Large-Scale Pattern

Large-scale patterns of primary productivity and the distribution of zooplankton and fishes often directly overlap climatic regions defined by the
physical features of oceanic gyres and constitute faunal provinces similar to

---▷

Figure 14.2. Faunal provinces in the Pacific Ocean, defined from the distributions
of 175 species of zooplankton. Darker shading indicates that a higher percentage of
species characteristic of that province are present in samples. These provinces correspond directly to the major pattern of surface circulation in the Pacific. (Courtesy
of Dr. J. McGowan.)

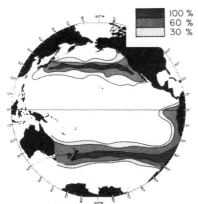

ESTIMATED PERCENT OF TRANSITION
ZONE FAUNA PRESENT

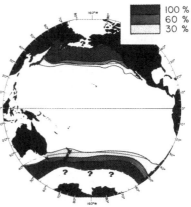

ESTIMATED PERCENT OF SUBARCTIC
OR SUBANTARCTIC FAUNA PRESENT

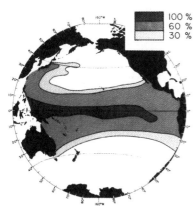

ESTIMATED PERCENT OF EQUATORIAL
FAUNA PRESENT

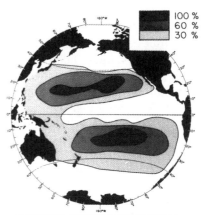

ESTIMATED PERCENT OF CENTRAL
FAUNA PRESENT

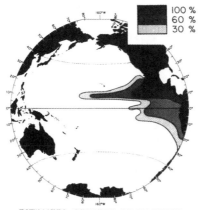

ESTIMATED PERCENT OF EASTERN
TROPICAL PACIFIC FAUNA PRESENT

the biomes of terrestrial communities (McGowan, 1971, 1974; Bachus et al., 1977; Reid et al., 1978). These patterns are well documented from the Pacific Ocean (Brinton, 1962; McGowan, 1971, 1986; Hayward and Mc-Gowan, 1979). Based on the distribution of hundreds of species of zooplankton, McGowan (1971) defined several faunal provinces (Fig. 14.2). Although these zoogeographic patterns are related to physical features, the critical factors controlling species abundances are unknown; it is unclear whether the limited dispersal of larvae and propagules, physiological tolerances of species, or interspecific biotic interactions along province boundaries are the major forces producing these patterns (Mackas et al., 1985). Beyond the simple maintenance of these patterns, McGowan stated that the processes that controlled their evolution also are unknown but must be related to physical processes. Faunal provinces also have been defined for coastal habitats. Along the western coast of North America, the large-scale distribution of benthos is most closely related to ocean temperature (Setchell, 1920; Hedgepeth, 1957; Newman, 1979a).

Benthic community patterns on such a large spatial scale are related to several factors, but in marine ocean basins they are most closely related to the depth and productivity of the overlying water column (Rowe, 1971, 1981; Mills, 1975); benthic biomass is low under central oceanic gyres and high under mid-to-high latitude gyres and continental shelf areas (Zenkevitch et al., 1971). Although the diversity of benthic megafauna generally increases to mesopelagic depths (1 to 3 km) then decreases toward the abyssal plain (4 km) (Haedrich and Rowe, 1977; Marshall, 1979; Haedrich et al., 1980), macrofaunal biomass shows a constant decrease with depth (Sanders and Hessler, 1969; Parsons et al., 1984). Superimposed on this depth relation, the biomass of macrofauna in the deep sea also is related to the proximity of sources of organic input (vertical or horizontal). Sources close to allochthonous inputs of organic matter, such as near continental shelves, have relatively high macrofaunal biomass. Jumars and Hessler (1976) found unexpectedly that the macrofaunal biomass of the Aleutian trench was unusually high, apparently owing to nearby sources of organic material.

Small Scale to Mesoscale Features

There also is close correspondence between patterns of physical and biological structure on scales of tens to hundreds of kilometers (Owen, 1981). Although mesoscale features are found throughout the oceans, boundary currents in particular are highly heterogeneous on this spatial scale. The well developed physical structure of coastal systems is evident from satellite images of sea surface temperature of boundary currents. In addition, the tight coupling of productivity to physical processes is shown by the frequently similar pattern of sea surface temperature and chlorophyll pigments. Higher trophic levels respond to patterns in primary productivity by behavioral aggregation and higher growth rates. Marine bird communities are aggregated on nearly all of the scales shown for plankton (see review by Hunt and Schneider, 1987).

Gulf Stream Rings

Mesoscale eddies have been intensively studied during the 1980s. In the North Atlantic, meanders of the Gulf Stream frequently separate from the main flow and form closed eddies (rings) around a core of either cold continental slope water or warm Sargasso Sea water from the south. Waters in the core of the ring retain the fauna and flora characteristic of their source. The rings drift away from the Gulf Stream and persist for weeks to years, changing in physical and biological character as they drift, until finally degrading or merging with the Gulf Stream. As isolated systems, these eddies are ideally suited for studying the progression of oceanographic and biological changes from the birth to the death of a ring. Wiebe et al., (1976) and The Ring Group (1981) summarized much of the early research on cold-core rings of the Gulf Stream, and Joyce and Wiebe (1983) presented an overview of warm-core rings. Similar eddies are found in other western boundary currents of most if not all of the world's oceans.

Eastern Boundary Current Turbulence

Eddies of various dimensions also are common in eastern boundary currents throughout the world and are affiliated with the distribution patterns of species from several trophic levels. The most well studied, the California current, has been sampled intensively since the late 1940s by the California Cooperative Oceanic Fisheries Investigations (Marine Research Committee, 1957). This study resulted in the compilation of an extensive time series (1949 to the present) of oceanographic characteristics and zooplankton abundance (zooplankton displacement volume) over a grid of stations covering thousands of kilometers.

The highly variable abundance and distribution of phytoplankton, zooplankton, and fish reflects the complex hydrodynamic structure of the California current (Hickey, 1979). Biotic abundances vary dramatically in the California current, especially when compared to the central gyre of the Pacific ocean, a larger-scale feature that is relatively homogeneous and stable (Haury, 1976; Hayward and McGowan, 1979; Chelton et al., 1982; McGowan and Walker, 1985; Brinton and Reid, 1986). McGowan and Walker (1985) compared the stability of pelagic zooplankton communities in the central Pacific gyre with those of the California current. Over thousands of kilometers and almost two decades the dominance hierarchy of the zooplankton community in the central gyre varied little, whereas in the California current zooplankton species shifted dramatically over small temporal and spatial scales (1 day, tens of kilometers). Interannual variation in the southward transport of the California current was directly related to temporal patterns of zooplankton biomass (Bernal, 1980; Chelton et al., 1982). Although McGowan and Walker (1979, 1985) attributed delineation of faunal provinces in the ocean to physical properties, they stated that the structure of the zooplankton community in the central gyre must be maintained by biological processes. The change from physical and biological homogeneity of the central gyre to marked heterogeneity for physical and well as biological pattern in the California current, however, suggests that physical factors exert strong control over biological patterns.

Coastal Eddies

Coastal eddies of various size affect the productivity of epipelagic and coastal planktonic and benthic communities. Mesoscale eddies can impinge along coastlines and modify phytoplankton productivity by advecting high nutrient waters into coastal areas (Smith, 1978b; Yentsch and Phinney, 1986). For example, Tont's (1976, 1981) analyses of coastal diatom blooms indicated that the approximately 5-week periodicity and duration of coastal blooms was related to California current eddies moving through the region. Vertical advection within and around eddies can increase the input of nutrients to the euphotic zone, particularly in the center of the eddy (Freeland and Denman, 1982).

Mesoscale eddies entrain the larvae of several nearshore species and provide mechanisms to increase their rate of return to coastal habitats via closed gyral circulation, especially in areas with stationary eddies (Johannes, 1978; Sale, 1980; Owen, 1981; Haury et al., 1986; Lobel and Robinson;1986). On smaller scales, several coral reef fish spawn at locations and times that result in larvae being swept off the reef into the open ocean where offshore eddies can entrap the larvae (Emery, 1972; Johannes, 1978; Lobel, 1978). Studies of the recruitment patterns of *Thalassoma bifasciatum*, a coral reef wrasse, on Caribbean reefs corroborate this idea, indicating that the patch size of larvae in the plankton was on the order of mesoscale features (Victor, 1984).

Within the southern California eddy, a persistent mesoscale feature of the southern California bight, recruitment rates of red sea urchins (*Strongylocentrotus franciscanus* and *Strongylocentrotus purpuratus*) are regular and less variable than to the north, outside of the eddy (Tegner and Barry, ms). The sheephead (*Semicossyphus pulcher*), a resident fish in coastal kelp beds, exhibits a similar pattern (Cowen, 1985). Compelling evidence for the importance of this process is shown during periods when the intensity or extent of such eddies is modified. The recruitment patterns of several fish in the southern California bight were changed dramatically during the 1982–1984 El Niño event; age structure analyses and recruitment densities indicate that recruitment of sheephead to San Nicolas Island, normally outside the influence of the southern California eddy, occurred only during years when the eddy pattern was disrupted and northward flow was intensified (Cowen, 1985).

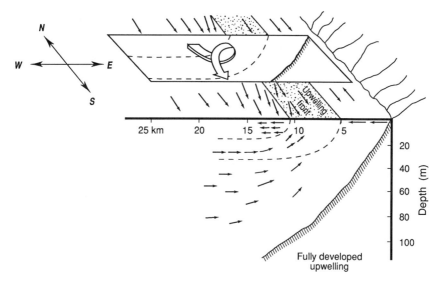

Figure 14.3. Conceptual illustration of fully developed coastal upwelling off the coast of Oregon. Arrows indicate the direction of flow. Dashed lines indicate lines of constant density. (From Barber and Smith, 1981.)

Coastal Upwelling

Coastal upwelling is a process by which nutrient-rich water from below the euphotic zone is brought to the surface as a consequence of horizontal and vertical advection along coastlines bordering the eastern margins of large anticyclonic atmospheric gyres over the ocean basins; a simplified diagram of upwelling is shown in Figure 14.3. Ekman drift, the net movement of water toward 90° to the left in the northern hemisphere and 90° to the right in the southern hemisphere, is thought to be responsible for the offshore drift of surface waters in the major coastal upwelling centers in the world. The offshore Ekman drift of surface waters, driven by wind stress parallel to the coastline, is replaced by the upwelling of subsurface waters, usually most intensively about 10 to 20 km from the coast, which are then entrained by the Ekman drift and transported offshore. The Ekman layer is on the order of tens of meters deep and is separated from a still deeper subsurface layer that flows shoreward to replace upwelled water. To complete the circulation cell, a convergence front usually forms at the outer edge of upwelling zones, up to several hundred kilometers offshore. This process is driven by the seasonal position and intensity of anticyclonic atmospheric gyres and strengthens when wind stress is maximal along the coast during the spring. Details of the physical dynamics of upwelling are not completely understood, and models including one-, two-, and three-celled systems have been proposed (see Richards, 1981; Hutchings et al., 1986).

Rather than a continuous coastal band, upwelling regions are physically heterogeneous in both time and space. These events are generated by wind stress variations caused by changes in large-scale atmospheric patterns. Events occur about four to six times per season (Huyer, 1976), and inter-event relaxation of upwelling is thought to be related to changes in the wind-field caused by coastally trapped atmospheric Kelvin waves (Beardsley et al., 1987; Send et al., 1987). Upwelling events usually occur at specific locations (upwelling centers) along a coastline and are characterized by large plumes (also termed filaments, jets, and squirts) of water that flow at high velocities (ca. 0.5 m/second up to several hundred kilometers away from the coast for several days to weeks (Flament et al., 1985). Filaments often form eddy-like features, with fronts along at least some of its boundaries. Research in upwelling systems (Coastal Ocean Dynamics Experiment) (Beardsley et al., 1987) has shown that the interaction of upwelling-favorable winds and coastal topography (changes in coastline orientation) can produce highly elevated and channelized surface winds, leading to locally high wind stress on the ocean surface (Winant et al., 1987, 1988) and suggesting that jets and squirts are generated under these conditions.

The advection of high nutrient waters into the euphotic zone and the presence of a two-layered flow that generates and maintains the vertical stability of the water column allow phytoplankton populations to grow rapidly without being mixed below euphotic depths. Because of large spatial and temporal variations, the dynamics of biological productivity in upwelling systems are complex. In general, however, an event includes a successional sequence with a period of high primary and secondary production followed by a decay period. Rapid phytoplankton growth is usually followed by an increase in zooplankton productivity, leading to high grazing rates, although in some cases, potentially high grazing rates by zooplankton early in the sequence or other unknown factors prevent the development of high phytoplankton standing stocks (Barber and Smith, 1981).

The often extreme patchiness in phytoplankton communities in upwelling areas is caused by several factors and is intimately tied to physical processes. Plankton abundances are related to the age of the upwelled water; upwelling communities go through a succession of stages from newly up-welled water with a small but rapidly growing phytoplankton assemblage and low zooplankton densities to higher standing stocks of both groups (see Barber and Vinogradov, 1980). Because of the transient and turbulent nature of upwelling systems, many species have difficulty remaining in suitable habitats and are patterned primarily by a combination of physical processes and differential growth and mortality. Other species (mainly zooplankton) are able to use vertical migrations to use the two-layer flow of upwelling to maintain their position in upwelling regions (Peterson et al., 1979).

Nekton and higher vertebrates in upwelling zones control their distributions through behavioral processes but are also distributed according to patterns ultimately derived from physical processes. Briggs and Chu (1987) showed that the abundances of several seabirds along the California coast were greatest near areas of intense upwelling, and their reproductive patterns were timed to coincide with the peak of the upwelling season. Upwelling frontal zones, in particular, have high concentrations of prey for several species of marine and coastal birds (Briggs et al., 1984; Briggs and Chu, 1987). Cushing (1982) cites whaling catch statistics for sperm whales showing that their distribution is related to the position of divergence zones at the margins of upwelling. The dependence of South American fur seals on upwelling productivity was dramatically illustrated during the 1983 El Niño when foraging success dropped, leading to abnormally high mortality of newborns and yearling pups (Trillmich et al., 1986). Even intertidal communities are affected by upwelling. Bosman et al. (1987) showed that organisms at all trophic levels in an intertidal community were influenced by the nutrient enhancement and high primary productivity of coastal upwelling; intertidal algae had higher production rates, leading to high productivity and standing stock of higher trophic levels. Along the California coast upwelling centers are usually located near points and form stationary, gyre-like circulation cells. Recruitment rates of sea urchins to nearby habitats are patchy on horizontal scales consistent with gyre-like flow; rates of recruitment are low inshore of the upwelling plume and higher to the south near the shoreward return (Ebert and Russell, 1988).

Benthic community patterns undoubtedly are strongly influenced by upwelling processes but are less well studied than planktonic patterns. The vertical flux of organic carbon sinking to benthic sedimentary communities in upwelling regions is high relative to other habitats owing to the shallow depth (usually over the continental shelf or slope) and to the high productivity of these systems (Suess, 1980; Dunbar, 1981). High sedimentation rates of organic material result in high food availability for benthic organisms. The organic carbon content of sediments in upwelling zones are generally much higher than comparable depths of nonupwelling areas (5 to 26% versus 0.5 to 2.0%) Diester-Haass, 1978; Thiel, 1978), and the biomass of benthic macrofauna reflects the high productivity of the overlying water. In some upwelling areas, however, high oxygen demand for bacterial degradation can result in low oxygen levels in the water column and sediments, particularly where the oxygen minimum impinges on the bottom, leading to anaerobic sediments and reduced macrofaunal densities (Rowe, 1971; Thiel, 1978).

Thorson (1950) was aware of the problem of larval retention and predicted that populations in regions with intense upwelling may experience catastrophic larval mortality due to offshore transport. For some species Thorson's prediction has been quantitatively documented. Parrish et al. (1981) noted that coastal fish in areas where offshore Ekman transport is greatest along the western North American coast (upwelling zones of California) spawn during the winter when surface drift is generally toward the coast, even though the most productive months, when larvae food levels are greatest, occur during the spring and summer. In fact, many of the coastal species of upwelling centers are migrating species that spawn in the California bight, where larvae are retained in the Southern California eddy. Roughgarden et al. (1988) showed that interannual variation in the rate of recruitment of intertidal barnacles along the central California coast, as well as the spatial distributions of their larvae in the plankton, were directly related to the Bakun index of coastal upwelling, which is indicative of offshore transport. During 1983 upwelling was negligible and recruitment rates were nearly two orders of magnitude greater than during 1985, when (during the 5-year study) upwelling was at its peak. Further evidence that larvae were advected offshore is shown by the abundance of larvae during the same period; when upwelling was weak (1983), barnacle larvae were found only up to 5 miles from the shore compared to at least 125 miles during other years. Although larval transport may be negatively affected by upwelling, high benthic productivity (catch rate) is correlated with more intense upwelling for some species (*Cancer magister, Siliqua patula, Pandalus jordani*) of nearshore environments (Peterson, 1972; Peterson and Miller, 1975).

Coastal upwelling has undergone intensive study since the 1970s, motivated in part by the considerable economic importance of fisheries of upwelling areas such as the Peruvian anchovy fishery. Although upwelling ecosystems account for only 0.1% of the ocean surface, 50% of the world fish catch are taken from these areas (Ryther, 1969). Several overviews of the physical and biological dynamics and patterns óf upwelling systems are available (Cushing, 1971; Barber and Vinogradov, 1980; Barber and Smith, 1981; Richards, 1981; Denman and Powell, 1984).

Fronts

Oceanic fronts are regions of contact between water masses that differ in the value of some parameter, such as temperature, salinity, or nutrient concentration; the parameter in question generally changes dramatically over short horizontal scales, usually owing to patterns of vertical circulation. This term has been used to describe a variety of structures from small scales (several meters) to features as large as the boundaries of oceanographic provinces covering several degrees of latitude. Many types of front exist in the ocean, and they may form at any depth near coastal regions and in midoceanic areas (see reviews by Bowman and Esaias, 1978; Owen, 1981; Denman and Powell, 1984).

Fronts form under a variety of conditions. Along coastlines, river or estuarine flows create salinity fronts as small as a few meters across. Shelf break fronts are located at the boundary between continental slope or oceanic waters and colder, less saline, continental shelf waters. Upwelling fronts form at the edges of upwelled water masses and are particularly evident in coastal upwelling sites. Shallow sea fronts occur when turbulent mixing is introduced to a stratified water column in shallow areas with fairly large tidal velocities; fronts form at the region where bottom friction becomes greater than the stratification of the water column. Seaward, the water column remains stratified; shoreward the sea is well mixed (Simpson and Pingree, 1978). Deep sea fronts develop at the junction of current systems and at the transition zones between westerly and easterly dominated atmospheric circulation (Owen, 1981).

The longevity of fronts varies according to oceanographic conditions, ranging from a few hours to being nearly permanent. Larger-scale features are generally more persistent; for example, the Antarctic convergence is a permanent temperature front. Shallow sea fronts are usually seasonal and depend on spring and summer warming to increase water column stratification prior to their formation, but they may vary considerably in extent and permanence (Pingree, 1978; Simpson and Pingree, 1978).

Primary productivity is enhanced near tidal fronts, particularly on the stratified edge of the front, where water clarity is high and nutrients are frequently upwelled into the euphotic zone by frontal hydrodynamic processes (upwelling, eddies) (Pingree et al., 1976, 1977; Simpson and Pingree, 1978). The abundance of plankton also is often higher at or adjacent to frontal zones (Owen, 1981; Nihoul, 1981, 1986; Denman and Powell, 1984; and references therein), but the mechanisms producing this pattern are sometimes obscure. Scrope-Howe and Jones (1985) (cited in Haury and Peiper, 1988) showed that even though zooplankton abundance was normally lower at a tidal front in the Irish Sea than the surrounding area, there were episodes of high zooplankton abundance at the front. These episodes were dominated by nauplii of calanoid copepods rather than adults and were associated with high levels of surface chlorophyll. Smith et al. (1986) also found higher concentrations of copepod nauplii in frontal zones of upwelling plumes off California. Moreover, these nauplii had feeding rates that were at least twice that outside the frontal zone, suggesting that higher production and population growth rates, rather than hydrodynamic mechanisms, were responsible for the higher abundances at or near the fronts. Consumer groups also aggregate near or in tidal fronts, probably in response to high food densities. Pingree et al. (1974) found that crustacean (copepod, euphausiid, meroplankton) densities in the frontal region of the English Channel were 74.5 times greater than outside the front, and several species of birds were feeding near the surface, particularly in downwelling convergences near the front. Marine mammals also utilize these tidal fronts as feeding grounds (Gaskin, 1968, 1976). Tidal fronts form at the boundary of stratified and mixed water masses on continental shelves, according to the intensity of tidal flow, water column stability, and water depth.

Although the relation between benthic community patterns and tidal fronts has been poorly documented, it is clear that hydrodynamic processes exert a strong influence on the structure of benthic shelf communities. The presence of tidal fronts, and especially the potential scouring effect of tidal currents, are directly responsible for the distribution of sedimentary types and benthic faunal associations in some areas. Bottom sediments are differentially sorted according to the intensity of currents, resulting in a range of sediment types from fine sediments and muds in sluggish waters to coarse or no sediments in energetic tidal environments. Warwick and Uncles (1980) demonstrated that sediments and benthic faunal assemblages in the Bristol Channel (UK) were directly related to tidal current stresses. In areas of high tidal flow, sediments were completely scoured and the bottom was dominated by a mussel community. Where sediments were present, tidal flows were directly correlated with grain size and were an important determinant of community structure (Warwick and Davies, 1977).

On the Bering Sea shelf, the presence of tidal (and other) fronts determine the distribution of nutrients, primary production, and the advection of plankton, leading to well defined patterns of planktonic and benthic faunal distributions (Grebmeier et al., 1988, 1989). Iverson et al. (1980), Goering and Iverson (1981), and others showed that the presence of outer, middle, and inner Bering Sea shelf fronts result in different schedules of seasonal phytoplankton blooms that differ in species composition. An oceanic group of larger species that can consume large chain-forming diatoms dominates the herbivorous zooplankton seaward of the middle shelf front, and a middle shelf community composed of species too small to handle larger diatoms are dominant inshore of the front. This disparity in grazing ability and the presence of the middle shelf front, which acts as a barrier to movement of the two grazer assemblages, leads to the accumulation of phytoplankton and high sedimentation rates of organic carbon in the middle shelf zone. This benthic food resource increases the production and standing stock of benthic macrofauna (fish and crabs) and deep-diving sea birds (Schneider and Hunt, 1982; Woodby, 1984). In contrast, herbivorous zooplankton on the outer shelf have higher grazing rates and productivity and thus a larger, more productive pelagic food chain, evidenced by the large stocks of birds (surface feeding), mammals, and pelagic fish (Schneider and Hunt, 1984; Schneider and Piatt, 1986). Radio-tag tracking of northern fur seal foraging trips from the Pribilof Islands show most individuals to be feeding on or seaward of the outer shelf (Loughlin et al., 1987).

It is important to note that time scales for biological and physical phenomena are not always the same. Near coastal upwelling plumes, fronts persist on a scale of several days to weeks, but population growth rates for phytoplankton and zooplankton range from days to years (Denman and Powell, 1984). Thus although phytoplankton populations may have ample time to respond to even short-lived fronts, zooplankton species usually have generation times much longer that the persistence of the front. Such events may nevertheless be critical to the dynamics of some populations. Failure of frontal development during an important stage in the life of a copepod or other zooplankter may result in year class failure due to low food concentrations or other factors. In response to the highly variable nature of upwelling systems, species typically found in upwelling zones apparently have evolved adaptations that allow successful feeding and reproduction (Barber and Smith, 1981; Haury and Pieper, 1988).

Internal Waves

Just as wind waves and swell travel across the air–sea interface of oceans and lakes, internal waves travel along the density interfaces or gradients under the sea surface. Internal waves are common throughout the oceans, especially on continental shelves where they are generated by tidal and wind stress forces and range in period from minutes to several hours (Pond and Pickard, 1983; Fu and Holt, 1984). If the water column is well stratified, internal waves travel horizontally along the pycnocline and, owing to the orbital motions of water associated with wave propagation, cause the formation of convergence and divergence zones at the sea surface; these circulation cells can aggregate some species of phytoplankton and zooplankton (Zeldis and Jillett, 1982; Haury et al., 1983) and result in the concentration of higher trophic levels (Brown, 1980; Haney, 1987). Surface convergence slicks associated with internal waves are potentially much more important to coastal species; plankton that are buoyant or that swim toward the surface can become aggregated in surface slicks found at convergence zones and can be carried with the wave as it propagates shoreward (Shanks, 1983; Kingsford and Choat, 1986). This mechanism is a potentially important one by which the larvae of benthic and marine species return to nearshore habitats. Because wave refraction affects internal waves as it does surface waves, the effect of coastal bathymetry on wave refraction patterns also is an important determinant of the shoreward flux of larvae by internal waves (Shanks, 1987; Shanks and Wright, 1987).

Langmuir Circulation

Langmuir cells—rows of helical circulation cells oriented parallel to the wind that may be several kilometers long but only about a meter in width and depth—are thought to be generated by interactions of the wind and surface drift associated with wave fields (Barstow, 1983). Depending on their swimming rates and behavior, plankton can become concentrated within langmuir cells (Fig. 14.4). Buoyant species aggregate in convergence zones near the air–sea interface; and sinking species may become concentrated near divergence zones (Parsons et al., 1984). Pelagic larvae and adults of many marine organisms, as well as other items, including micro- and macroalgae, zooplankton, leaves, foam, bird feathers, ice, oil, and tar, are found in aggregations in Langmuir cell convergence zones

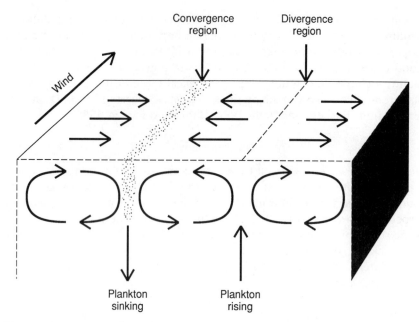

Figure 14.4. Conceptual drawing of Langmuir circulation showing the concentration of plankton in surface convergence zones. (From Nybakken, 1988.)

(Alldrege, 1982; Harding et al., 1982; Barstow, 1983; Jillett and Zeldis, 1985; Hamner and Schneider, 1986; Haney, 1986; Shanks, 1987).

Microscale Hydrodynamic Features

On even smaller scales, biological patchiness can be generated by the response of species to microscale structure in the oceans. In plankton communities, horizontal and vertical patchiness of the order of centimeters to a meter have been observed for several species (Cassie, 1959, 1963; Emery, 1968; Clutter, 1969; Haury et al., 1978; Silver et al., 1978). For example, marine snow (aggregates of marine organic detritus derived from several sources) is a site of locally high bacterial and photosynthetic activity (Silver et al., 1978; Knauer et al., 1982; Angel, 1984). Even on the air–sea interface, bacterial, phytoplankton, and microheterotrophic abundances are typically high (Sorokin, 1981).

Even though the time scales of microscale hydrodynamic features are usually short, they can have important biological consequences. Microscale concentration of planktonic prey is thought to increase the availability of food for the early larvae of some pelagic fish (Vlymen, 1977; Lasker, 1981a,b). Periods of low wind stress, which permit microstratification of the surface waters, lead to concentrations of food that exceed the minimum threshold for larval anchovy and result in higher rates of growth and survival (Lasker, 1981b; Peterman and Bradford, 1987).

On a still smaller scale, the excretory plumes of zooplankton may be an important source of limiting nutrients for phytoplankton in some areas (McCarthy and Goldman, 1979; Angel, 1984). As yet, it is unclear if these plumes can contribute to the level of primary production (Eppley and Petersen, 1979; Williams and Muir, 1981).

Polar Environments

The presence of a polar ice sheet on the ocean creates several features that contribute to heterogeneity in planktonic and benthic communities not found in lower latitude ecosystems. Seasonal variations in the extent of the polar ice sheets are enormous, particularly around the Antarctic, where sea ice cover can change from 20×10^6 km^2 during winter to 3×10^6 km^2 in summer (Gordon, 1981); the cover of Arctic Sea ice does not undergo such large seasonal change. This difference leads to different patterns for local biota for each polar region. In the north, surface feeding species are excluded from some regions year-round and are limited to polynyas (ice-free locations) and the relatively stationary ice edge, whereas in the Antarctic both producers and consumers have distributions that vary with the large seasonal shifts in ice cover (see review by Smith, 1987).

Physical oceanographic processes near the margins of polar ice sheets influence the structure and productivity of planktonic and benthic communities. Phytoplankton blooms in high latitudes, unlike low latitude systems; they are rarely nutrient-limited and, instead, are controlled by light and water column stability. As polar ice sheets degrade during spring, localized upwelling and, more importantly, the input of low-salinity sea ice meltwater into the surface layer of the ocean leads to increased water column stability, shoaling of the mixed layer, and initiation of spring phytoplankton blooms (Smith and Nelson, 1985, 1986; Smith, 1987). The high productivity of phytoplankton concentrates individuals from all levels of the food chain; zooplankton, fish, birds, and mammals aggregate near regions of high chlorophyll (Ainley and Jacobs, 1981; Fraser and Ainley, 1986; Nelson and Smith, 1986; Wilson et al., 1986). In particular, frontal regions are important feeding areas for apex predators such as marine mammals and birds (see review in Ainley and Demaster, in press).

The interaction of ice cover and the regional current field can also generate heterogeneous patterns of phytoplankton populations, leading to similar patterns for benthic algae and invertebrates. In eastern McMurdo Sound, Antarctica, for example, currents from the north carry plankton-rich waters to nearby benthic communities that have extremely high standing stocks of macrofaunal invertebrates. Less that 40 km away, benthic populations are one to two orders of magnitude less abundant, likely because of the low phytoplankton content of currents originating under the Ross Ice Shelf (Dayton et al., 1986; Barry, 1988a; Barry and Dayton, 1988).

Sea ice formation also affects the distribution of species on small to micro scales. Because "anchor ice" (ice that forms on the sea floor at shallow depths, less than 30 m) limits the survival of several species at these depths, there is a marked depth zonation in these subtidal benthic communities (Dayton et al., 1969, 1970). Even within sea ice, changes in light levels due to variations in snow and ice thickness create differences in the distribution and abundance of sea ice microbial communities (Palmisano and Sullivan, 1983; Garrison et al., 1986; Gosselin et al., 1986). Underneath sea ice the formation of platelet ice can provide refuge space or simply an attachment surface for microalgae, leading to heightened productivity on higher trophic levels.

Hydrothermal Vents

The fauna of hydrothermal vents, distributed along fissures and edges of tectonic plates, are a notable exception to the pattern of most benthic communities. Unlike other benthic communities below euphotic depths, these isolated systems are not dependent on the allochthonous input of organic matter. Instead, chemolithotrophic bacteria act as the primary producers on which consumers depend for food. These relatively recently discovered marine communities and similar communities found at some deep sea cold seeps deserve a chapter of their own but are not considered here; they have been reviewed by Jones (1985) and have yielded several important insights into the evolution of symbiotic relationships. Much more importantly, they furnish habitats for several mesozoic relics (Newman, 1985), which include the most primitive living sessile barnacle (Newman, 1979b; Newman and Hessler, 1989).

Effects of Topography and Substratum

On almost any spatial scale, modification of the flow field by bottom topography can result in patchiness in biotic communities (Owen, 1981; Denman and Powell, 1984). On a global scale, wind-driven oceanic gyres exist because of the deflection of currents by continents; these closed gyral circulation cells define faunal provinces in the ocean (see above). A classic example of the effect of the interaction of currents and topography on plankton communities is the formation of tidal fronts in areas with strong tidal currents (see above).

Current Intensification

The interaction of currents with pinnacles and seamounts also generate heterogeneous community patterns. Genin et al. (1986) analyzed the distribution of macrofauna on a deep seamount and related their abundance patterns to variations in current speeds over the seamount. Currents were intensified near peaks, and the densities of benthic filter feeders were much higher than at similar depths on the relatively flat slope of the seamount. Plankton communities near seamounts have long been known to be rich, compared to nearby regions (Genin et al., 1988). Genin and Boehlert (1985) tested the hypothesis that topographically induced upwelling above a seamount could lead to higher rates of primary productivity in the surface waters. Although their data did not show elevated production rates, they hypothesized that such upwelling events are episodic and might result in higher productivity in the down-current region near the seamount.

Topographic influences can be an important determinant of community structure on small as well as large spatial scales. In addition to modifying the distribution of larvae settling on the substratum, changes in water motion induced by topography can result in differential feeding rates for sessile benthic invertebrates (Schmidt and Warner, 1984). Sebens (1984, 1986) showed that even though zooplankton densities were similar throughout his study areas, the growth rates of anthozoans were related to their position on vertical or horizontal walls. Individuals higher on the wall where water flow was greater or in areas with greater wave motion had higher growth rates due to higher prey fluxes past feeding tentacles despite homogeneous zooplankton densities. Differential growth caused by variation in the flow field introduced patchiness to the structure of the benthic community.

On even smaller spatial scales, small changes in the speed and direction of currents in the immediate vicinity of the feeding tentacles can modify the prey capture rates of the crinoid *Oligometra serripinna* (Holland et al., 1987; Leonard et al., 1988). Current speeds found to be most efficient for feeding were similar to those measured in the habitat typical for these crinoids, indicating that microhabitat selection is strongly linked to the local microscale current field.

Wave Effects

Waves and wave-borne objects often affect the distribution, growth, and survival of benthic species in shallow water communities (Kitching, 1941; Southward and Orton, 1954; Jones and Demetropoulos, 1968; Harger, 1970; Dayton, 1971, 1973; Menge, 1976; Seapy and Littler, 1978; Denley and Underwood, 1979; Paine, 1979; Velimirov and Griffiths, 1979; Foster, 1982; Kastendiek, 1982; McQuaid and Branch, 1984, 1985; Denny, 1985; Witman, 1987). Although increased water motion can elevate the feeding rates of some species (Sebens, 1984), sessile species must be able to withstand the force of waves without being damaged or dislodged. On exposed coastlines, resistance to wave forces is likely a strong selective force determining the morphology as well as some of the life history characteristics of coastal species. Denny (1988) presented an extensive review of the physics of wave motion and its effects on intertidal organisms, clearly illustrating mechanisms by which water motion can determine the distribution of coastal species. Spatial and temporal variation in the frequency and intensity of damage from wave-borne objects such as rocks and rolling boulders can affect the structure of coastal populations and interrupt competitive monopolies. Algal assemblages on small boulders usually are maintained in early successional stages compared to larger boulders, which are overturned less frequently by waves (Sousa, 1979; Littler and Littler, 1984). In the Pacific northwest, damage to intertidal communities by logs tossed ashore by waves increase the mortality of exposed individuals (Dayton, 1971). On many coastlines rolling boulders and even pebbles affect the demography of shallow water species. Limpets in wave-exposed intertidal boulder fields have higher mortality rates than those in more protected habitats, leading to modified size frequency distributions and age structures (Shanks and Wright, 1986).

Substratum Effects

Much of the ocean bottom is covered with sediments that vary according to characteristics of the overlying water; organic-rich muds are typical below highly productive waters near continental margins; fine-grained deposits (oozes) low in organic content are most abundant in open ocean benthos. The diversity and abundance of all sizes of benthic faunal assemblages are patterned according to characteristics of sediments including grain size, sorting, and organic content (Petersen, 1913; Jones, 1950; Thorson, 1957; Rhoads and Young, 1970; Gray, 1974; Rhoads, 1974; Coull et al., 1977; Marshall, 1979; Grebmeier et al., 1988, 1989). Hughes et al. (1972) used multivariate techniques to show that substratum characteristics accounted for 46% of the variance in the frequency of occurrence of polychaete and echinoderms in St. Margaret's Bay, Nova Scotia.

In the deep sea and other sedimentary communities that are not disturbed by water motion, sediment oxygen profiles show strong gradients with aerobic conditions on the surface and a reduced sulfide zone of anaerobic conditions only millimeters to centimeters below. The vertical distributions of many species of meiofauna and microfauna are determined by the position and steepness of this oxygen gradient (see Fenchel, 1969, for review). In euphotic depths, even the penetration of light into sediments influences the vertical distribution of benthic photosynthetic bacteria due to changes in intensity and wavelength with increasing depth in the sediments over millimeters of depth (Perkins, 1963; Fenchel and Straarup, 1971).

Similar patterns arise from variation in the quality of hard substrata. Even subtle differences in the quality of the substratum, particularly for shallow water species, are capable of causing large changes in community structure. Raimondi (1988) showed that the rate of barnacle settlement was higher on granite than on basalt substrata in an intertidal community, perhaps related to differences in the thermal capacity of these substrata. Regardless of potential variations in the rate of settlement to alternative substrata, postsettlement survivorship may vary according to substratum effects (Caffey, 1982; Barry, 1988b; see review in Crisp, 1984).

Physical characteristics of the substratum such as contour, grain size, porosity, adsorbed chemicals, and many others can cause the nonrandom distribution of newly settled larvae and other propagules (Jones, 1950; Gray, 1974). The literature concerning settlement behavior and preference for benthic species is enormous. For example, several species of barnacles respond to chemical stimuli or other substratum characteristics, leading to highly aggregated patterns (Crisp and Meadows, 1963; Crisp, 1974; Barnett and Crisp, 1979; Denley and Underwood, 1979; Grossberg, 1982; Burke, 1983; Hudon et al., 1983; Yule and Crisp, 1983; Wethey, 1984; Connell, 1985; Chabot and Bourget, 1988; Le Tourneux and Bourget, 1988; Raimondi, 1988). Similarly, *Phragmatopoma lapidosa californica*, a tube-building polychaete from intertidal and shallow subtidal benthic habitats, has highly specific responses to chemical compounds from the tubes of conspecifics, resulting in a highly aggregated distribution (Jensen and Morse, 1984; Pawlik, 1986).

Until recently, active settlement choice was considered the primary mechanism by which larvae settling from the plankton became heterogeneously distributed. Irrespective of larval behavior, microscale topography can cause settling larvae to become highly patterned, even for passive species (Eckman, 1983; Hannon, 1984; Bhaud et al., 1985). Fine to microscale patterns of flow are now being recognized as determinants of the distributions of larvae and adults. Butman (1987) presented an excellent review of the literature and the evidence for both active selection and hydrodynamic hypotheses, concluding that both processes are important. In addition, hydrodynamic processes control the distribution of larvae over the bottom, causing the aggregation of passive and swimming larvae on several spatial scales (Parrish et al., 1981; Cameran and Rumrill, 1982; Eckman, 1983; Hannon, 1984; Jackson, 1986). This process is responsible for a large component of the microscale distribution of some barnacle larvae (Wethey, 1986).

Biogenic Environmental Heterogeneity

In addition to physical cues, biotic interactions and (in particular) biogenic environmental changes may be responsible for much of the microscale horizontal patchiness in benthic communities, especially in soft bottom habitats of the deep sea (Jumars, 1975, Levin et al., 1986). Adult–adult and adult–larval interactions within and between species contribute to microscale patchiness through competitive interactions or simply by settlement inhibition (Rhoads and Young, 1970; Gray, 1974; Rhoads, 1974; Levin, 1981, 1982). Tube-building species are an example of the potentially major modification of the quality of the sedimentary habitats for other species. Tube mats can bind sediment and stabilize the sea floor. One species, *Owenia fusiformis*, by selecting particular shapes of sand grains for its tube, can concentrate the mineral hornblende by 25-fold or more. Together with an anemone, *Zaolutus actius*, *O. fusiformis* can stabilize the sediments such that there are several other species of animals and plants that are associated with these areas in the midst of otherwise shifting sediments (Fager, 1964; but see Eckman et al., 1981).

On a smaller scale, *Molpadia oolitica*, a holothuroid of Cape Cod Bay, Massachusetts, ingests fine-grained particles from below the surface of the sediment and deposits feces at the surface in a fecal mound. This microheterogeneity, in the form of topographic relief, provides a suitable site for the settlement and growth of suspension feeders, which are not present between the mounds (Rhoads and Young, 1971). Jumars (1975) showed that one of the most significant microscale features affecting the distribution of benthic paraonid polychaetes was the presence of mudballs created by cirratulid polychaetes. Marine plants such as sea grasses and algae also stabilize sediments and alter environmental quality for several other species (Petersen, 1986). Even in wave-disturbed habitats, increased sedimentation in macroalgal turfs results in changes in the associated faunal and floral assemblages (Barry, 1988b). Rhoads (1974) reviewed the literature on organisms–sediment relations of benthic sedimentary communities and discussed several important environmental consequences of biotic processes, including the effects of tube mat formation, sediment sorting, and bioturbation leading to the mixing and transport of particles, water, nutrients, and dissolved gases. In particular, Rhoads pointed out the importance of the formation and deposition of fecal pellets to the size distribution of particles and its effect on the topography, porosity, and aeration of sediments.

Biological Interactions and Heterogeneity in Marine Communities

The importance of biotic interactions in many benthic communities in shallow water habitats are well known and are only briefly mentioned here. Predation intensity is known to have dramatic effects in several communities, particularly for species with the capability of broad control over other subordinate species in the community (e.g., Connell, 1970; Dayton, 1971, 1984; Paine, 1974, 1980; Estes and Palmisano, 1974; Estes et al., 1978). Competitive interactions are similarly important (Connell, 1961; Paine, 1966, 1980, 1984; Dayton, 1971; Dayton et al., 1974. The roles of biological interactions in plankton communities are less well documented but are nevertheless considered to be important. Even though the megascale faunal provinces of plankton communities in the oceans are bounded by hydrodynamic features, the processes that control the structure of these communities are, by lack of contrary evidence, thought to be biological (McGowan and Walker, 1985).

Refugia

We mention biological processes here mainly to emphasize the role of physical heterogeneities in regulating the intensity of these interactions. Variation in the quality of the substratum can contribute to heterogeneous species distributions by modifying physiological processes linked to settlement, growth, and reproduction, as well as by mediating biological interactions that can control recruitment and survivorship.

Woodin (1978) outlined five types of refugia related to disturbance, most of which are well documented in benthic communities. Temporal refugia can include periods affected and unaffected by a disturbing factor. Nocturnally active prey have refuge from diurnal predators (Nelson and Vance, 1979). "Fugitive species" utilize life history strategies or rapid maturation and high fecundity to persist within the effective temporal period of disturbances (e.g., Dayton, 1973; Paine, 1979). Spatial refuges are defined by two types: refugia outside the limit of the disturbance and physical heterogeneities within the range of the disturbance. Connell's (1961) classic study of barnacle zonation illustrated the former spatial refuge, as do several other studies (e.g., Paine, 1974; Menge, 1978; Hay, 1981; Kastendiek, 1982; Littler et al., 1983; D'Antonio, 1986). Crevices or small pits in the substratum provide some species with refugia by scour from boulders, ice, and other disturbances (Emson and Faller-Fritsch, 1976; Birkeland and Randall, 1981; Garrity and Levings, 1981; Menge and Lubchenco, 1981; Lubchenco, 1983; Menge et al., 1985; Bergeron and Bourget, 1986; Shanks and Wright, 1986). Biogenic refugia are noteworthy in that the interaction of physical and biological processes leads to increased habitat quality for associated species. *Diopatra*, by stabilizing the sediment, creates a spatial refuge from disturbance for several species of benthic invertebrates (Woodin, 1978). These sorts of refugia can be produced in many ways; the increase in substratum texture due to the presence of barnacles leads to lower competitive stress on *Patelloida*, a small limpet outcompeted for space on smooth surfaces (Creese, 1982). The defense of nesting territory by damselfish creates a spatial refuge for benthic invertebrates, which escape from the higher nearby predation rates (Lobel, 1980). Finally, some individuals can attain refuge in size, such that they are not vulnerable to some predators (e.g., Dayton, 1971; Paine, 1974; Tegner and Levin, 1983).

Physical-Biological Coupling and Community Patchiness

Individuals of any species encounter a profusion of environmental stresses, including those related strictly to physical features and those associated with biological entities or processes. We have presented several examples of physical processes or patterns that increase the heterogeneity of marine populations, and have reviewed some of the literature concerning the role of biogenic modification of environmental quality that increases the patchiness of associated species. In addition to heterogeneity produced by physical or biogenic changes in the environment, the structure of communities can be determined by the effects of physical features on biotic interactions. Here we use examples from two coastal communities to show how physical patterns of the environment can mediate biotic interactions.

Substratum Effects in a Rocky Intertidal Community

Substratum characteristics have a dual role in controlling the demography of epilithic species in rocky intertidal habitats in southern California by modifying the rates of physical and biological disturbances (Barry, 1988b). In the mid to upper intertidal zones of intertidal headlands composed of a mosaic of soft friable sandstone interlaced with much less friable (hard) sandstone, the abundances of invertebrates and algae differ significantly between substrata (Fig. 14.5). Two processes contribute to this pattern. First, the abundances of sessile invertebrates on either rock type are related to rates of settlement and survival. Mussels recruit to both rock types in the middle intertidal zone but are removed from soft sandstone more frequently and intensively by large storm waves owing to the lower adhesive strength of mussels on soft, versus hard, rock (Fig. 14.6). Barnacle densities also are lower on soft sandstone because of its more rapid rate of erosion. Second, substratum type mediates competitive interactions for space between sessile invertebrates (mainly barnacles) and limpet grazers. On soft rock the effects of grazing by limpets are much more severe than on hard rock, leading to low barnacle abundance on soft rock in the presence of limpets (Fig 14.7). Thus although limpet grazing alone can decrease barnacle survival and abundance, changes in habitat quality exacerbate this biotic disturbance for barnacles on soft sandstone.

Kelp Beds

Spatial and temporal variations in the structure of kelp bed communities provide another example of the interacting roles of physical and biological processes. Kelp beds thrive in shallow rocky habitats of temperate coastlines with cool, nutrient rich waters. On a global scale, patterns of sea surface temperature and the availability of nutrients determine their distribution. On a regional scale, substratum type and water clarity generally determine their longshore and cross-shore distribution; most kelps (actually

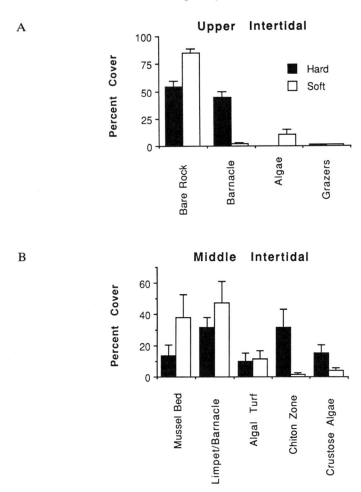

Figure 14.5. Percentage cover of intertidal biota on friable (soft) and unfriable (hard) sandstones along the coast of San Diego, California (Barry, 1988b). (A) The upper intertidal zone is dominated by barnacles (*Chthamalus* spp.) and bare rock. (B) The middle intertidal zone is dominated by mussels (*Mytilus* spp.), a limpet/barnacle assemblage similar to the upper intertidal, macroalgal turf, and crustose algae.

genera from both the Laminariales and Fucales) occur only on hard substrata, though several species are capable of colonizing and anchoring themselves in soft sediments (Dayton, 1985a). Not all hard substrata are alike, however, and mudstone or soft sandstone can increase the severity of disturbance by large storms as well as decrease water clarity by erosion (Cowen et al., 1982; Dayton, 1985a). Light attenuation with depth and

A

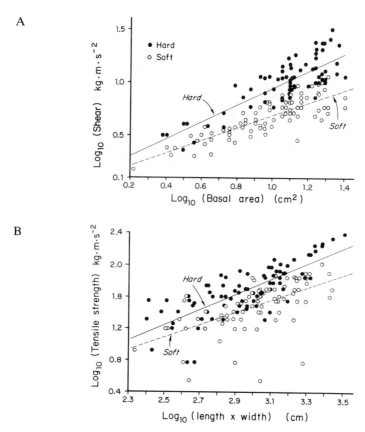

B

Figure 14.6. Adhesive strengths of barnacles (*Chthamalus* spp.) and mussels (*Mytilus* spp.) on hard and soft sandstone. (A) *Chthamalus*. (B) *Mytilus*.

turbidity restrict kelp to shallow habitats (see review in Drew, 1983). Near their distributional boundaries, periods of nutrient limitation may devastate local populations (Gerard, 1982; Zimmerman and Robertson, 1985), and in these areas the injection of nutrients from deeper waters by internal waves is important to kelp growth (Zimmerman and Kremer, 1984). On smaller scales, biological interactions appear critical to the structure of kelp beds, just as in some terrestrial forests competition for light leads to recruitment inhibition and local competitive exclusion (Dayton, 1975a,b; Dayton et al., 1984; Reed and Foster, 1984).

The interaction of physical and biological processes causes spatial and temporal changes in the distribution of species within and around kelp beds. The presence of kelp beds modifies coastal oceanographic patterns and provides habitats for several species of planktivores that utilize kelp as a habitat from which to forage. Coastal currents, waves, and associated pro-

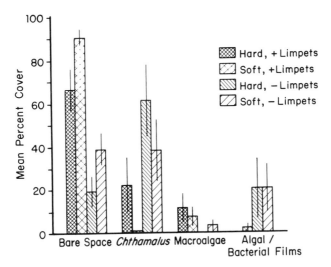

Figure 14.7. Community structure in the upper intertidal zone on hard and soft sandstone with and without the presence of grazers. Note the negative effect of grazers (limpets) on barnacle cover for both substrata and the near absence of barnacles on soft rock when limpets were present.

cesses are modified when they encounter the kelp bed, leading to dramatic changes in the behavior and distribution of several species. The drag by kelp beds on coastal currents causes the current to sweep around the bed, with some of the flow diffusing into and through the bed (Jackson and Winant, 1983); internal waves, which can transport larvae and neuston shoreward, are also disrupted by kelp beds (Jackson, 1984). Some species, particularly planktivorous fish, aggregate near the edge of kelp beds where larvae and zooplankton may accumulate owing to changes in advective patterns (Bray and Ebeling, 1974; Hobson and Chess, 1976; Bray, 1981). Sluggish currents within the bed and the attenuation of internal waves by the kelp bed, coupled with high predation rates at the edge and within the kelp bed, reduce the shoreward flux of larvae toward intertidal habitats, thereby increasing the horizontal patchiness within these communities (Bernstein and Jung, 1979; Gaines and Roughgarden, 1987). In this way interactions between kelp beds and physical patterns and processes determine, in part, the predation rates on planktonic larvae and zooplankton, with a consequence of even greater heterogeneity for inshore coastal communities.

Grazing pressure, particularly by sea urchins, has a significant effect in kelp bed communities, often leading to complete removal of all kelp. In South American kelp beds, grazing by urchins restricts kelp to shallow habitats, where wave surge inhibits grazing activities (Dayton, 1985b). Along North American coastlines the outcome of interactions between urchins and kelp varies in space and time and, in some cases, is determined by temporal variations in physical factors (Ebeling et al., 1985; Johnson and Mann, 1986; Witman, 1987). Intense winter storms remove the detrital material that normally rains to the bottom of kelp communities, feeds urchin populations, and leads to a change in urchin foraging behavior and heavy grazing pressure on any remaining plants. Severe storms can cause reversals of community structure from an urchin-dominated barren ground with little or no kelp to a luxuriant kelp bed with low urchin densities, or back, depending on the initial conditions (Ebeling et al., 1985).

Temporal Heterogeneity

Although the bulk of this chapter has dealt with spatial variation, temporal changes in physical stress and the intensity of biological interactions are similarly important and merit a chapter of their own; they are only briefly explored here. Heterogeneous patterns in marine communities can be found on almost any temporal scale. Here we focus on examples from two scales, partly due to personal bias and partly because of recent general interest in episodic events and climate change.

El Niño Southern Oscillation Events

The effects of the 1982–1984 El Niño Southern Oscillation (ENSO) event were so striking that it was more intensively investigated than earlier ENSO events. In addition, more comprehensive oceanographic and atmospheric data were available during this event, allowing better documentation of their relation to biological changes. Although the processes that lead to ENSO events are poorly understood, it appears that a decrease in the intensity of easterly trade winds and the relaxation and reversal of equatorial surface currents lead to a warming of surface waters in the eastern equatorial Pacific (Halpern et al., 1983; Luther et al., 1983). Several other oceanographic and biological phenomena are associated with extreme ENSO events, including high latitude changes in oceanographic and atmospheric flow (Barber and Chavez, 1983; Wooster and Fluharty, 1985). These events are periodic and occur approximately once every 5 to 6 years, with intense events about each 12 years.

During the 1982–1984 event, a few coastal communities of California were strongly affected. The normally consistent southerly flow of nutrient-rich water in the California current deceased and even reversed, leading to warmer, nutrient-poor conditions in southern California. The depth of the surface layer of warm, nutrient-poor water increased (Simpson, 1983), leading to extremely low levels of nutrients, chlorophyll *a*, and macrozooplankton biomass. These conditions were reflected in all trophic levels of pelagic food chains by depressed productivity (Fiedler, 1984; Fiedler et al.,

1986; McGowan, 1985). In addition to poor conditions for primary production, winter storms were more intense and frequent owing to a deeper than normal Aleutian low pressure center, leading to several extreme wave events that severely damaged coastal biotic communities (Dayton and Tegner, 1984a; Seymour et al., 1984; Ebeling et al., 1985).

The effects on benthic communities were less dramatic that in pelagic communities and were most pronounced in kelp bed communities (Dayton and Tegner, 1990). The combination of low nutrient conditions and severe winter storms devastated coastal kelp beds. Following this event, however, the kelp recruited rapidly, and most beds recovered to pre-ENSO status within 1 to 2 years. Several kelp-associated animals were also affected and had anomalous recruitment patterns shortly after the event, but this variability was rapidly attenuated as the kelp beds recovered. ENSO-related changes in intertidal communities were less evident (Gunnill, 1985; Paine, 1986), except for the disturbance effects of waves (Barry, 1989).

Episodic phenomena such as ENSO events generate heterogeneity in biological communities via their effects on spatial and temporal patterns of species abundances. The extreme storms associated with ENSO events can cause persistent changes in the structure of kelp communities, as replacement species may form stable associations in areas earlier occupied by a different suite of species (Dayton and Tegner, 1984a; Ebeling et al., 1985). For long-lived species, disastrous disturbance events (sensu Harper, 1977) are encountered frequently enough to contribute to the evolution of these species and thus provide a mechanism by which populations may respond to episodic events. For some species of brown algae, dispersal tactics appear to be related to water motion associated with storms (Reed et al., 1988). Even far below surface waters, large storms affect benthic habitats. Time series measurements of various parameters show that the effects of large storms and hurricanes can penetrate the deep ocean, causing benthic storms (Gardner and Sullivan, 1981).

Some species have adaptations to recover from mortality caused by episodic, but important, environmental events. *Phragmatopoma lapidosa californica*, a tube-building marine annelid, lives in large aggregations that form small reefs along the Pacific coast of North America; it appears to respond to extreme storms with increased reproductive output. Damage to colonies caused by waves and wave-borne objects induce the worms to shed gametes into the water column, a reproductive response analogous to seed dehiscence in some fire-adapted plants of terrestrial communities. Between 1983 and 1987, the recruitment rates of *Phragmatopoma* to intertidal reefs near San Diego, California were highly correlated with recent wave height, related to the potential for disturbance to the colonies (Barry, 1989) (Fig. 14.8). Recruitment of *P.1. californica* to intertidal habitats was dense throughout southern California following the extreme storms of the 1982–1983 winter but was low during the succeeding 2 years with mild conditions. Recruitment occurred again during spring 1986, following the large

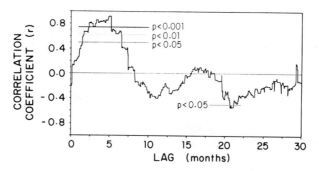

Figure 14.8. Correlation between the percentage cover of *Phragmatopoma lapidosa californica* on intertidal reefs in San Diego, California, with recent wave disturbance events (maximum wave height during the preceding 45 days). The highest correlation occurs with a lag of *P. l. californica* cover 2 to 5 months after wave disturbances. Percentage cover increased after intense wave events, apparently due to increased reproductive output.

storms of the previous winter. This species appears to have evolved a reproductive tactic by which it increases its reproductive output according to environmental cues that indicate an increased potential for successful recruitment (free space created by storms) and impending mortality (damage to colonies).

Climatic Variation

Fisheries statistics often provide the longest time series of biological data available for analysis of climatic variations. Cushing (1971, 1982) has presented several sources of evidence to show that climatic variations over years to decades result in changes in the distribution of fish stocks and, in some areas, a dramatic shift in a suite of community species. Mysak et al. (1982) analyzed 40 to 80 years of coastal sea level, temperature, and salinity data from the northeastern Pacific Ocean and found cycles for periods of 2.3, 3, 5 to 6, and 11 years. The 5- to 6-year period was related to changes in the composition of the sockeye salmon and herring catch at several sites; the average weight and the total catch of sockeye salmon were coherent with the 5- to 6-year signal that propagated northward at a rate similar to that predicted for a coastally trapped Kelvin wave. The role of environmental forcing in regulating interannual changes in groups of species or entire communities is shown by the similar patterns of recruitment of marine fish in the northeastern Pacific Ocean. Hollowed et al. (1987) found that extreme years in a 30-year series of year-class strength occurred simultaneously throughout the region studied for 59 stocks from 28 species of fish.

The Russell cycle (Cushing and Dickson, 1976; Southward, 1980) is a classic example of decadal and longer-period cycles in community patterns that are rarely studied because of the long time series requirement. During the mid to late 1920s, the abundance of several species in the English Channel essentially disappeared and were replaced by other species, corresponding to a drop in the magnitude of the winter phosphate maximum. The quantity of macroplankton, the arrow worms *Sagitta elegans and Sagitta setosa*, herring, and nonclupeid fish decreased or disappeared between 1925 and 1935. At the same time pilchard abundance increased. This pattern persisted until the mid-1960s, when the number of fish larvae increased. By the mid-1970s winter phosphorus, as well as the abundances of most species that had disappeared during the 1930s, had returned to their former levels.

Discussion

Our goal in this chapter has been to review some of the important processes that increase the heterogeneity of benthic and planktonic communities in marine environments. Although still recognizing the power of biological interactions, we have tried to highlight the fundamental role of physical processes and point out the importance of biogenic environmental heterogeneity produced by the interaction of biotic and abiotic factors.

Relevance of Perspective and Scale

Ecologists often have utilized one of two perspectives when studying ecosystem patterns and processes. The functional viewpoint of systems ecologists focuses on energy flow through systems, with little regard to individuals, populations, or species. In contrast, population and community ecologists have, in a strict sense, concentrated on biological interactions and deemphasized or ignored the importance of physical patterns and processes in order to understand community patterns and dynamics. Neither approach is unjustified, and enormous insights into natural patterns have emerged from both. Nevertheless, neither approach provides an understanding of roles of physical and biological properties in creating and maintaining patterns on all spatial and temporal scales in natural communities. If we view the regulation of ecosystem patterns and processes as spatiotemporal hierarchies (sensu O'Neill et al., 1986), abiotic environmental factors must be viewed as the highest level of organization within which other factors are subordinate. On the largest scales, physical patterns define the distributional boundaries of groups of species—the faunal provinces of the oceans. Progressively smaller scale features are more transient, yet are coupled to heterogeneity in planktonic and benthic marine communities. Thus ecologists must consider the potential effects of physical processes, as they set the stage for other levels of population control. This fact by no means implies that other processes (e.g., competition) are unimportant—rather, they should be considered in the context of the physical setting and the relevant scales.

Because the scale of observation alone determines, in large part, the type of question, methodology, and results of most projects (Dayton and Tegner, 1984b), research questions should, whether the question is related to topics such as nutrient cycling or population dynamics, include consideration of the level of organization within a hierarchy of ecosystem processes. O'Neill et al. (1986) suggested that functional and community perspectives form dual hierarchies that encompass ecosystem patterns at different levels of organization or scale, with community studies more appropriate for lower-level or smaller-scale studies.

The appropriate scale of study may vary considerably depending on the question or approach. Systems ecologists must be aware of spatial and temporal variability on the scale of the system or compartment of interest. In contrast, the emphasis for community ecologists must relate to processes that control the structure of the community (recruitment, growth, and reproduction) rather than energy flow, and the relevant scale may differ even in the same system. For questions relevant to ecological time frames, consideration must be given to processes responsible for temporal and spatial variations in the recruitment of important species. In contrast, ecologists interested in evolutionary processes must consider scales appropriate to address genome patchiness or population boundaries.

Returning to the Stommel diagram of space and time scales in the ocean (Fig 14.1), we can see that observations of any marine system over a particular scale limits our ability to measure variability on other scales. Over short time frames, large-scale temporal cycles are relatively invariant, whereas short-term fluctuations may be undetectable by observations on a larger scale. For example, seasonal changes in the flux of organic carbon to the sea floor would be undetectable in a week-long study of benthic patterns, but these seasonal cycles would be seen as noise in studies of similar processes from sediment cores encompassing hundreds to thousands of years. Thus by virtue of a limited perspective, processes active at the scale of the study are likely to be related to most of the variability in any measure. If we are aware of the scale dependence of our research, we are more likely to gain a broader understanding of the limitations and relevance of a study.

In addition to consideration of the appropriate temporal and spatial scales, how can ecologists incorporate the concept of hydrodynamic biological oceanography into a research program? First, knowledge of the myriad patterns that can be generated by physical processes in the ocean furnishes a broader perspective. Second, study of biological responses to hydrodynamic or other physical processes requires measurement of the appropriate biological parameters on the relevant scales. It may require new techniques, particularly when attempting to measure rates of variation for biological properties (e.g., biomass or abundance) rather than simply the level of the variable (Legendre and Demers, 1984).

Coupling the Scales of Physical and Biological Processes

The importance of coupling the scales of physical and biological processes is poorly understood. For some systems, such as the dynamics of phytoplankton productivity in coastal upwelling ecosystems, the physical and biological scales are similar and result in a highly responsive physical and biotic coupling. Zooplankton production operates on a different scale (seasonal) but still matches a relevant temporal scale of upwelling systems. Other scales of climatic variations, such as ENSO events, can disrupt ecological patterns; but it is not clear how tightly coupled, if at all, biological processes are to such events. Not all scales of physical processes produce strong responses in biotic systems, and the role of environmental forcing is not well understood, even for species important for human exploitation (e.g., commercial fishery species).

Recognition of the crucial importance of recruitment processes to adult abundance, especially for species with long-lived pelagic larvae, increases the need to couple population studies to hydrodynamic processes on the appropriate scales. Although the suggestion that variable rates of recruitment are the primary regulator of the abundances of many marine organisms is long-standing (Hjort, 1914, 1926; Thorson, 1950; Coe, 1956; Loosanoff, 1964), ecologists, particularly those studying benthic communities, have since concentrated on small-scale biotic interactions, primarily between adults. There has been a return, however, to "supply-side ecology"—the idea that the dynamics of adult populations are more related to the arrival rates of larvae than to postsettlement processes (Underwood and Denley, 1984; Roughgarden et al., 1987; Underwood and Fairweather, 1989); many studies in the rocky intertidal zone have corroborated Thorson's ideas (e.g., Gaines et al., 1985; Gaines and Roughgarden, 1985, 1987; Sutherland, 1987; Roughgarden et al., 1988).

In hopes of explaining interannual variability in recruitment and the apparent lack of a relation between adult stock size and recruitment for several coastal fishery species, population models have increased the emphasis on hydrodynamic processes that control regional phytoplankton blooms and larval retention. Hjort's (1914) critical phase hypothesis predicted that interannual variations in oceanic circulation play a role in recruitment variation for herring populations. Cushing's (1975) match/mismatch hypothesis related fish spawning success to the coincidence of phytoplankton blooms and the location of drifting larvae. The herring hypothesis (Iles and Sinclair, 1982) has implicated oceanographic processes even more strongly, postulating that herring stocks are determined by geographically stable larval retention areas defined by topographic features and circulation patterns. Sinclair (1988) regarded the regulation of marine species with planktonic larvae primarily as a function of physical processes that regulate the rate of loss of individuals (vagrancy) from geographically suitable areas; other processes are thought to be active but less important.

The recognition of physical features and processes as potentially controlling interannual variation in population growth also has led to a greater awareness of the importance of these processes in the evolutionary history of these species. Because individual fitness may be related to selection for responses to environmental characteristics, species undoubtedly have undergone such selection or perished. Sinclair (1988) termed it "life cycle selection," where selection is strong to remain a "member" of a spatially defined population.

Benthic, Pelagic, and Terrestrial Comparisons

Pelagic systems appear to be organized by processes different from those of benthic or terrestrial communities (Dayton, 1984; McGowan and Walker, 1985). Benthic communities are strongly affected by biological processes that regulate the utilization of space. On rocky substrata in particular, one or a few apecies often are able to monopolize space [e.g., barnacles (Connell, 1961) and mussels (Paine, 1966)], and heterogeneity may arise only by disturbance events via physical processes [e.g., wave-related effects (Dayton, 1971)] or biological interactions [e.g., predation on mussels by sea stars (Paine, 1966, 1974)]. Species with disproportionately large effects (usually predators) often are characteristic of systems with competitive monopolies. In sediment communities the roles of biological interactions are less clear; competitive monopolies occur, e.g., tube-mat forming worms (Woodin, 1978), but the foraging activities of large demersal predators can disturb benthic sediments and disrupt monopolies (VanBlaricom, 1982; Oliver et al., 1983a,b; Fukuyama and Oliver, 1985).

In planktonic communities, the strength of biological interactions in regulating the distribution and abundance of species appears much weaker than in the benthos, even when physical factors are relatively benign. Zooplankton communities in the central gyre of the Pacific Ocean are not chaotic and exhibit highly stable structure over many years and thousands of kilometers; yet no evidence of "keystone" species or competitive dominance has been discovered (McGowan and Walker, 1985). McGowan and Walker nevertheless postulated that biotic interactions must be important owing to the fruitless search for physical factors that could generate such diversity and stability.

Differences in community regulation between benthic and pelagic communities are analogous to differences between marine and terrestrial systems. Long-term environmental records have indicated that environmental variability is greater in terrestrial systems than in the ocean, particularly on short to medium time scales (Steele, 1985). The inherent variability of terrestrial systems may constrain terrestrial biota to particular periodicities, whereas in the ocean species are less closely tied to these scales of variation (Steele, 1985).

Conclusions

The concept of hydrodynamic biological oceanography, simply put, states that variations in the distribution and abundance of organisms in marine communities are intimately tied to physical processes, either directly through obvious constraints on the dispersal and survival of individuals or indirectly by affecting the intensity of other biological and physical controls. Knowledge of the inherent variability of the marine environment provides a framework on which studies of marine ecosystem dynamics can be based. Rather than ignoring the importance of biological patterns or interactions, this perspective considers potential sources of environmental heterogeneity that are generated by physical pattern, the interaction of physical and biological factors, and their subsequent effects on individuals.

To best understand the dynamics of communities and ecosystems, we as ecologists must be aware of the fundamental role of physical patterns, particularly on the appropriate spatial and temporal scales. Biotic interactions must be considered in the context of the inherent physical variability of the system in question. Abiotic gradients on the appropriate scale can alter the intensity of biotic interactions and indirectly modify the outcome of interactions. Furthermore, although competition, predation, and other biotic interactions dominate the regulation of community patterns on some scales, the outcome of these interactions often is linked to physical variability and the interaction of physical and biological factors. Incorporation of this perspective into ecological research, rather than restricting studies to particular formats or questions, allows a broader knowledge of the relative importance of physical and biological factors and the scale dependence of these processes.

Acknowledgments. We thank several people for their assistance, for either editing early versions of the manuscript or providing illustrations: L. Haury, J. McGowan, J. Nybakken, J. Pineda, E. Portillo, D. Reed, and W. Wakefield. Because of its broad scope, the theme of this chapter has been considered, at least in part, by several other authors. For planktonic systems, we have benefited from the outstanding reviews prepared during the past several years that have provided a wealth of information. In particular, the reviews of Denman and Powell (1984) and Haury and Pieper (1988) have been useful.

References

Ainley DG, Demaster DP (1990) The upper trophic levels in polar ecosystems. In Smith WO (ed) *Polar Oceanography*. Academic press, Orlando, pp 599–630

Ainley DG, Jacobs SS (1981) Sea-bird affinities for ocean and ice boundaries in the Antarctic. Deep Sea Res 28A:1173–1185

Alldredge AL (1982) Aggregation of spawning appendicularians in surface windrows. Bull Mar Sci 32:250–254

Angel MV (1984) Deep-water biological processes in the northwest region of the Indian Ocean. Deep Sea Res 31:935–950

Bachus RH, Craddock JE, Haedrich RL, Robinson BH (1977) Atlantic mesopelagic zoogeography. In *Fishes of the Western North Atlantic*. Part 7. Sears Foundation for Marine Research. Yale University, New Haven, pp 266–287

Bainbridge R (1957) The size, shape, and density of marine phytoplankton concentrations. Biol Rev 32:91–115

Barber RT, Chavez FP (1983) Biological consequences of El Nino. Science 222:1203–1210

Barber RT, Smith RL (1981) Coastal upwelling ecosystems. In Longhurst (ed) *Analysis of Marine Ecosystems*. Academic Press, Orlando, pp 31–68

Barber RT, Vinogradov ME (eds) (1980) *Productivity of Upwelling Ecosystems*. Elsevier, Amsterdam

Barnett BE, Crisp DJ (1979) Laboratory studies of gregarious settlement in *Balanus balanoides* and *Elminius modestus* in relation to competition between these species. J Mar Biol Assoc UK 59:581–590

Barry JP (1988a) Hydrographic patterns in McMurdo Sound, Antarctica and their relationship to local benthic communities. Polar Biol 8:377–391

Barry JP (1988b) Pattern and process: patch dynamics in a rocky intertidal community in Southern California. Dissertation, University of California, San Diego

Barry JP (1989) Reproductive response of a marine annelid to winter storms: an analog to fire adaptation in plants? Mar Ecol Prog Ser 54:99–107

Barry JP, PK Dayton (1988) Current patterns in McMurdo Sound, Antarctic and their relationship to local biotic communities. Polar Biol 8:367–376

Barstow SF (1983) The ecology of Langmuir circulation: a review. Mar Environ Res 9:211–236

Beardsley RC, Dorman CE, Friehe CA, Rosenfeld LK, Winant CD (1987) Local atmospheric forcing during the coastal ocean dynamics experiment 1. A description of the marine boundary layer and atmospheric conditions over a northern California upwelling region. J Geophys Res 92:1467–1488

Bergeron P, Bourget E (1986) Shore topography and spatial partitioning of crevice refuges by sessile epibenthos in an ice disturbed environment. Mar Ecol Prog Ser 28:129–145

Bernal PA (1980) Large-scale biological events in the California current: the low frequency response to the epipelagic ecosystem. Dissertation, University of California, San Diego

Bernstein BB, Jung N (1979) Selective pressures and coevolution in a kelp canopy community in southern California. Ecol Monogr 49:335–355

Bhaud M, Aubin D, Duhamel G (1985) Benthic recruitment of invertebrate larvae: role of hydrodynamics. Oceanis 7:97–113

Birkeland C, Randall RH (1981) Facilitation of coral recruitment by echinoid excavations. In *Proceedings of the Fourth International Coral Reef Symposium*, Manila, pp 695–698

Bosman AL, Hockey PAR, Seigfried WR (1987) The influence of coastal upwelling on the functional structure of rocky intertidal communities. Oecologia 72:226–232

Bowman MJ, Esaias WE (1978) *Oceanic Fronts in Coastal Processes*. Springer-Verlag, New York

Bray RN (1981) Influence of water currents and zooplankton densities on daily foraging movements of blacksmith, *Chromis punctipinnia*, a planktivorous reef fish. Fish Bull 78:829–841

Bray RN, Ebeling AW (1974) Food, activity and habitat of three picker-type microcarnivorous fishes in the kelp forests off Santa Barbara, California. Fish Bull 731:815–829

Briggs KT, Chu EW (1987) Trophic relationships and food requirements of California seabirds: updating models of trophic impact. In Croxall JP (ed) *Seabirds: Feeding, Ecology and Role in Marine Ecosystems*. Cambridge University Press, Cambridge

Briggs KT, Dettman KF, Lewis DB, Tyler WB (1984) Phalarope feeding in relation to autumn upwelling off California. In Nettleship DN, Sanger GA, Springer PF (eds) *Marine Birds: Their Feeding Ecology and Commercial Fisheries Relationships*. Minister of Supply and Services, Ottawa, pp 51–62

Brinton E (1962) The distribution of Pacific Euphausiids. Bull Scripps Inst Oceanogr 8:51–270

Brinton E, Reid JL (1986) On the effects of interannual variations in circulation and temperature upon the euphausiids of the California current. In *Pelagic Biogeography*. UNESCO Technical Papers on Marine Science 49, pp 25–34

Brown RGB (1980) Seabirds as marine animals. In Burger J, Olla BL, Winn HE (eds) *Behavior of Marine Animals*. Vol 4. Plenum Press, New York, pp 1–39

Burke RD (1983) The induction of metamorphosis of marine invertebrate larvae: stimulus and response. Can J Zool 61:1701–1719

Butman CA (1987) Larval settlement of soft-sediment invertebrates: the spatial scales of pattern explained by active habitat selection and the emerging role of hydrodynamical processes. Oceanogr Mar Biol Annu Rev 24:113–166

Caffey HM (1982) No effect of naturally-occurring rock types on settlement or survival in the intertidal barnacle, *Tessoropora rosea* (Krauss). J Exp Mar Biol Ecol 63:119–132

Cameran RA, Rumrill SS (1982) Larval abundance and recruitment of the sand dollar *Dendraster excentricus* in Monterey Bay, California. Mar Biol 71:197–202

Cassie RM (1959) Micro-distribution of plankton. NZ J Sci 2:398–409

Cassie RM (1963) Microdistribution of plankton. Oceanogr Mar Biol Annu Rev 1:223–252

Chabot R, Bourget E (1988) Influence of substratum heterogeneity and settled barnacle density on the settlement of cypris larvae. Mar Biol 97:45–56

Chelton DB, Bernal PA, McGowan JA (1982) Large-scale interannual physical and biological interaction in the California current. J Mar Res 40:1095–1125

Chesson PL, Case TJ (1986) Overview: nonequilibrium community theories: chance, variability, history, and coexistence. In Case TJ, Diamond J, Roughgarden J, Schoener T (eds) *Community Ecology*, Harper & Row, New York, pp 229–239

Clements FE, Shelford VE (1939) *Bioecology*. Wiley, New York

Clutter RI (1969) The microdistribution and social behavior of some pelagic mysid shrimps. J Exp Mar Biol Ecol 3:125–155

Coe WR (1956) Fluctuations in populations of marine invertebrates. J Mar Res 15:212–232

Connell JH (1961) The influence of interspecific competition and other factors on the distribution of the barnacle *Chthamalus stellatus*. Ecology 42:710–723

Connell JH (1970) A predator-prey system in the marine intertidal region. 1. *Balanus glandula* and several predatory species of *Thais*. Ecol Monogr 40:49–78

Connell JH (1985) The consequences of variation in initial settlement vs post-settlement mortality in rocky intertidal communities. J Exp Mar Biol Ecol 93:11–45

Coull BC, Ellison RL, Fleeger JW, Higgins RP, Hope WD, Hummon WD, Rieger

RM, Sterrer WE, Thiel H, Tietjen JH (1977) Quantitative estimates of the meiofauna from the deep sea off North Carolina, USA. Mar Biol 39:233–240

Cowen RK (1985) Large scale pattern of recruitment by the labrid, *Semicossyphus pulcher*: causes and implications. J Mar Res 43:719–742

Cowen RK, Agegian CR, Foster MS (1982) The maintenance of community structure in a central California giant kelp forest. J Exp Mar Biol Ecol 64:189–201

Creese RG (1982) Distribution and abundance of the acmaeid limpet, *Patelloida lagistrigata*, and its interaction with barnacles. Oecologia 52:85–96

Crisp DJ (1974) Factors influencing the settlement of marine invertebrate larvae. In Grant PT, Mackie AM (eds) *Chemoreception in Marine Organisms*. Academic Press, Orlando, pp 177–265

Crisp DJ (1984) Overview of research on marine invertebrate larvae, 1940–1980. In Costlow JD, Tipper RC (eds) *Marine Biodeterioration: An Interdisciplinary Study*. Naval Institute Press, Annapolis, pp 103–126

Crisp DJ, Meadows PS (1963) Adsorbed layers: the stimulus to settlement in barnacles. Proc R Soc Bull 158:364–387

Cushing DH (1971) Upwelling and the production of fish. Adv Mar Biol 9:255–334

Cushing DH (1975) *Marine Ecology and Fisheries*. Cambridge University Press, Cambridge

Cushing DH (1982) *Climate and Fisheries*. Academic Press, Orlando

Cushing DH, Dickson RR (1976) The biological response in the sea to climatic changes. Adv Mar Biol 14:1–122

D'Antonio CM (1986) Role of sand in the domination of hard substrata by the intertidal alga *Rhodomela larix*. Mar Ecol Prog Ser 27:263–275

Dayton PK (1971) Competition, disturbance and community organization: the provision and subsequent utilization of space in a rocky intertidal community. Ecol Monogr 41:351–389

Dayton PK (1973) Dispersion, dispersal and persistence of the annual intertidal alga *Postelsia palmaeformis* Ruprecht. Ecology 54:433–438

Dayton PK (1975a) Experimental studies of algal canopy interactions in a sea otter-dominated kelp community at Amchitka Island, Alaska. US Natl Mar Fish Serv Bull 73:230–237

Dayton PK (1975b) Experimental evaluation of ecological dominance in a rocky intertidal algal community. Ecol Monogr 45:137–159

Dayton PK (1984) Processes structuring some marine communities: are they general? In Strong DR, Simberloff D, Abele LG, Thistle AB (eds) *Ecological Communities: Conceptual Issues and the Evidence*. Princeton University Press, Princeton, pp 181–197

Dayton PK (1985a) Ecology of kelp communities. Annu Rev Ecol Syst 16:215–245

Dayton PK (1985b) The structure and regulation of some South America kelp communities. Ecol Monogr 55:447–468

Dayton PK, Tegner MJ (1984a) Catastrophic storms, El Nino, and patch stability in a southern California kelp community. Science 224:283–285

Dayton PK, Tegner MJ (1984b) The importance of scale in community ecology: a kelp forest example with terrestrial analogs. In Price PW, Slobodchikoff CM, Gaud WS (eds) *A New Ecology: Novel Approaches to Interactive Systems*. Wiley, New York, pp 457–481

Dayton PK, Tegner MJ (1990) Bottoms beneath troubled waters: benthic impacts of the 1982–1984 El Niño in the temperate zone. In Glynn PW (ed) *Ecological Consequences of the 1982–1983 El Niño to marine life*. Elsevier Oceanography Series, Amst, pp. 433–472

Dayton PK Robilliard GA, DeVries AL (1969) Anchor ice formation in McMurdo Sound, Antarctica, and its biological effects. Science 163:273–274

Dayton PK, Robilliard GA, Paine RT (1970) Benthic faunal zonation as a result of anchor ice at McMurdo Sound, Antarctica. In Holdgate MV (ed) *Antarctic Ecology*. Vol 1. Academic Press, Orlando, pp 244–258

Dayton PK, GA Robilliard, RT Paine, Dayton LB (1974) Biological accommodation in the benthic community at McMurdo Sound, Antarctica. Ecol Monogr 44:105–128

Dayton PK, Currie V, Gerrodette T, Keller BD, Rosenthal R, Ven Tresca D (1984) Patch dynamics and stability of some California kelp communities. Ecol Monogr 54:253–289

Dayton PK, Watson D, Palmisano A, Barry JP, Oliver JS, Rivera D (1986) Distribution patterns of benthic microalgal standing stock at McMurdo Sound, Antarctica. Polar Biol 6:207–213

DeAngelis DL, Waterhouse JC (1987) Equilibrium and nonequilibrium concepts in ecological models. Ecol Monogr 57:1–21

Denley EJ, Underwood AJ (1979) Experiments on factors influencing settlement, survival, and growth of two species of barnacles in New South Wales. J Exp Mar Biol Ecol 36:269–293

Denman KL, Powell TM (1984) Effects of physical processes on planktonic ecosystems in the coastal ocean. Oceanogr Mar Biol Annu Rev 22:125–168

Denny MW (1985) Wave forces on intertidal organisms: a case study. Limnol Oceanogr 30:1171–1187

Denny MW (1988) *Biology and the Mechanics of the Wave-swept Environment*. Princeton University Press, Princeton

Diester-Haass L (1978) Sediments as indicators of upwelling. In Boje R, Tomczak M, (eds) *Upwelling Ecosystems*. Springer-Verlag, New York, pp 261–281

Drew EA (1983) Light. In Earll R, Erwin DG (eds) *Sublittoral Ecology: The Ecology of the Shallow Sublittoral Benthos*. Clarendon Press, Oxford, pp 10–57

Dunbar RB (1981) Sedimentation and the history of upwelling and climate in high fertility areas of the northeastern Pacific Ocean. Dissertation, University of California, San Diego

Ebeling AW, Laur DR, Rowley RJ (1985) Severe storm disturbances and reversal of community structure in a southern California kelp forest. Mar Biol 84:287–294

Ebert TA, Russell MP (1988) Latitudinal variation in size structure of the West Coast purple urchin: a correlation with headlands. Limnol Oceanogr 33:286–294

Eckman JE (1983) Hydrodynamic processes affecting benthic recruitment. Limnol Oceanogr 28:241–257

Eckman JE, Nowell ARM, Jumars PA (1981) Sediment destabilization by animal tubes. J Mar Res 39:361–374

Emery AR (1968) Preliminary observations on coral reef plankton. Limnol Oceanogr 13:293–303

Emery AR (1972) Eddy formation from an oceanic island: ecological effects. Carib J Sci 12:121–128

Emson RH, Faller-Fritsch RJ (1976) An experimental investigation into the effect of crevice availability on abundance and size-structure in a population of Littorina rudis (Maton): Gastropoda: Prosobranchia. J Exp Mar Biol Ecol 23:285–297

Eppley RW, Peterson BJ (1979) Particulate organic matter flux and planktonic new production in the deep ocean. Nature 282:677–680

Estes JA, Palmisano JF (1974) Sea otters: their role in structuring nearshore communities. Science 185:1058–1060

Estes JA, Smith NS, Palmisano JS (1978) Sea otter predation and community organization in the western Aleutian Islands, Alaska. Ecology 59:822–833

Fager EW (1964) Marine sediments: effects of a tube-building polychaete. Science 143:356–359

Fasham MJR (1977) The application of some stochastic processes to the study of plankton patchiness. In Steele JH (ed) *Spatial Pattern in Plankton Communities*. Plenum Press, New York, pp 131–156

Fasham MJR (1978) The statistical and mathematical analysis of plankton patchiness. Oceanogr Mar Annu Rev 16:43–79

Fenchel T (1969) The ecology of marine microbenthos. IV. Structure and function of the benthic ecosystem, its chemical and physical factors and the microfauna communities with special reference to the ciliated protozoa. Ophelia 6:1–182

Fenchel T, Straarup BJ (1971) Vertical distribution of photosynthetic pigments and the penetration of light in marine sediments. Oikos 22:172–182

Fiedler PC (1984) Satellite observations of the 1982–1983 El Nino along the U.S. Pacific coast. Science 224:1251–1254

Fiedler PC, Methot RD, Hewitt RP (1986) Effects of the California El Nino on the northern anchovy. J Mar Res 44:317–338

Flament P, Armi L, Washburn L (1985) The evolving structure of an upwelling filament. J Geophys Res 90:11765–11778

Foster MS (1982) The regulation of macroalgal associations in kelp forests. In *Synthetic and Degradative Processes in Marine Macrophytes*. de Gruyter, New York, pp 185–205

Fraser WR, Ainley DG (1986) Ice edges and seabird occurrence in Antarctica. Bioscience 36:258–263

Freeland HJ, Denman KL (1982) A topographically controlled upwelling center off southern Vancouver Island. J Mar Res 40:1069–1093

Fu L, Holt B (1984) Internal waves in the gulf of California: observations from a spaceborne radar. J Geophys Res 89:2053–2060

Fukuyama AK, Oliver JS (1985) Sea star and walrus predation on bivalves in Norton Sound, Bering Sea, Alaska. Ophelia 24:17–36

Gaines S, Brown S, Roughgarden J (1985) Spatial variation in larval concentrations as a cause of spatial variation in settlement for the barnacle, *Balanus glandula*. Oecologia 67:267–272

Gaines S, Roughgarden J (1985) Larval settlement rate: a leading determinant of structure in an ecological community of the marine intertidal zone. Proc Natl Acad Sci USA 82:3707–3711

Gaines SD, Roughgarden J (1987) Fish in offshore kelp forests affect recruitment to intertidal barnacle populations. Science 235:479–481

Gardner WD, Sullivan LG (1981) Benthic storms: temporal variability in a deep-ocean nepheloid layer. Science 213:329–331

Garrison DL, Sullivan CW, Ackley SF (1986) Sea ice microbial communities in Antarctica. Bioscience 36:243–250

Garrity SD, Levings SC (1981) A predator-prey interaction between two physically and biologically constrained tropical rocky shore gastropods: direct, indirect and community effects. Ecol Monogr 5:267–286

Gaskin DE (1968) Distribution of Delphinidae (Cetacea) in relation to sea surface temperatures off eastern and southern New Zealand. NZ J Mar Freshwat Res 2:727–534

Gaskin DE (1976) The evolution, zoogeography and ecology of Cetacea. Oceanogr Mar Biol Annu Rev 14:247–346

Genin A, Boehlert GW (1985) Dynamics of temperature and chlorophyll structures above a seamount: an oceanic experiment. J Mar Res 43:907–924

Genin A, Dayton PK, Lonsdale PF, Spiess FN (1986) Corals on seamount peaks provide evidence of current acceleration over deep-sea topography. Nature 322:59–61

Genin A, Haury L, Greenblatt P (1988) Interactions of migrating zooplankton with shallow topography: predation by rockfishes and intensification of patchiness. Deep Sea Res 35:151–175

Gerard VA (1982) In situ rates of nitrate uptake by giant kelp, *Macrocyctis pyrifera* (L.) C. Agardh: tissue differences, environmental effects, and predictions of nitrogen-limited growth. J Exp Mar Biol Ecol 62:211–224

Goering JJ, Iverson RL (1981) Phytoplankton distribution on the southeastern Bering Sea shelf. In Hood DW, Calder JA (eds) *The Eastern Bering Sea Shelf: Oceanography and Resources*. Vol 1. University of Washington Press, Seattle, pp. 933–946

Gordon AL (1981) Seasonality of southern ocean sea ice. J Geophys Res 86:4193–4197

Gosselin M, Legendre L, Therriault JC, Demers S, Rochet M (1986) Physical control of the horizontal patchiness of sea-ice microalgae. Mar Ecol Prog Ser 29:289–298

Gray JS (1974) Animal-sediment relationships. Oceanogr Mar Biol Annu Rev 12:223–261

Grebmeier JM, McRoy CP, Feder HM (1988) Pelagic-benthic coupling on the shelf of the northern Bering and Chukchi Seas. I. Food supply source and benthic biomass. Mar Ecol Prog Ser 48:57–67

Grebmeier JM, Feder HM, McRoy CP (1989) Relagoc-benthic coupling on the shelf of the northern Bering and Chukchi Seas. II. Benthic community structure. Mar Ecol Prog Ser 54:121–131

Grossberg RK (1982) Intertidal zonation of barnacles: the influence of planktonic zonation of larvae on vertical distribution of adults. Ecology 63:894–899

Gunnill FC (1985) Population fluctuations of seven macroalgae in southern California during 1981–1983 including effects of severe storms and an El Nino. J Exp Mar Biol Ecol 85:149–164

Haedrich RL, Rowe GT (1977) Megafaunal biomass in the deep sea. Nature 269:141–142

Haedrich RL, Rowe GT, Pollini PT (1980) The megabenthic fauna in the deep sea south of New England, USA. Mar Biol 57:165–179

Halpern D, Hayes SP, Leetmaa A, Hansen DV, Philander SGH (1983) Oceanographic observations of the 1982 warming of the tropical eastern Pacific. Science 221:1173–1175

Hamner WM, Schneider D (1986) Regularly spaced rows of medusae in the Bering Sea: role of langmuir circulation. Limnol Oceanogr 31:171–177

Haney JC (1986) Seabird patchiness in tropical oceanic waters: the influence of Sargassum "reefs." Auk 103:141–151

Haney JC (1987) Ocean internal waves as sources of small-scale patchiness in seabird distribution on the Blake Plateau. Auk 104:129–133

Hannon CA (1984) Planktonic larvae may act like passive particles in turbulent near-bottom flows. Limnol Oceanogr 29:1108–1116

Harding GC, Vass WP, Drinkwater KF (1982) Aspects of larval American lobster (*Homarus americanus*) ecology in St. Georges Bay, Nova Scotia. Can J Fish Aquat Sci 39:1117–1129

Harger JRE (1970) The effect of wave impact on some aspects of the biology of sea mussels. Veliger 12:401–414

Harper JL (1977) *Population Biology of Plants*. Academic Press, Orlando

Harris RP (1987) Spatial and temporal organization in marine plankton communities. In Gee JHR, Giller PS (eds) *Organization of Communities: Past and Present*. Blackwell, Oxford, pp 327–336

Haury LR (1976) A comparison of zooplankton patterns in the California current and north Pacific central gyre. Mar Biol 37:159–167

Haury LR, Pieper RE (1988) Zooplankton: scales of biological and physical events. In Soule DF, Kleppel GS (eds) *Marine Organisms as Indicators*. Springer-Verlag, New York, pp 35–72

Haury LR, McGovern JS, Wiebe P (1978) Patterns and processes in the time-space scales of plankton distributions. In Steele J (ed) *Spatial Pattern in Plankton Communities*. Plenum Press, New York, pp 277–327

Haury LR, Wiebe PH, Orr MH, Briscoe MG (1983) Tidally generated high-frequency internal wave packets and their effects on plankton in Massachusetts Bay. J Mar Res 41:65–112

Haury LR, Simpson JJ, Pelaez J, Koblinsky C, Wiesenhahn D (1986) Biological consequences of a persistent eddy off Point Conception, California. J Geophys Res 91:12937–12956

Hay ME (1981) Herbivory, algal distribution, and the maintenance of between-habitat diversity on a tropical fringing reef. Am Nat 118:520–540

Hayward TL, McGowan JA (1979) Pattern and structure in an oceanic zooplankton community. Am Zool 19:1045–1055

Hedgpeth JW (1957) Classification of marine environments. Mem Geol Soc Am 67:17–28

Hickey BM (1979) The California current system—hypotheses and facts. Prog Oceanogr 8:191–279

Hjort J (1914) Fluctuations in the great fisheries of northern Europe reviewed in the light of biological research. Rapp Proces Verb Cons Int Exp Mer 20:1–228

Hjort J (1926) Fluctuations in the year classes of important food fishes. J Conseil 1:1–38

Hobson ES, Chess JR (1976) Trophic interactions among fishes and zooplankters near shore at Santa Catalina Island, California. Fish Bull 74:567–598

Holland ND, Leonard AB, Strickler JR (1987) Upstream and downstream capture during suspension feeding by *Oligometra serripinna* (Echinodermata: Crinoidea) under surge conditions. Biol Bull 173:552–556

Hollowed AB, Bailey KM, Wooster WS (1987) Patterns in recruitment of marine fishes in the northeast Pacific Ocean. Biol Oceanogr 5:99–131

Horne EPW, Platt T (1984) The dominant space and time scales of variability in the physical and biological fields on continental shelves. Rapp Proces Verb Cons Int Explor Mer 183:9–19

Hudon C, Bourget E, Legendre P (1983) An integrated study of the factors influencing the choice of the settling site of *Balanus crenatus* cyprid larvae. Can J Fish Aquat Sci 40:1186–1194

Hughes RN, Peer DL, Mann KH (1972) Use of multivariate analysis to identify functional components of the benthos in St. Margaret's Bay, Nova Scotia. Limnol Oceanogr 17:111–121

Hunt GL, Schneider DC (1987) Scale-dependent processes in the physical and biological environment of marine birds. In Croxall JP (ed) *Seabirds: Feeding, Ecology and Role in Marine Ecosystems*. Cambridge University Press, Cambridge, pp 7–41

Hutchings L, Armstrong DA, Mitchell-Innes BA (1986) The frontal zone in the southern Benguela Current. In Nihoul JCJ (ed) *Marine Interfaces Ecohydrodynamics*. Elsevier, New York, pp 67–94

Huyer A (1976) A comparison of upwelling events in two locations: Oregon and Northwest Africa. J Mar Res 34:531–546

Iles TD, Sinclair M (1982) Atlantic herring: stock discreteness and abundance. Science 215:627–633

Iverson RL, Coachman LK, Cooney RT, English TS, Goering JJ, Hunt GL, Macauley MC, McRoy CP, Reeburg WS, Whitledge TH (1980) Ecological significance of fronts in the southeastern Bering Sea. In *Ecological Processes in Coastal and Marine Systems*. Plenum Press, New York, pp 437–468

Jackson GA (1984) Internal wave attenuation by coastal kelp stands. J Phys Oceanogr 14:1300–1306

Jackson GA (1986) Interaction of physical and biological processes in the settlement of planktonic larvae. Bull Mar Sci 39:202–212

Jackson GA, Winant CD (1983) Effect of a kelp forest on coastal currents. Continental Shelf Res 2:75–80

Jensen RA, Morse DE (1984) Intraspecific facilitation of larval recruitment: gregarious settlement of the polychaete *Phragmatopoma californica* (Fewkes). J Exp Mar Biol Ecol 83:107–126

Jillett JB, Zeldis JR (1985) Aerial observations of surface patchiness of a planktonic crustacean. Bull Mar Sci 37:609–619

Johannes RE (1978) Reproductive strategies of coastal marine fishes in the tropics. Environ Biol Fish 3:65–84

Johnson CR, Mann KH (1986) Diversity, patterns of adaptation, and stability of Nova Scotian kelp beds. Ecol Monogr 58:129–154

Jones ML (1985) (ed) Hydrothermal vents of the eastern Pacific: an overview. Bull Biol Soc Wash 6.

Jones NS (1950) Marine bottom communities. Biol Rev 25:283–313

Jones WE, Demetropoulos A (1968) Exposure to wave action: measurements of an important ecological parameter on rocky shores on Anglesey. J Exp Mar Biol Ecol 2:46–63

Joyce T, Wiebe P (1983) Warm-core rings of the Gulf Stream. Oceanus 26:34–44

Jumars PA (1975) Environmental grain and polychaete species' diversity in a bathyal benthic community. Mar Biol 30:253–266

Jumars PA, Hessler RA (1976) Hadal community structure: implications from the Aleutian trench. J Mar Res 34:547–560

Kastendiek J (1982) Factors determining the distribution of the sea pansy, *Renilla kollikeri*, in a subtidal sand-bottom habitat. Oecologia 52:340–347

Kingsford MJ, Choat JH (1986) The influence of surface slicks on the distribution and onshore movement of small fish. Mar Biol 91:161–171

Kitching JA (1941) Studies in sublittoral ecology III. *Laminaria* forest on the west coast of Scotland: a study of zonation in relation to wave action and illumination. Biol Bull Woods Hole Mar Biol Lab 80:324–327

Knauer GA, Hebel D, Cipriano F (1982) Marine snow: major site of primary production in coastal waters. Nature 300:630–631

Lasker R (1981a) Factors contributing to variable recruitment of the northern anchovy (*Engraulis mordax*) in the California current: contrasting years, 1975 through 1978. Rapp Proces Verb Cons Int Explor Mer 178:375–388

Lasker R (1981b) The role of a stable ocean in larval fish survival and subsequent recruitment. In Lasker R (ed) *Marine Fish Larvae—Morphology, Ecology, and Relation to Fisheries*. Washington Sea Grant Program, Seattle

Legendre L (1981) Hydrodynamic control of marine phytoplankton production: the paradox of stability. In Nihoul JCJ (ed) *Ecohydrodynamics*. Elsevier, New York, pp 191–208

Legendre L, Demers S (1984) Towards dynamic biological oceanography and limnology. Can J Fish Aquat sci 41:2–19

Leonard AB, Strickler JR, Holland ND (1988) Effects of current speed on filtration during suspension feeding in *Oligometra serripinna* (Echinodermata: Crinoidea). Mar Biol 97:111–125

Le Tourneux F, Bourget E (1988) Importance of physical and biological settlement cues used at different spatial scales by the larvae of *Semibalanus balanoides*. Mar Biol 97:57–66

Levin LA (1981) Dispersion, feeding behavior and competition in two spionid polychaetes. J Mar Res 39:99–117

Levin LA (1982) Interference interactions among tube-dwelling polychaetes in a dense infaunal assemblage. J Exp Mar Biol Ecol 65:107–119

Levin LA, DeMaster DJ, McCann LD, Thomas CL (1986) Effects of giant protozoans (class: Xenophyophorea) on deep-seamount benthos. Mar Ecol Prog Ser 29:99–104

Littler MM, Littler DS (1984) Relationships between macroalgal functional form groups and substrata stability in a subtropical rocky-intertidal system. J Exp Mar Biol Ecol 74:13–34

Littler MM, Martz DR, Littler DS (1983) Effects of recurrent sand deposition on rocky intertidal organisms: importance of substrate heterogeneity in a fluctuating environment. Mar Ecol Prog Ser 11:129–139

Lobel PS (1978) Diel, lunar and seasonal periodicity in the reproductive behavior of the Pomacanthid fish, *Centropyge potteri* and some other reef fishes in Hawaii. Pacific Sci 32:193–207

Lobel PS (1980) Herbivory by damselfishes and their role in coral reef community ecology. Bull Mar Sci 30:273–289

Lobel PS, Robinson AR (1986) Transport and entrapment of fish larvae by ocean mesoscale eddies and current in Hawaiian waters. Deep Sea Res 33:483–500

Loosanoff VL (1964) Variations in time and intensity of setting of the starfish, *Asterias forbesi*, in Long Island Sound during a twenty-five-year period. Biol Bull 126:423–429

Loughlin TR, Bengtson JL, Merrick RL (1987) Characteristics of feeding trips of female northern fur seals. Can J Zool 65:2079–2084

Lubchenco J (1983) *Littorina* and *Fucus*: effects of herbivores, substratum heterogeneity and plant escapes during succession. Ecology 64:1116–1123

Luther DS, Harrison DE, Knox RA (1983) Zonal winds in the central equatorial pacific and El Nino. Science 222:327–329

Mackas DL, Denman KL, Abbott MR (1985) Plankton patchiness: biology in the physical vernacular. Bull Mar Sci 37:652–674

Marine Research Committee (1957) The Marine Research Committee, 1947–1955. California Cooperative Oceanic Fish Investion Report 1953–1955, pp 7–9 Scripps Inst. Oceanography, La Jolla

Marshall NB (1979) *Deep-Sea Biology: Developments and Perspectives.* Garland STPM Press, New York

McCarthy JJ, Goldman JC (1979) Nitrogenous nutrition of marine phytoplankton in nutrient-depleted waters. Science 203:670–672

McGowan JA (1971) Oceanic biogeography of the Pacific. In Funnell BM, Riedel WR (eds) *The Micropalaeontology of the Oceans.* Cambridge University Press, Cambridge, pp 3–74

McGowan JA (1974) The nature of oceanic ecosystems. In Miller CB (ed) *The Biology of the Oceanic Pacific.* Oregon State University Press, Corvallis, pp 9–28

McGowan JA (1985) El Niño 1983 in the southern California bight. In Wooster WS, Fluharty DL (eds) *El Niño North, Niño Effects in the Eastern Subarctic Pacific Ocean.* Washington Sea Grant Program, University of Washington, Seattle, pp 166–184

McGowan JA (1986) The biogeography of pelagic ecosystems: pelagic biogeography. UNESCO Technical Papers in Marine Science 49, pp 191–200

McGowan JA, Walker PW (1979) Structure in the copepod community of the north Pacific central gyre. Ecol Monogr 49:195–226

McGowan JA, Walker PW (1985) Dominance and diversity maintenance in an oceanic ecosystem. Ecol Monogr 55:103–118

McQuaid CD, Branch GM (1984) Influence of sea temperature, substratum and wave exposure on rocky intertidal communities: an analysis of faunal and floral biomass. Mar Ecol Prog Ser 19:145–151

McQuaid CD, Branch GM (1985) Trophic structure of rocky intertidal communities: response to wave action and implications for energy flow. Mar Ecol Prog Ser 22:153–161

Menge BA (1976) Organization of the New England rocky intertidal community:

role of predation, competition and environmental heterogeneity. Ecol Monogr 46:355–393

Menge BA (1978) Predation intensity in a rocky intertidal community: relation between predator foraging activity and environmental harshness. Oecologia 34:1–16

Menge BA, Lubchenco J (1981) Community organization in temperate and tropical rocky intertidal habitats: prey refuges in relation to consumer pressure gradients. Ecol Monogr 51:429–450

Menge BA, Lubchenco J, Ashkenas LR (1985) Diversity, heterogeneity and consumer pressure in a tropical rocky intertidal community. Oecologia 65:394–405

Merriam CH (1898) Life zones and crop zones of the United States. Bull US Biol Surv 10:1–79

Mills EL (1975) Benthic organisms and the structure of marine ecosystems. J Fish Res Board Can 32:1657–1663

Mysak LA, Hsieh WW, Parsons TR (1982) On the relationship between interannual baroclinic waves and fish populations in the northeast Pacific. Biol Oceanogr 2:63–103

Nelson BV, Vance RR (1979) Diel foraging patterns of the sea urchin *Centrostephanus coronatus* as a predator avoidance strategy. Mar Biol 51:251–258

Nelson DM, Smith WO (1986) Phytoplankton bloom dynamics of the western Ross Sea ice edge. II. Mesoscale cycling of nitrogen and silicon. Deep Sea Res 33: 1389–1412

Newman WA (1979a) A Californian Transition zone: significance of short-range endemics. In Gray J, Boucot AJ (eds) *Historical Biogeography Plate Tectonics and the Changing Environment*. Oregon State University Press, Corvallis, pp 399–416

Newman WA (1979b) A new scalpellid (Cirripedia); a mesozoic relic living near an abyssal hydrothermal spring. Trans San Diego Soc Nat Hist 19:153–167

Newman WA (1985) The abyssal hydrothermal vent invertebrate fauna: a glimpse of antiquity? Bull Biol Soc Wash 6:231–242

Newman WA, Hessler RR (1989) A new abyssal hydrothermal verrucomorphan (Cirripedia; Sessilia): the most primitive living sessile barnacle. Trans San Diego Soc Nat Hist 21:259–273

Nihoul JCJ (1981) (ed) *Ecohydrodynamics*. Proceedings of the 12th International Liege Colloquium on Ocean Hydrodynamics. Elsevier, New York

Nihoul JCJ (1986) (ed) *Marine Interfaces Ecohydrodynamics*. Elsevier, New York

Nybakken JW (1988) *Marine Biology: An Ecological Approach*. 2nd ed. Harper & Row, New York

Oliver JS, Slattery PN, Silberstein MA, O'Connor EF (1983a) A comparison of gray whale, *Eschrichtius robustus*, feeding in the Bering Sea and Baja California. Fish Bull 81:513–522

Oliver JS, Slattery PN, O'Connor EF, Lowry LF (1983b) Walrus, *Odobenus rosmarus*, feeding in the Bering Sea: a benthic perspective. Fish Bull 81:501–512

O'Neill RV, DeAngelis DL, Waide JB, Allen TFH (1986) *A Hierarchical Concept of Ecosystems*. Princeton University Press, Princeton

Owen RW (1981) Fronts and eddies in the sea: mechanisms, interactions and biological effects. In Longhurst AR (ed) *Analysis of Marine Ecosystems*. Academic press, Orlando, pp 197–233

Paine RT (1966) Food web complexity and species diversity. Am Nat 100:65–75

Paine RT (1974) Intertidal community structure: experimental studies on the relationship between a dominant competitor and its principal predator. Oecologia 15:93–120

Paine RT (1979) Disaster, catastrophe, and local persistence of the sea palm *Postelsia palmaeformis*. Science 205:685–687

Paine RT (1980) Food webs: linkage, interaction strength and community infrastructure. J Anim Ecol 49:667–685

Paine RT (1984) Ecological determinism in the competition for space. Ecology 65:1339–1348

Paine RT (1986) Benthic community-water column coupling during the 1982–1983 El Nino: are community changes at high latitudes attributable to cause or coincidence? Limnol Oceanogr 31:351–360

Palmisano AC, Sullivan CW (1983) Sea ice microbial communities (SIMCO) 1. Distribution, abundance, and primary production of ice microalgae in McMurdo Sound, Antarctica in 1980. Polar Biol 2:171–177

Parrish RH, Nelson CS, Bakun A (1981) Transport mechanisms and reproductive success of fishes in the California current. Biol Oceanogr 1:175–203

Parsons TR, Takahashi M, Hargrave B (1984) *Biological Oceanographic Processes.* 3rd ed. Pergamon Press, New York

Pawlik JR (1986) Chemical induction of larval settlement and metamorphosis in the reef-building tube worm *Phragmatopoma californica* (Sabellariidae: Polychaeta). Mar Biol 91:59–68

Pearson TH, Rosenberg R (1987) Feast and famine: structuring factors in marine benthic communities. In Gee JHR, Giller PS (eds) *Organization of Communities: Past and Present.* Blackwell, Oxford, pp 373–395

Perkins EJ (1963) The penetration of light into littoral soils. J Ecol 51:687–692

Peterman RM, Bradford MJ (1987) Wind speed and mortality rate of a marine fish, the northern anchovy (*Engraulis mordax*). Science 235:354–355

Petersen CGJ (1913) Valuation of the sea. II. The animal communities of the sea bottom and their importance for marine zoogeography. Rep Dan Biol Stn 21: 1–44

Peterson CH (1986) Enhancement of *Mercenaria mercenaria* densities in seagrass beds: is pattern fixed during settlement season or altered by subsequent differential survival? Limnol Oceanogr 31:200–205

Peterson WT (1972) Upwelling indices and annual catches of dungeness crab, *Cancer magister*, along the west coast of the United States. Fish Bull US 71:902–910

Peterson WT, Miller CB (1975) Year-to-year variations in the planktology of the Oregon upwelling zone. Fish Bull US 73:642–653

Peterson WT, Miller CB, Hutchinson A (1979) Zonation and maintenance of copepod populations in the Oregon upwelling zone. Deep Sea Res 26A:467–494

Pingree RD (1978) Mixing and stabilization of phytoplankton distributions on the northwest European continental shelf. In Steele J (ed) *Spatial Pattern in Plankton Communities.* Plenum Press, New York, pp 181–220

Pingree RD, Forster GR, Morrison GK (1974) Turbulent convergent tidal fronts. J Mar Biol Assoc UK 54:469–479

Pingree RD, Holligan PM, Mardell GT, Head RN (1976) The influence of physical stability on spring, summer and autumn phytoplankton blooms in the Celtic Sea. J Mar Biol Assoc UK 56:845–873

Pingree RD, Holligan PM, Head RN (1977) Survival of dinoflagellate blooms in the western English Channel. Nature 265:266–269

Pond S, Pickard GL (1983) *Introductory Dynamical Oceanography.* 2nd ed. Pergamon Press, New York

Raimondi PT (1988) Rock type affects settlement, recruitment, and zonation of the barnacle *Chthamalus anisopoma* Pilsbury. J Exp Mar Biol Ecol 123:253–268

Reed DC, Foster, M (1984) The effects of canopy shading on algal recruitment and growth in giant kelp (*Macrocystis pyrifera*) forest. Ecology 65:937–948

Reed DC, Laur DR, Ebeling AW (1988) Variation in algal dispersal and recruitment: the importance of episodic events. Ecol Monogr 58:321–335

Reid JL, Brinton E, Fleminger A, Venrick EL, McGowan JA (1978) Ocean circulation and marine life. In Charnock H, Deacon G (eds) *Advances in Oceanography*. Plenum Press, New York, pp 65–130

Rhoads DC (1974) Organism-sediment relations on the muddy sea floor. Oceanogr Mar Biol Annu Rev 12:263–300

Rhoads DC, Young DK (1970) The influence of deposit-feeding organisms on sediment stability and community trophic structure. J Mar Res 28:150–178

Rhoads DC, Young DK (1971) Animal-sediment relations in Cape Code Bay, Massachusetts. II. Reworking by *Molpadia oolitica* (Holothuroidea). Mar Biol 11:255–261

Richards FA (1981) (ed) *Coastal Upwelling. Coastal and Estuarine Sciences*. Vol 1. American Geophysical Union, Washington DC

Roughgarden J, Gaines SD, Pacala SW (1987) Supply side ecology: the role of physical transport processes. In Gee JHR, Giller PS (eds) *Organization of Communities: Past and Present*. Blackwell, Oxford, pp 491–528

Roughgarden J, Gaines S, Possingham H (1988) Recruitment dynamics in complex life cycles. Science 241:1460–1466

Rowe GT (1971) Observations on bottom currents and epibenthic populations in Hatteras Submarine Canyon. Deep Sea Res 18:569–581

Rowe GT (1981) The deep-sea ecosystem. In Longhurst AR (ed) *Analysis of Marine Ecosystems*. Academic Press, Orlando, pp 235–226

Ryther JH (1969) Photosynthesis and fish production in the sea. Science 166:72–76

Sale PF (1980) The ecology of fishes on coral reefs. Oceanogr Mar Biol Annu Rev 18:367–421

Sanders HL, Hessler RR (1969) Ecology of deep-sea benthos. Science 163:1419–1424

Schmidt GH, Warner GF (1984) Effects of caging on the development of a sessile epifaunal community. Mar Ecol Prog Ser 15:251–263

Schneider D, Hunt GL (1982) Carbon flux to seabirds in waters with different mixing regimes in the southeastern Bering Sea. Mar Biol 67:337–344

Schneider D, Hunt GL (1984) A comparison of seabird diets and foraging distribution around the Pribilof Islands, Alaska. In Nettleship DN, Sanger GA, Springer PF (eds) *Marine Birds: Their Feeding Ecology and Commercial Fisheries Relationships*. Minister of Supply and Services, Ottawa, pp 86–95

Schneider DC, Piatt JF (1986) Scale dependent correlation of seabirds with schooling fish in a coastal ecosystem. Mar Ecol Prog Ser 32:237–246

Scrope-Howe S, Jones DA (1985) Biological studies in the vicinity of a shallow-sea tidal mixing front. V. Composition, abundance and distribution of zooplankton in the western Irish Sea, April 1980 to November 1981. Philos Trans R Soc Lond [Biol] 310:501–519

Seapy RR, Littler MM (1978) The distribution, abundance, community structure, and primary productivity of macroorganisms from two central California rocky intertidal habitats. Pacific Sci 32:293–314

Sebens KP (1984) Water flow and coral colony size: interhabitat comparisons of the octocoral *Alcyonium siderium*. Proc Natl Acad Sci USA 81:5473–5477

Sebens KP (1986) Community ecology of vertical rock walls in the Gulf of Maine USA: small-scale processes and alternative community states. In Moore PG, Seed R (eds) *The Ecology of Rocky Coasts*. Columbia University Press, New York, pp 346–371

Send URC, Beardsley RC, Winant CD (1987) Relaxation from upwelling in the coastal ocean dynamics experiment. J Geophys Res 92:1683–1698

Setchell WA (1920) The temperature interval in the geographic distribution of marine algae. Science 52:187–190

Seymour RJ, Strange RR III, Cayan DR, Nathan RA (1984) Influence of El Niños on California's wave climate. In Billy LE (ed) *Proceedings of the International*

Conference (19th Coastal Engineering Conference). American Society of Civil Engineers, New York, pp 577–592

Shanks AL (1983) Surface slicks associated with tidally forced internal waves may transport pelagic larvae of benthic invertebrates and fishes shoreward. Mar Ecol Progr Ser 13:311–315

Shanks AL (1987) The onshore transport of an oil spill by internal waves. Science 235:1198–1199

Shanks AL, Wright WG (1986) Adding teeth to wave action: the destructive effects of wave-borne rocks on intertidal organisms. Oecologia 69:420–428

Shanks AL, Wright WG (1987) Internal-wave-mediated shoreward transport of cyprids, megalopae, and gammarids and correlated longshore differences in the settling rates of intertidal barnacles. J Exp Mar Biol Ecol 114:1–13

Silver MW, Shanks AL, Trent JD (1978) Marine snow: microplankton habitat and source of small-scale patchiness in pelagic populations. Science 201:371–373

Simpson JJ (1983) Large-scale thermal anomalies in the California current during the 1982–1983 El Niño. Geophys Res Lett 10:937–940

Simpson JH, Pingree RD (1978) Shallow sea fronts produced by tidal stirring. In Bowman MJ, Esias WE (eds) *Oceanic Fronts in Coastal Processes.* Springer-Verlag, New York, pp 29–42

Sinclair M (1988) *Marine Populations: An Essay on Population Regulation and Speciation.* Washington Sea Grant Program, University of Washington Press, Seattle

Smith PE (1978a) Biological effects of ocean variability: time and space scales of biological response. Rapp Proces Verb Cons Int Explor Mer 173:117–127

Smith PC (1978b) Low-frequency fluxes of momentum, heat, salt, and nutrients at the edge of the Scotian shelf. J Geophys Res 83:4079–4096

Smith SL, Jones BH, Atkinson LP, Brink KH (1986) Zooplankton in the upwelling fronts off Pt. Conception, California. In Nihoul JCJ (ed) *Marine Interfaces Ecohydrodynamics.* Elsevier, New York, pp 195–214

Smith WO (1987) Phytoplankton dynamics in marginal ice zones. Oceanogr Mar Biol Annu Rev 25:11–38

Smith WO, Nelson DM (1985) Phytoplankton bloom produced by a receding ice edge in the Ross Sea: spatial coherence with the density field. Science 227:163–166

Smith WO, Nelson DM (1986) Importance of ice edge phytoplankton production in the southern ocean. Bioscience 36:251–257

Sorokin YI (1981) Microheterotrophic organisms in marine ecosystems. In Longhurst AR (ed) *Analysis of Marine Ecosystems.* Academic Press, London, pp 293–342

Sousa WP (1979) Disturbance in marine intertidal boulder fields: the nonequilibrium maintenance of species diversity. Ecology 60:1225–1240

Southward AJ (1980) The western English Channel—an inconstant ecosystem. Nature 285:361–366

Southward AJ, Orton JH (1954) The effects of wave-action on the distribution and numbers of the commoner plants and animals living on the Plymouth breakwater. J Mar Biol Assoc UK 33:1–19

Steele JH (1985) A comparison of terrestrial and marine ecological systems. Nature 313:355–358

Suess E (1980) Particulate organic carbon flux in the oceans—surface productivity and oxygen utilization. Nature 288:260–263

Sutherland JP (1987) Recruitment limitation in a tropical intertidal barnacle: *Tetraclita panamensis* (Pilsbury) on the Pacific coast of Costa Rica. J Exp Mar Biol Ecol 113:267–282

Tegner MJ, Levin LA (1983) Spiny lobsters and sea urchins: analysis of a predator-prey interaction. J Exp Mar Biol Ecol 73:125–150

Tegner MJ, Barry JP Ms. Oecanographic effects on larval dispersal affect recruitment and survival of red sea urchins (*Strongylocentrotus franciscanus*) in the southern California bight. Submitted to J. Exp. Mar. Biol. Ecol.

Tett P, Edwards A (1984) Mixing and plankton: an interdisciplinary theme in oceanography. Oceanogr Mar Biol Annu Rev 1984 22:99–123

The Ring Group (1981) Gulf Stream cold-core rings: their physics, chemistry, and biology. Science 211:1091–1099

Thiel H (1978) Benthos in upwelling regions. In Boje R, Tomczak M (eds) *Upwelling Ecosystems*. Springer-Verlag, New York, pp 124–138

Thorson G (1950) Reproductive and larval ecology of marine bottom invertebrates. Biol Rev 25:1–45

Thorson G (1957) Bottom communities (sublittoral or shallow shelf). Geol Soc Am Mem 67:461–534

Tont SA (1976) Short-period climatic fluctuations: effects on diatom biomass. Science 194:942–944

Tont SA (1981) Temporal variations in diatom abundance off southern California in relation to surface temperature, air temperature and sea level. J Mar Res 39:191–201

Trillmich F, Kooyman GL, Majluf P, Sanchez-Grinan M (1986) Attendance and diving behavior of South American fur seals during El Nino in 1983. In Gentry RL, Kooyman GL (eds) *Fur Seals: Maternal Strategies on Land and at Sea*. Princeton University Press, Princeton, pp 153–167

Underwood AJ, Denley EJ (1984) Paradigms, explanations, and generalizations in models for the structure of intertidal communities on rocky shores. In Strong DR, Simberloff D, Abele LG, Thistle AB (eds) *Ecological Communities: Conceptual Issues and the Evidence*. Princeton University Press, Princeton, pp 151–180

Underwood AJ, Fairweather PG (1989) Supply-side ecology and benthic marine assemblages. Trends Ecol Evol 4:16–20

VanBlaricom GR (1982) Experimental analyses of structural regulation in a marine sand community exposed to oceanic swell. Ecol Monogr 52:283–305

Velimirov B, Griffiths CL (1979) Wave-induced kelp movement and its importance for community structure. Bot Mar 22:169–172

Victor BC (1984) Coral reef fish larvae: patch size estimation and mixing in the plankton. Limnol Oceanogr 29:1116–1119

Vlymen WJ (1977) A mathematical model of the relationship between larval anchovy (*Engraulis mordax*) growth, prey microdistribution, and larval behavior. Environ Biol Fish 2:211–233

Warwick RM, Davies JR (1977) The distribution of sublittoral macrofauna communities in the Bristol Channel in relation to the substrate. Estuar Coast Mar Sci 5:267–288

Warwick RM, Uncles RJ (1980) Distribution of benthic macrofauna associations in the Bristol Channel in relation to tidal stress. Mar Ecol Progr Ser 3:97–103

Wethey DS (1984) Spatial pattern in barnacle settlement: day to day changes during the settlement season. J Mar Biol Assoc UK 64:687–698

Wethey DS (1986) Ranking of settlement cues by barnacle larvae: influence of surface contour. Bull Mar sci 39:393–400

Wiebe PH, Hulbert EM, Carpenter EJ, Jahn AE, Knapp GP, Boyd SH, Ortner PB, Cox JL (1976) Gulf Stream cold core rings: large-scale interaction sites for open ocean plankton communities. Deep Sea Res 23:695–710

Williams PJL, Muir LR (1981) Diffusion as a constraint on the biological importance of microzones in the sea. In Nihoul JCJ (ed) *Ecohydrodynamics*. Elsevier, New York, pp 209–218

Wilson DL, Smith WO, Nelson DM (1986) Phytoplankton bloom dynamics of the western Ross Sea ice edge. I. Primary productivity and species-specific production. Deep Sea Res 33:1375–1387

Winant CD, Friehe CA, Dorman CE, Beardsley RC (1987) Hydraulic control of the marine layer over coastal upwelling zones. In *Proceedings of the 3rd International Symposium on Density-Stratified Flows*. California Institute of Technology, Pasadena, pp 1–19

Winant CD, Dorman CE, Friehe CA, Beardsley RC (1988) The marine layer off northern California: an example of supercritical channel flow. J Atm Sci 45:3588–3605

Witman JD (1987) Subtidal coexistence: storms, grazing, mutualism, and the zonation of kelps and mussels. Ecol Monogr 57:167–187

Woodby DA (1984) The April distribution of murres and prey patches in the southeastern Bering Sea. Limnol Oceanogr 29:181–188

Woodin SA (1978) Refuges, disturbance, and community structure: a marine soft-bottom example. Ecology 59:274–284

Wooster WS, Fluharty DL (1985) (eds) *El Niño North: Niño Effects in the Eastern Subarctic Pacific Ocean*. Washington Sea Grant, University of Washington Press, Seattle

Yentsch CS, Phinney DA (1986) The role of streamers associated with mesoscale eddies in the transport of biological substances between slope and ocean waters. In Nihoul JCJ (ed) *Marine Interfaces Ecohydrodynamics*. Elsevier, New York, pp 153–163

Yule AB, Crisp DJ (1983) Adhesion of cypris larvae of the barnacle, Balanus balanoides, to clean and arthropodin treated surfaces. J Mar Biol Assoc UK 63:261–271

Zeldis JR, Jillett JB (1982) Aggregation of pelagic *Munida gregaria* (Fabricius) (Decapoda, Anomura) by coastal fronts and internal waves. J Plankton Res 4:839–857

Zenkevitch LA, Filatova A, Belyaev GM, Lukyanova TS, Suetova IA (1971) Quantitative distribution of zoobenthos in the world ocean. Bull Mosk Gen Naturforsch Biol 76:27–33

Zimmerman RC, Kremer JN (1984) Episodic nutrient supply to a kelp forest ecosystem in southern California. J Mar Res 42:591–604

Zimmerman RC, Robertson DL (1985) Effects of the 1983 El Niño on growth of giant kelp, *Macrocystis pyrifera*, at Santa Catalina Island. Limnol Oceanogr 30:1298–13024

Index